Building the New Man

CEU Press Studies in the History of Medicine

Volume III

Series Editor: Marius Turda

PUBLISHED IN THE SERIES:

Svetla Baloutzova
Demography and Nation
Social Legislation and Population Policy in Bulgaria, 1918–1944

Christian Promitzer · Sevasti Trubeta · Marius Turda, eds.
Health, Hygiene and Eugenics in Southeastern Europe to 1945

Building the New Man

Eugenics, Racial Science and Genetics in Twentieth-Century Italy

Francesco Cassata

Translated by Erin O'Loughlin

CEU PRESS

Central European University Press
Budapest—New York

This is an extended version of the 2006 Italian edition
Molti, sani e forti. L'eugenetica in Italia, by Bollati Boringhieri Editore, Torino

Published in 2011 by
Central European University Press
An imprint of the Central European University Limited Company

Nádor utca 11, H-1051 Budapest, Hungary
Tel: +36-1-327-3138 or 327-3000 · *Fax:* +36-1-327-3183
E-mail: ceupress@ceu.hu
Website: www.ceupress.com

400 West 59th Street, New York NY 10019, USA
Tel: +1-212-547-6932 · *Fax:* +1-646-557-2416
E-mail: mgreenwald@sorosny.org

The translation of this work has been funded by SEPS
SEGRETARIATO EUROPEO PER LE PUBBLICAZIONI SCIENTIFICHE

Via Val d'Aposa 7 - 40123 Bologna - Italy
seps@seps.it · www.seps.it

ISBN 978-963-9776-83-8 cloth

ISSN 2079-1119

Library of Congress Cataloging-in-Publication Data

Cassata, Francesco.
[Molti, sani e forti. English]
Building the new man : eugenics, racial science and genetics in twentieth-century Italy / Francesco Cassata ; translated by Erin O'Loughlin.
p. cm. -- (CEU Press studies in the history of medicine ; v. 3)
Includes bibliographical references and index.
ISBN 978-9639776838 (hbk.)
1. Eugenics--Italy--History. I. O'Loughlin, Erin. II. Title.
HQ755.5.I8C3713 2010
363.9'2--dc22
2010039399

Printed in Hungary by Akadémiai Nyomda, Martonvásár

To my parents,
Adele and Letterio

CONTENTS

ACKNOWLEDGEMENTS

Research and writing for this book has been supported by many institutions and individuals. My first debt of gratitude is to Marius Turda, for his support and guidance during the process of revision, translation and publication of the book. Discussions with colleagues at seminars and conferences where I presented my work helped in clarifying many of my arguments: I would like to mention in particular the workshop on *Regards croisés sur l'hérédité pathologique et l'essor de la génétique humaine au XXe siècle. Une histoire comparée Italie-France*, organized by the CERMES (CNRS-INSERM-EHESS) and the Institute for the History of Medicine in Rome (5 December 2005); the conference on *Eugenics, Modernization and Biopolitics*, organized by the Institute of History of the Polish Academy of Science, in cooperation with the Working Group on the History of Race and Eugenics, Oxford Brookes University (17–19 April 2008); and the workshop on *Human Heredity in the Twentieth Century (A Cultural History of Heredity V)*, organized by the ESRC Research Centre for Genomics in Society (University of Exeter), in collaboration with the Centre for Medical History of the University of Exeter and the Max-Planck-Institute for the History of Science in Berlin (2–4 September, 2010).

Research on the book would not have been possible without the help of many librarians and archivists: for their substantial help with bibliographic and archival research I would like to thank the staff of the Fondazione Luigi Einaudi in Turin; the Institute for the History of Medicine in Rome; the Archivio Centrale dello Stato in Rome; the American Philosophical Society in Philadelphia.

Many people have made comments and criticisms of the arguments developed here. I owe special thanks to Luc Berlivet, Mauro Capocci, Gilberto Corbellini, Giovanni Favero, Massimo Moraglio, Claudio Pogliano, Sandro Rinauro, Paul J. Weindling.

Anna Treves, until her untimely death, was a friend and a mentor. This book owes a great deal to her understanding of population studies in twentieth century Italy.

Francesco Cassata
March 2011

INTRODUCTION

Francis Galton's gospel was quickly spread around the world. In 1924, a report of the International Commission of Eugenics published in *Eugenical News* listed fifteen countries in which eugenics had assumed an institutional form: England, Germany, the United States, Italy, France, Belgium, Switzerland, Holland, Denmark, Sweden, Czechoslovakia, Norway, Argentina, Cuba and Russia; countries that were cooperating with the International Commission included Brazil, Canada, Colombia, Mexico, Venezuela, Australia and New Zealand.[1] In the same year, a bibliography dedicated to eugenic issues already counted 7,500 titles, including monographs and articles.[2]

It therefore seems most appropriate to approach eugenics as a cultural, social and political phenomenon with a broad international relevance. As Frank Dikötter put it, eugenics should be considered as "a fundamental aspect of some of the most important cultural and social movements of the twentieth century, intimately linked to ideologies of 'race,' nation and sex, inextricably meshed with population control, social hygiene, state hospitals, and the welfare state."[3] Initially focused on the cases of Great Britain, the United States[4] and

1 Mark B. Adams, ed., *The Wellborn Science: Eugenics in Germany, France, Brazil and Russia* (New York: Oxford University Press, 1990), 5.

2 Samuel J. Holmes, *A Bibliography of Eugenics* (Berkeley: University of California Press, 1924).

3 Frank Dikötter, "Race Culture: Recent perspectives on the history of eugenics," *The American Historical Review* 103, no. 2 (April 1998): 467. See also Marius Turda, "New Perspectives on Race and Eugenics," Historical Journal 51, no. 4 (2008): 1115–24.

4 Publications on eugenics in Great Britain and United States are too numerous to list here exhaustively. See, in particular, Lindsay Andrew Farrall, *The Origins and Growth of the English Eugenics Movement 1865–1925* (New York: Garland Pub., 1965); Daniel J. Kevles, *In the Name of Eugenics. Genetics and the Uses of Human Heredity*, rev. ed. (Cambridge, MA: Harvard University Press, 1995); Richard A. Soloway, *Demography and*

Germany,[5] the historiography on eugenics has recently assumed a more open and varied comparative perspective, following the pioneering suggestions offered by Mark B. Adams in 1990:

> Even if our ultimate goal is to comprehend the "essence" of eugenics as a phenomenon, or to find the invariant laws or processes underlying the character of knowledge, or even to ascertain what is unique or atypical in a given movement or development, we cannot hope to do so without comparative studies. And this is as true for eugenics and the history of the sciences generally as it is for embryology, molecular biology, or linguistics.[6]

Nowadays, the general interpretative framework seems extremely fresh and stimulating.

First of all, eugenics no longer appears as a homogenous movement, coherent within itself and essentially reducible to the Anglo-Saxon matrix. Instead, it could be described as a "multiform archipelago," composed of multiple national styles:[7] the Scandinavian coun-

Degeneration. Eugenics and the Declining Birthrate in Twentieth Century Britain (Chapel Hill: North Carolina University Press, 1990); Pauline M. H. Mazumdar, *Eugenics, Human Genetics and Human Failings: the Eugenics Society, its Source and its Critics in Britain* (London–New York: Routledge, 1992); Garland E. Allen, "The Misuse of Biological Hierarchies: the American Eugenics Movement, 1900–1940," *History and Philosophy of the Life Sciences* 2, no. 5 (1983): 105–128; Mark H. Haller, *Eugenics: Hereditarian Attitudes in American Thought* (New Brunswick, NJ: Rutgers University Press, 1984); Edward J. Larson, *Sex, Race, and Science: Eugenics in the Deep South* (Baltimore: Johns Hopkins University Press, 1995); Ian Robert Dowbiggin, *Keeping America Sane: Psychiatry and Eugenics in the US and Canada, 1880–1940* (Ithaca, NY: Cornell University Press, 1997); Philip R. Reilly, *The Surgical Solution: a History of Involuntary Sterilization in the United States* (Baltimore: Johns Hopkins University Press, 1991); Wendy Kline, *Building a Better Race: Gender, Sexuality, and Eugenics from the Turn of the Century to the Baby Boom* (Berkeley and London: University of California Press, 2001); Edwin Black, *War Against the Weak. Eugenics and America's Campaign to Create a Master Race* (New York: Four Walls Eight Windows, 2003); Alexandra Minna Stern, *Eugenic Nation: Faults and Frontiers of Better Breeding in Modern America* (Berkeley: University of California Press, 2005); Alison Bashford and Philippa Levine, eds., *The Oxford Handbook of the History of Eugenics* (New York: Oxford University Press, 2010); Marius Turda, *Modernism and Eugenics* (Basingstoke: Palgrave Macmillan, 2010).

5 On German eugenics, see: Gisela Bock, *Zwangssterilisation im Nationalsozialismus* (Opladen: Westdeutscher Verlag, 1986); Robert Proctor, *Racial Hygiene: Medicine under the Nazis* (Cambridge, MA: Harvard University Press, 1988); Peter Weingart, Jürgen Kroll, and Kurt Bayertz, *Rasse Blut und Gene: Geschichte der Eugenik und Rassenhygiene in Deutschland* (Frankfurt a. M.: Suhrkamp, 1988); Paul J. Weindling, *Health, Race and German Politics between National Unification and Nazism, 1870–1945* (Cambridge: Cambridge University Press, 1989); Paul J. Weindling, "The 'Sonderweg' of German Eugenics: Nationalism and Scientific Internationalism," *The British Journal for the History of Science* 22, no. 3 (September 1989): 321–33; Sheila F. Weiss, "The Race Hygiene Movement in Germany, 1904–1945," in Adams, ed., *The Wellborn Science*, 8–68; Michael Burleigh and Wolfgang Wipperman, *The Racial State: Germany 1933–1945* (Cambridge: Cambridge University Press, 1991).

6 Adams, ed., *The Wellborn Science*, 6.

7 Peter Weingart, "Science and Political Culture: Eugenics in Comparative Perspective," *Scandinavian Journal of History* 24, no. 2 (June 1999): 163–177. On international networks, see, in particular, Stefan Kühl, *The Nazi Connection: Eugenics, American Racism, and German National Socialism*, (Oxford: Oxford University Press,

tries,[8] Central and Eastern Europe,[9] Latin America,[10] but also China, India, and Japan are among the regions and countries most recently studied.[11]

Secondly, on a theoretical level, next to Mendelism, which was dominant in the British and North American contexts, neo-Lamarckism has been identified as one of the constitutive elements of the eugenic discourse, above all in several nations, such as France, Russia and Brazil.[12] In

1994) and *Die Internationale der Rassisten. Aufstieg und Niedergang der internationalen Bewegung für Eugenik und Rassenhygiene im 20. Jahrhundert* (Frankfurt a. M.: Campus Verlag, 1997). See also Alison Bashford, *Internationalism, Cosmopolitanism, and Eugenics*, in Bashford and Levine, eds., *The Oxford Handbook of the History of Eugenics*, 154–72.

8 On eugenics in Scandinavia, see: Gunnar Broberg and Nils Roll-Hansen, eds., *Eugenics and the Welfare State: Sterilization Policy in Denmark, Sweden, Norway and Finland* (East Lansing, MI: Michigan State University Press, 1996); Dorothy Porter, "Eugenics and the sterilization debate in Sweden and Britain before World War II," *Scandinavian Journal of History* 24, no. 2 (1999): 145–162; Alain Drouard, "Concerning Eugenics in Scandinavia. An Evaluation of Recent Research and Publications," *Population: an English Selection* 11 (1999): 261–70; Mattias Tydén, *The Scandinavian States: Reformed Eugenics applied*, in Bashford and Levine, eds., *The Oxford Handbook of the History of Eugenics*, 363–76.

9 Maria Bucur, *Eugenics and Modernization in Interwar Romania* (Pittsburgh: Pittsburgh University Press, 2002); Brigitte Fuchs, *'Rasse', 'Folk', 'Geschlecht'. Anthropologische Diskurse in Österreich, 1850–1960* (Frankfurt a. M.: Campus Verlag, 2003); Kamila Uzarczyk, *Podstawy ideologiczne higieny ras i ich realizacja na przykładzie Śląska w latach 1924–1944* (Toruń: Wydawnictwo Adam Marszałek, 2003); Magdalena Gawin, *Rasa i nowczesność. Historia polskiego ruchu eugenicznego, 1880–1952* (Warsaw: Wydawnicwo Neriton, 2003); Heinz Eberhard and Wolfgang Neugebauer, eds., *Vorreiter der Vernichtung? Eugenic, Rassenhygiene und Euthanasie in der österreichischen Discussion vor 1938* (Vienna: Böhlau Verlag, 2005); Gerhard Baader, Veronika Hofer and Thomas Mayer, eds., *Eugenik in Österreich: Biopolitischer Methoden und Strukturen vor 1900–1945* (Vienna: Czernin Verlag, 2007); Marius Turda and Paul J. Weindling, eds., *Blood and Homeland: Eugenics and Racial Nationalism in Central and Southeast Europe 1900–1940* (Budapest–New York: Central European University Press, 2007); Christian Promitzer, Sevasti Trubeta, Marius Turda, eds., *Health, Hygiene and Eugenics in Southeastern Europe to 1945* (Budapest–New York: Central European University Press, 2011).

10 Nancy Leys Stepan, *The "Hour of Eugenics": Race, Gender and Nation in Latin America* (Ithaca–London: Cornell University Press, 1991); Alexandra Minna Stern, "From Mestizophilia to Biotypology. Racialization and Science in Mexico, 1920–1960," in Nancy P. Appelbaum, Anne S. Macpherson, and Karin Alejandra Rosenblatt, eds., *Race & Nation in Modern Latin America* (Chapel Hill and London: University of North Carolina Press, 2003), 187–210.

11 Frank Dikötter, *Imperfect Conceptions: Medical Knowledge, Birth Defects and Eugenics in China* (New York: Columbia University Press, 1998); Patrick McGinn, "'Quality not quantity tells': The Eugenics Movement in India," unpublished manuscript; Sabine Frühstück, *Die Politik der Sexualwissenschaft: Zur Produktion und Popularisierung sexologischen Wissens in Japan 1908–1941* (Vienna: Institut Ostasienwissenschaften, 1997); Yuehtsen Juliette Chung, *Eugenics in China and Hong Kong: Nationalism and Colonialism*, in Bashford and Levine, eds., *The Oxford Handbook of the History of Eugenics*, 258–73; Jennifer Robertson, *Eugenics in Japan: Sanguinous Repair*, in Bashford and Levine, eds., *The Oxford Handbook of the History of Eugenics*, 430–48.

12 William H. Schneider, *Quality and Quantity. The Quest for Biological Regeneration in Twentieth-Century France* (Cambridge: Cambridge University Press, 1990); Nancy Leys Stepan, "Eugenics in Brazil, 1917–1940," in Adams, ed., *The Wellborn Science*, 110–52; Gilberto Hochman, Nisia Trindade Lima, and Marcos Chor Maio, *The Path of Eugenics in Brazil: Dilemmas of Miscegenation*, in Bashford and Levine, eds., *The Oxford Handbook of the History of Eugenics*, 493–510; Mark B. Adams, "Eugenics in Russia, 1900–1940," in Adams, ed., *The Wellborn Science*, 153–216. ; Nikolai Krementsov, "From 'Beastly Philosophy' to Medical Genetics: Eugenics in Russia and the Soviet Union," *Annals of Science* 68, no. 1 (January 2011): 61–92. On Lamarckian eugenics, see also: Peter J. Bowler, "E. W. MacBride's Lamarckian Eugenics and Its Implications for the Social Construction of Scientific Knowledge," *Annals of Science* 41, no. 3 (May 1984): 245–60.

parallel, "Nordic" eugenics has been coupled with "Latin" eugenics, widespread in Catholic countries such as Italy, France, Spain, Belgium and some Latin American nations.[13]

Thirdly, the definition of eugenics as a "pseudo-science" is being progressively substituted by an analysis that is more conscious of the relationships of eugenics to genetics and other scientific disciplines, such as demography, statistics and psychology.[14]

Finally, the myth of eugenics as an essentially reactionary ideology, mostly linked to sexist, racist, anti-Semitic and generally right-wing movements, has been replaced with an historically more mature evaluation, which is more knowledgeable about the fascination exercised by the eugenic thinking also in the left-wing milieu: from the first British feminists to German and Swedish social-democrats; from Spanish anarchists to French communists.[15]

---

13 See, in particular, Marisa Miranda and Gustavo Vallejo, eds., *Darwinismo social y eugenesia en el mundo latino* (Buenos Aires: Siglo Veintiuno de Argentina Editores, 2005); Armando García González, Raquel Álvarez Peláez, *En busca de la raza perfecta. Eugenesia e higiene en Cuba (1898–1958)* (Madrid: CSIC, 1998); "Dossier: Estudios sobre eugenesia," ed. Raquel Álvarez Peláez, special issue, *Asclepio* 51, no. 2 (1999): 5–148; Patience A. Schell, *Eugenics Policy and Practice in Cuba, Puerto Rico, and Mexico*, in Bashford and Levine, eds., *The Oxford Handbook of the History of Eugenics*, 477–92.

14 On genetics and eugenics, see: Kevles, *In the Name of Eugenics*; Diane B. Paul, *Controlling Human Heredity: 1865 to the Present* (Atlantic Highlands, NJ: Humanities Press, 1995); Jan Sapp, "The Struggle for Authority in the Field of Heredity, 1900–1932: New perspectives on the Rise of Genetics," *Journal of the History of Biology* 16, no. 3 (1983): 311–42; Jonathan Harwood, "Geneticists and the Evolutionary Synthesis in Interwar Germany," *Annals of Science* 42, no. 3 (May 1985): 279–301; Paul Weindling, "Weimar Eugenics: The Kaiser Wilhelm Institute for Anthropology, Human Heredity and Eugenics in Social Context," *Annals of Science* 42, no. 3 (May 1985): 303–18; Garland E. Allen, "The Eugenics Record Office at Cold Spring Harbor, 1910–1940. An Essay in Institutional History," *Osiris*, 2nd series 2 (1986): 225–64; Nils Roll-Hansen, "Geneticists and the Eugenics Movement in Scandinavia," *The British Journal for the History of Science* 22, no. 3 (September 1989): 335–46; David Barker, "The Biology of Stupidity: Genetics, Eugenics and Mental Deficiency in the Inter-War Years," *The British Journal for the History of Science* 22, no. 3 (September 1989): 347–75; Hans-Peter Kröner, *Von der Rassenhygiene zur Humangenetik* (Munich: Urban & Fischer, 1998); Nathaniel Comfort, "'Polyhybrid Heterogeneous Bastards': Promoting Medical Genetics in America in the 1930s and 1940s," *Journal of the History of Medicine and Allied Sciences* 61, no. 4 (October 2006): 415–55. On eugenics and demography, see: Edmund Ramsden, "Carving up Population Science: Eugenics, Demography and the Controversy over the 'Biological Law' of Population Growth," *Social Studies of Science*, 32, no. 5–6 (October–December 2002): 857–99; Edmund Ramsden, "Social Demography and Eugenics in the Interwar United States," *Population and Development Review*, 29, no. 4 (December 2003): 547–93; Edmund Ramsden, "Eugenics from the New Deal to the Great Society: Genetics, Demography and Population Quality," *Studies in History and Philosophy of Biological and Biomedical Sciences* 39 (2008): 391–406.

15 See: Donald K. Pickens, *Eugenics and the Progressives* (Nashville: Vanderbildt University Press, 1968); Michael Freeden, "Eugenics and Progressive Thought: a Study in Ideological Affinity," *Historical Journal* 22 (1979): 645–71; Loren R. Graham, "Science and Values: the Eugenic Movement in Germany and Russia in 1920s," *American Historical Review* 82, no. 5 (December 1977): 1133–1964; Diane B. Paul, "Eugenics and the Left," *Journal of the History of Ideas* 45 (1984): 567–90; Kevin Repp, "'More Corporeal, More Concrete': Liberal Humanism, Eugenics and German Progressives at the Last Fin de Siècle," *Journal of Modern History* 72

In the context of this fertile comparative approach, the Italian case—notwithstanding its crucial importance from an international point of view, due to the role of Fascism and of the Catholic Church—has long been neglected, or has been studied in an incomplete manner, as a component of the fascist population policy or state racism.[16]

Based on previously unexplored archival documentation, this book offers a first general overview of the history of Italian eugenics, not limited to the decades of the fascist regime, but instead ranging from the beginning of the 1900s to the first half of the seventies.

The word *eugenica* (or, less frequently, *eugenìa* and *eugenetica*) began to spread in Italy in 1912, in the wake of the First International Congress of Eugenics, held in London, under the presidency of Leonard Darwin. Even recalling the intense proto-eugenic debate existing in Italy from the final decades of the nineteenth century, the Italian participation at the London Congress not only stimulated a process of institutionalization of Italian eugenics—through the constitution in 1913 of the first Italian Committee of Eugenic Studies—but also demonstrated from the beginning the particular originality of the Italian approach to eugenics. Neo-Lamarckian theoretical influences; Pareto's theory of the elite and social exchange; the anthropology of racial breeding and migrations; and the Lombrosian connection between genius and degeneration, all created a scientific and intellectual framework that made Italian eugenics inassimilable to the Anglo-Saxon model.

The First World War, which is addressed in chapter 2, represented an important moment of development for Italian eugenics. Interpreted as dramatic "counter-selection"; or, vice-versa, as a means of biological optimi-

(2000): 683–730; M. Schwartz, *Sozialistische Eugenik. Eugenische Sozialtechnologien in Debatten und Politik der deutschen Sozialdemokratie, 1890–1993* (Bonn: Dietz, 1995); Richard Cleminson, *Anarchism, Science and Sex: Eugenics in Eastern Spain, 1900–1937* (Oxford–Bern: Peter Lang 2000); Richard Sonn, "'Your body is Yours': Anarchism, Birth Control, and Eugenics in Interwar France," *Journal of the History of Sexuality* 14, no. 4 (October 2005): 415–32; Richard Cleminson, "'A Century of Civilization under the Influence of Eugenics': Dr. Enrique Diego Madrazo, Socialism and Scientific Progress," *Dynamis* 26 (2006): 221–51.

16 On eugenics and fascist population policy, see: David Horn, *Social Bodies. Science, Reproduction and Italian Modernity* (Princeton: Princeton University Press, 1994); Carl Ipsen, *Dictating Demography: The Problem of Population in Fascist Italy* (Cambridge: Cambridge University Press, 1996); Maria Sophia Quine, *Population Politics in Twentieth Century Europe: Fascist Dictatorships and Liberal Democracies* (London: Routledge, 1996). On eugenics and racism in fascist Italy, see: Roberto Maiocchi, *Scienza italiana e razzismo fascista* (Florence: La Nuova Italia, 1999); Giorgio Israel, Pietro Nastasi, *Scienza e razza nell'Italia fascista* (Bologna: Il Mulino, 1998); Aaron Gillette, *Racial Theories in Fascist Italy* (New York: Routledge, 2002). Recent works have provided a more comprehensive approach. See, in particular, Claudia Mantovani, *Rigenerare la società. L'eugenetica in Italia dalle origini ottocentesche agli anni Trenta* (Soveria Mannelli: Rubbettino, 2004).

zation of the nation, the conflict provided eugenicists with important lessons: in particular, it demonstrated the relevance of a "unity of command" and the efficiency of direct state management, economically rational, of the biological resources of the nation.

Anxieties over national regeneration, technocratic ambitions and new social welfare-oriented policies, which, after the war, accompanied the crises of the last liberal governments and the progressive rise of fascism, favored the affirmation of eugenics as a part of social medicine and public health. In this context, eugenics was progressively seen as a paradigm of national efficiency, based on the subordination of individual liberty to superior collective interests for the "defense of society and the race." Such a technocratic and managerial conception of the population fascinated the Italian political elite in this period, the left as much as the right, ranging from nationalism to reformist socialism, and of course fascism. It was in these years—as discussed in chapter 3—that Italian eugenics was institutionalized, with the constitution of the Institute of Public Welfare and Assistance (*Istituto di Previdenza e Assistenza Sociale,* IPAS); the Italian Society for the Study of Sexual Questions (*Società Italiana per lo studio delle Questioni Sessuali,* SISQS); the Italian Society for Genetics and Eugenics (*Società Italiana di Genetica e Eugenica,* SIGE); and the Italian League of Hygiene and Mental Prophylaxis (*Lega Italiana di Igiene e Profilassi Mentale,* LIPIM). In the same period, the eugenic debate went through a season of extreme richness and variety, exploring the fundamental issues of birth control, premarital certification, sterilization and mental hygiene.

The orthodoxy based on the binomial "quantitative" eugenics—pronatalist population policy was imposed officially and definitively in 1927. The turning point was above all political, and it was sanctioned by the alliance between fascist natalist policy, inaugurated in May 1927 with Mussolini's famous Ascension Day Speech, and Catholic sexual morals, reaffirmed by the Holy See in December 1930, with the encyclical *Casti Connubii* [On Christian marriage]. SIGE's leadership mirrored this ideological and political fusion: the president was the demographer and statistician Corrado Gini, who contemporaneously managed also ISTAT and CISP; the vice-president was Agostino Gemelli, founder and dean of the Milan Catholic University, and principle exponent of Italian Catholic eugenics.

On a more specifically scientific level, starting from the second half of the 1920s, the theoretical paradigm that fascist eugenics was based on was constituted by the convergence between Corrado Gini's "integral" demography—synthesis of demography, biology, anthropology, economy, sociology and, obviously, eugenics—and constitutional biotypological medicine. The latter was represented above all by the endocrinologist Nicola Pende, close to the Catholic environments. Both Gini's "regenerative" eugenics and Pende's biotypological "orthogenesis" opposed the "Nordic" Anglo-Germanic and Scandinavian model.

This opposition—scientific, ideological and political all at the same time—was expressed at an institutional level by Italy's exit from the IFEO, and the constitution in 1935 of the Latin Federation of Eugenic Societies: an alternative model, the birth of which coincided not surprisingly with the most critical phase of diplomatic relationships between fascist Italy and Nazi Germany.

Starting from 1936, and in particular in 1938 with the introduction of state racism in fascist Italy, the ideological and political convergence of fascism and national socialism also influenced the relationship between eugenics and racism, feeding new tensions and oppositions. This issue is analyzed in chapter 5. Between 1938 and 1943 the nature/nurture debate became the battleground for the clash between the different racisms of fascism: "biological" (Telesio Interlandi, Guido Landra, etc.) and "esoteric-traditionalist" racism (Julius Evola, Giovanni Preziosi, etc.) adopted the negative Nazi eugenic model, while "nationalist" and "Mediterranean" racism (Giacomo Acerbo, Nicola Pende, etc.) remained faithful to the "Latin" model, environmentalist and neo-Lamarckian. The two positions were opposed in their definition of Italian racial identity, but converged in their discrimination of racial enemies, in particular the half-caste and the "Jew."

The end of the Second World War and the discovery of the tragic consequences of National Socialist racism did not signal the definitive end of eugenics. In the 1950s and 1960s, eugenics in Italy was not stigmatized as taboo, but it was progressively redefined, passing through a sort of *no man's land,* in which struggles and oppositions occurred on different levels. Institutionally and academically, the statisticians and demographers of SIGE clashed with the geneticists (Adriano Buzzati-Traverso, Giuseppe Montalenti, Claudio Barigozzi), who decided, in 1953, to constitute a new autonomous association (*Associazione Italiana di Genetica,* AGI). Instead, the physicians (Carlo Foà, Luigi Gedda, Luisa Gianferrari) in 1951 constituted the first Italian Society of Medical Genetics (*Società Italiana di Genetica Medica*), opposed to both Gini's SIGE and the AGI. Politically, mainline Italian eugenics, after the Second World War, became an important component of international scientific racism, expressed by the IAAEE and the *Mankind Quarterly,* encountering the anti-fascist and anti-racist components of the reform and new Italian eugenics.

Finally, from an ideological point of view, Catholic, familial and natalist eugenics, supported above all by Luigi Gedda's "Gregorio Mendel" Institute, opposed secular eugenics, which advocated birth control and family planning policies. The history of eugenics and genetics in Italy after the Second World War is covered in chapters 6 and 7.

The book concludes in the second half of the 1970s, with the introduction in Italy of prenatal diagnosis in 1975, followed in 1978 with the approval of Law 194 on the legalization of abortion: the eugenic debate entered in a new phase—that of so-called *new eugenics*—which in Italy even today feeds an intense, and at times lacerating, political and cultural debate.[17]

[17] On "new" eugenics, see Diane B. Paul, "Eugenic Anxieties, Social Realities, and Political Choices," *Social Research* 59, no. 3 (1992): 663–83; Jean Gayon and Daniel Jacobi, eds., *L'éternel retour de l'eugénisme* (Paris: Presses Universitaires de France, 2006); Alison Bashford, *Where Did Eugenics Go?,* in Bashford and Levine, eds., *The Oxford Handbook of the History of Eugenics,* 539–58.

CHAPTER I

BETWEEN LOMBROSO AND PARETO

THE ITALIAN WAY TO EUGENICS

The First International Eugenics Congress was held in London between 24 and 31 July 1912, under the presidency of Leonard Darwin. The large Italian delegation included some of the most relevant figures of positivist science: jurist Raffaele Garofalo (1851–1934), anthropologists Giuseppe Sergi (1841–1936) and Vincenzo Giuffrida-Ruggeri (1872–1921), psychiatrists Enrico Morselli (1852–1929) and Antonio Marro (1840–1913), economist Achille Loria (1857–1943), sociologist Roberto Michels (1876–1936), and statisticians Alfredo Niceforo (1876–1960) and Corrado Gini (1885–1965). From a disciplinary point of view, it was a heterogenous group, and also contained a reasonable cross-section of political orientations, from the socialism of Loria and Niceforo to the nationalism of Gini.

In the history of Italian eugenics, the First International Eugenics Congress was a defining moment, from two points of view. First, the London congress contributed to the process of organization and institutionalization of the eugenic movement. Before 1912, the Italian scientific and cultural context had seen some debate that centered around the problems of the biological regeneration of the nation. The hygienist utopia of Paolo Mantegazza, professor of the first chair of anthropology in Italy, physician and scientific popularizer of extraordinary success;[1] the eighteenth-century development of social medicine;[2] and the brief appearance, between

[1] See, for example, Paolo Mantegazza, *L'anno Tremila – Sogno* (2nd ed.) (Milan: Treves, 1897); Paolo Mantegazza, *Un giorno a Madera. Una pagina dell'igiene dell'amore* (Florence: Salani, 1910).

[2] See, in particular, Gaetano Bonetta, *Corpo e nazione. L'educazione ginnastica, igienica e sessuale nell'Italia liberale* (Milan: Franco Angeli, 1990) and Mantovani, *Rigenerare la società*, 87–114.

1910 and 1913, of a neo-Malthusian movement,[3] clearly demonstrate the presence of a sort of Italian *proto-eugenics.* But it was only after the London Congress that the term "eugenics" (in Italian, "eugenia," "eugenica" or "eugenetica") became diffused in the scientific press and amongst the wider public. In 1912, Serafino Patellani was assigned the first university course of "social eugenics," and in 1913 an Italian Committee of Eugenic Studies was instituted at the Roman Society of Anthropology, with Giuseppe Sergi[4] nominated as president.

Secondly, the reconstruction of the scientific paths of the most important members of the delegation allows the identification of a set of problems at the origins of Italian eugenics: these included the notion of atavism, the relationship between genius and degeneration, the anthropological heterogeneity of the Italian population, and the demographic dynamic of social exchange. All these issues reveal the intellectual influence on Italian eugenics of two intellectual figures of extreme relevance in the history of social sciences: the anthropologist and criminologist Cesare Lombroso, and the economist and statistician (not to mention sociologist) Vilfredo Pareto.

The specificity of Italian eugenics in the international context, including its opposition—as much theoretical as ideological and political—to the Anglo-Saxon mainstream, developed from the singular convergence of these two different and conflicting streams of thought.

1. Lombroso's Way: the Problem of Degeneration

The Lombrosian path to eugenics can be first of all identified in the particular meanings that *dégénérescence* assumes in the theoretical production of the well-known Italian criminologist. A great deal has been written on the importance of the concept of degeneration in the genesis of the eugenic discourse.[5] Nevertheless, the degeneration–eugenics nexus varies notably according to the cultural reference scenarios.

---

3 See Bruno Wanrooij, *Storia del pudore. La questione sessuale in Italia* (Venice: Marsilio, 1990); Giorgio Rifelli, *Per una storia dell'educazione sessuale* (Florence: La Nuova Italia, 1991).

4 See "Notizie," *Rivista di antropologia* 18 (1913): 289.

5 See, in particular, Daniel Pick, *Faces of Degeneration: A European Disorder* (Cambridge: Cambridge University Press, 1989); Soloway, *Demography and Degeneration.* The information available on Cesare

In particular, the theories of atavism and born criminal do not seem to share the pessimistic belief in the omnipresence and dissemination of degeneration that induced Francis Galton, in 1883,[6] to coin the term *eugenics* to indicate a program of planning and rationalization of human reproduction, aimed at the biological improvement of the species. Italian criminal anthropology identified the base cause of innatism to crime as arrested development. Therefore, the primary objective of the discipline was not to intervene in the reproductive process, but rather to isolate dysgenic types (antisocial delinquents) and segregate them from the rest of society. As a consequence, Lombroso's "new criminal therapy" outlined a large reformist project of social control, developed from a complex anthropological and psychiatric taxonomy: the regulation of migratory flows and a rapid repressive justice, the segregation of habitual criminals and the control of "honest, but weak" citizens, taxes on alcohol and a protracted surveillance of youth and derelicts through "voluntary" or "compulsory asylums" and "industrial schools." For born criminals and the criminally insane, measures were different and more serious: "life segregation," forced work, criminal asylums and, finally, the death penalty. It was above all in relation to this latter measure that the eugenic intent was explicit:

> While it is correct to consider that the roots of certain evils cannot be overcome with the death of a few felons, it is however true that crime has diminished in intensity and ferocity in the last centuries thanks in part to the death penalty. Distributed so widely and with much publicity, if it has contributed to a share of new crimes with a spirit of imitation and ferocious public spectacle, it must also have diminished many others, preventing every evasion, every relapse and heredity in criminals, doing that which nature does in the selection of the species, when, from inferior beings, it gives us the grand dominators of the globe.[7]

The theory of born criminals was the subject of numerous criticisms, but the *cordon sanitaire* of social defense theorized by Lombroso, with its sequence

---

Lombroso is vast: for a recent overview, see Silvano Montaldo and Paolo Tappero, eds., *Cesare Lombroso cento anni dopo* (Turin: UTET, 2009). See also Mary Gibson, *Born to Crime. Cesare Lombroso and the Origins of Biological Criminology* (London: Praeger, 2002).

6 Francis Galton, *Inquiries into Human Faculty and its Development* (London: Macmillan, 1883).

7 Cesare Lombroso, *Troppo presto. Appunti al nuovo progetto di codice penale con appendici* (1888; repr., Turin: Bocca, 1889), 23–4.

of prevention, socio-economic utilization, segregation and—only as a last resort—elimination of the dysgenic elements, had a lasting influence on Italian eugenics, defining its specific position in the international context.[8] Given this framework, it is not surprising that it was above all the anthropologists of the Lombrosian school who first called for a preventive recording of the population, in the conviction that the availability of data and numbers constituted the most rational techno-bureaucratic management of human material. The idea of a "biographical card" in fact came to be suggested multiple times: for the military, as desired by the medical captain Salvatore Guida in 1879; for criminals and workers, as hoped in the first decades of the 1900s by the legal physician Salvatore Ottolenghi, student of Lombroso and founder of the School of Scientific Police; and for students, according to the theories of Alfredo Niceforo in 1913.[9]

The atavistic model, which assumed a biological predisposition to evil, and the principle of social defense, based on the institutional practices of segregation, prevention and control, still fell within the Lombrosian theoretical scheme, with its belief in an evolutionary dimension of degeneration. From this point of view, Lombroso's reflections on genius assume fundamental importance. Although deeming Galton's *Hereditary Genius* a "valuable work," Lombroso challenged its statistical data, declaring, in opposition to Galton, the weaker "hereditary action" of genius, compared to insanity.[10]

Therefore, while on one hand a "fatal parallelism" existed between genius and degeneration, on the other, genius represented, in Lombroso's views, a progressive anomaly *par excellence*: the genius action was innovating and could change the world, and degeneration could produce progress. While Galton maintained that natural selection needed to be reinforced with an artificial eugenic selection, for Lombroso, eugenics was a part of the same evolutionary mechanisms of natural selection, even in its degenerative aspects. It was not by chance that genius—carrier of degeneration, but innovator and creator of progress—represented only one aspect of the positive transgression of the norms theorized by Lombroso: revolutionary

8 See Francesco Cassata, "Dall'Uomo di genio all'eugenica," in Montaldo and Tappero, eds., *Cesare Lombroso cento anni dopo*, 175–84.

9 On the projects of recording, see Mantovani, *Rigenerare la società*, 50–51.

10 Cesare Lombroso, *Genio e follia in rapporto alla medicina legale, alla critica e alla storia* (Turin: Bocca, 1882).

spirit, modern evolutionary criminality, and the social function of crime were others.

Even this second dimension of Lombroso's anthropology exercised a lasting influence on nineteenth-century Italian eugenics. On many occasions, the refusal of negative eugenics (above all, sterilization) was inspired by the Lombrosian idea that biological degeneration could in reality generate genius; that the deformed or epilepsy sufferers could be hiding a Leopardi or a Manzoni in their midst.

In 1880 Lombroso founded the journal *Archivio di psichiatria, scienze penali ed antropologia criminale* [Journal of psychiatry, penal science and criminal anthropology]. In these pages, it is possible to notice the distinctive Lombrosian interpretation, and the attention with which the development of the international eugenic debate was followed is also evident. From its inception, the *Archivio* dealt with eugenics, informing its readers about the legislative initiatives on sterilization and castration introduced, in those years, in the United States and Europe.[11] The principle source was the *Eugenical News*, while the most perceptive editor seemed to be Prospero Mino, voluntary assistant of the medical clinic at the University of Turin, and author, in the 1920s, of a highly informative essay on "hereditary illnesses and their etiology."[12]

After the death of Lombroso in 1909, Mario Carrara, his son-in-law and successor to the direction of *Archivio* and the Institute of Legal Medicine in Turin, oriented the periodical towards a synthesis between biology and legal medicine, in which eugenics assumed a significant role. Author of several statistical-genealogical analyses on the intelligence of "men of genius,"[13] Carrara was strictly influenced by Lombroso's theories. He was convinced that the principle of "social defense" needed to be founded on the concept of "social danger," which came as much from an "originally deviant psychophysiological constitution" as from a constitution deviated by an "acquired

11 See "Selezione artificiale," *Archivio di antropologia criminale, psichiatria e medicina legale*, 34, (1913): 468; "Sterilizzazione di criminali in America," *Archivio di antropologia criminale, psichiatria e medicina legale* 34 (1913): 613.

12 For a critique of the concept of degeneration in a Mendelian framework, see in particular Prospero Mino, "Sulle malattie ereditarie e sulla loro etiologia (continuazione e fine)," *Archivio di antropologia criminale, psichiatria e medicina legale* 43 (1923): 5.

13 Salvatore Ottolenghi and Mario Carrara, "Perioptometria e psicometria di uomini geniali," *Archivio di psichiatria, scienze penali ed antropologia criminale* 13 (1892).

postnatal illness."[14] On these premises, in 1911, Carrara rejected sterilization as a "scientific *boutade*," which "everyone feels can have no practical importance,"[15] although he did not exclude the adoption of that practice—with the necessary precautions and guarantees—for a very limited number of extreme cases.[16] Instead he favored other measures of a eugenic nature, above all therapeutic abortion, for which he repeatedly requested decriminalization, and the "permanent segregation" of recidivist criminals.[17]

The 1912 First International Eugenics Congress undoubtedly marked a turning point for the *Archivio*'s coverage of eugenic themes. For the transition of the Lombrosian school to eugenics, the London Congress had a double importance. In the first place, the Italian delegation was a synthesis of those disciplines—anthropology, psychiatry, criminology, legal medicine—on which Lombroso had exerted a powerful influence. A glance through the names reveal intellectual figures—such as, in particular, Giuseppe Sergi, Raffaele Garofalo, Alfredo Niceforo and Enrico Morselli—whose scientific and personal links with Lombroso are well-known. Sergi broadly shared the Lombrosian position on atavism and the biological inferiority of females; in 1880, together with Lombroso, Garofalo was co-founder of the above-mentioned *Archivio*; Niceforo had been controversially labeled by Napoleone Colajanni as "the latest Lombrosian" for his statistical-anthropological investigation on the "cursed race" of Southern Italy;[18] while Morselli was particularly interested in Lombroso's innovations during his early years, although this interest never translated into open adherence, and was replaced in later years by a position of complete distance. It is also worth remembering the numerous exponents of legal

14 Mario Carrara, "La difesa sociale nel Diritto private," *Archivio di antropologia criminale, psichiatria e medicina legale* 44 (1924): 7; see also Carrara, *Lezioni di medicina legale* (Turin: Litografia A. Viretto, 1913); Carrara, "Influenze della biologia sulle leggi," *La Parola* (September 1925) offprint.

15 Mario Carrara, "Il VII Congresso Internazionale d'Antropologia Criminale," *Archivio di antropologia criminale, psichiatria e medicina legale* 32 (1911): 664.

16 [Mario Carrara], review of L. Altmann, *Die Fruchtabtreibung* (Vienna: Hölder-Pichler-Tempsky, 1926), *Archivio di antropologia criminale, psichiatria e medicina legale* 46 (1926): 731; [Mario Carrara], review of G. Sampaio, *A estarilizaçäo eugenica e a deontologia medica* (1928), *Archivio di antropologia criminale, psichiatria e medicina legale* 49 (1929): 732; [Mario Carrara], review of O. Kankeleit, *Die Unfruchtbarmachung aus rassenhygienischen und sozialen Gründen* (1929), *Archivio di antropologia criminale, psichiatria e medicina legale* 50 (1930): 787.

17 [Mario Carrara], "Primo congresso di Eugenetica sociale," *Archivio di antropologia criminale, psichiatria e medicina legale* 45 (1925): 72.

18 For comments on the debate regarding the "two Italies" from an anthropological point of view, see Vito Teti, *La razza maledetta: origini del pregiudizio antimeridionale* (Rome: Manifestolibri, 1993); Claudia Petraccone, *Le due Italie: la questione meridionale tra realtà e rappresentazione* (Rome: Laterza, 2005).

and military medicine, inspired by Lombroso, who took part in 1913 in the first Italian Committee of Eugenic Studies, starting with Mario Carrara and Salvatore Ottolenghi, Lombroso's assistant in Turin from 1885 to 1893.

But what characterized "Lombrosian" eugenics at the London Congress? Senator Raffaele Garofalo did not present a specific paper, but appeared as an honorary member of the Congress, implicitly revealing how important eugenics was for the Italian positivist school of criminal law. From 1885 in fact, the jurist had loudly supported the custody of the perpetrators of crimes against people in criminal asylums for indeterminate periods. This was because from the "precedence of other crimes, hereditary degeneration or a complex of marked psychological and anthropological characteristics, we can assume that the criminal is either a moral imbecile or an instinctive criminal."[19] Garofalo believed above all in the need for eugenic protection, which justified the restoration of the death penalty to the penal code. In the past, the death penalty had had the merit of "rendering the reproduction of criminals impossible, and therefore leading to a lower number."[20]

At the London Congress, Alfredo Niceforo was president of the Italian Consultative Committee. For Niceforo, eugenics was a theoretical corollary of his research on the anthropological causes—both hereditary and environmental—of the inferiority of the Italian "southern race" and the poor classes, which he had begun to investigate in the final years of the nineteenth century.[21] In Niceforo's view, biological weakness was the principle cause of socioeconomic inferiority: "The groups formed by individuals belonging to the lower classes present, in comparison with subjects of the higher classes, a lesser development of the figure, of the cranial circumference, of the sensibility, of the resistance to mental fatigue, a delay in the epoch when puberty manifests itself, a slowness in the growth, a larger number of anomalies and of cases of arrested development."[22]

19 Raffaele Garofalo, *Criminologia. Studio sul delitto, sulle sue cause e sui mezzi di repressione* (Turin: Bocca, 1885), 449–50.

20 Garofalo, *Criminologia*, 419.

21 On this theme, see in particular Bernardino Farolfi, "Antropometria militare e antropologia della devianza (1876–1906)," in Franco Della Peruta, ed., *Storia d'Italia. Annali*, vol. 7, *Malattia e medicina* (Turin: Einaudi, 1984), 1181–1222.

22 Alfredo Niceforo, "The cause of the inferiority of physical and mental characters in the lower social classes," in *Problems in Eugenics: Papers Communicated to the First International Eugenics Congress held at the University of London July 24th to 30th* (London: Eugenics Education Society, 1912), 187.

Therefore the bio-psychical characteristics of the individual were the subject of social exchange: those most endowed tended to be concentrated in the superior classes, while the weakest and defective inevitably descended the social scale. Niceforo understood eugenics as an "anthropology of the poorer classes" or "anthropology of social classes," which studied how to facilitate the natural circulation" of "social molecules": upwards for the superior who find themselves below, downwards for the inferior who find themselves above.[23]

Among the Italian delegates to London, Giuseppe Sergi, who later became president of the first Italian Committee of Eugenic Studies in 1913, was the only member to have personally met Francis Galton: in 1886, when the British scientist visited Rome. He was later a guest in the Galton's London house and met him again on the successive trips to Rome, the last of which was in 1903.[24] Sergi's approach to eugenics can be seen, in addition to his knowledge of the theories of Darwin and Galton, and in general to Anglo-Saxon scientific culture, in his specific treatment of the problem of degeneration, to which he dedicated a specific essay in 1889. In it, defining degeneration as a form of "inferior adaptation," a sort of residuum from the process of natural selection, Sergi described various categories of degenerates, which reproduced the usual positivist approach of social pathology: the insane, criminals, suicides, prostitutes, the "serfs and the servile," vagabonds, beggars and parasites.

In the face of this harvest of human degeneration, what sense could "regeneration" still have? With lengthy citations from Herbert Spencer, Sergi passionately denounced the dangerous effects of "sentimental altruism": protecting degenerates only increased their chances of reproducing. The "protection of the weak" could be useful for victims of misfortune or illness, but could not be extended to vagabonds, beggars and criminals.[25] Natural selection must therefore be supported by "artificial selection," with the aim of the "regeneration" of the stock. This artificial selection had to be characterized by a double objective: "prevent

23 Niceforo, "The cause of the inferiority of physical and mental characters in the lower social classes," 189.

24 Giuseppe Sergi, "Francis Galton," *Rivista di Antropologia*, 41, 1 (1911): 179–81. On Sergi's eugenics, see also Luca Tedesco, "'For a healthy, peace-loving and hardworking race': anthropology and eugenics in the writings of Giuseppe Sergi," *Modern Italy* 16, 1 (February 2011): 51–65.

25 Giuseppe Sergi, *Le degenerazioni umane* (Milan: Fratelli Dumolard, 1889), 204.

the increase of degenerates" and "diminish and make existing degenerates disappear."[26] The first aspect dealt with the protection of parents, guaranteeing them "useful nutrition," a job, "adequate rest" and the "necessary recreation." As for children, Sergi identified various categories. For the children of "serious degenerates" ("those in advanced states of tuberculosis, rachitis and scrofula") he hoped for "rapid elimination." For the children of less serious degenerates, it was necessary to distinguish between "criminal" or "pathological" characteristics of degeneration, and decide the treatment accordingly. For the "children of normal parents who may lack resistance," Sergi outlined a program of biosocial "regeneration," that included correct nutrition, "protection from the external environment" and, above all, education.

As for the second aspect—the diminution of existing degenerates—Sergi called for the abandonment of sentimentalism in the name of "prudent philanthropy." This signified the abolition of homeless night shelters and maternity shelters, condemnation to work through deportation to deserted isles, prohibition of marriage and prevention of illegitimate children.

In the first years of the twentieth century, Sergi's interest in the theories of hereditary transmission continued, opening up the pages of his journal *Rivista di Antropologia* [Anthropological review] to what could be considered the first steps of genetics in Italy.[27] In the nature/nurture debate, Sergi clearly opposed the Mendelian-Weismannian paradigm in the name of the Lamarckian principle of the hereditariness of acquired characteristics, attributing the role of prime motor in the modification of the germ plasm to environmental conditions (social, economic, etc.)[28] At the London Congress, Sergi contested Franz Boas' research on the role of the environment in the modification of the cephalic index of Italian immigrants in United States, but at the same time maintained that it was necessary to carry out "new and rigorous observations in order to be able to prove decisively that human heredity proceeds according to Mendel's theory."[29] Sergi's

26 Sergi, *Le degenerazioni umane*, 223.

27 See, for example, Cesare Artom, "Principi di genetica," *Rivista di antropologia* 19, 1–2 (1914): 281–410. On the initial phases of genetics in Italy, see Alessandro Volpone, *Gli inizi della genetica in Italia* (Bari: Cacucci, 2008).

28 Giuseppe Sergi, *Problemi di scienza contemporanea* (Milan: Remo Sandron Editore, 1904), 155

29 Giuseppe Sergi, "Variazione ed eredità nell'uomo," in *Problems in Eugenics*, 14.

skepticism regarding the risk of excessive "Mendelian" generalizations was connected to his definition of eugenics as a discipline suspended between biology and sociology, focused on the environmental role in hereditary transformations and on the centrality of "education."[30] The same positivist concept of progress was used to justify the eugenic power of education: "We must concede some value to educational power, if the education is rational and under the guidance of biology and that genetics of which we until now know very little and which has different interpretations according to different theories."[31]

Sergi's sociological environmentalism was devised however, not as an alternative, but as a complement to negative eugenics: "It is not enough to eliminate the human elements that carry hereditary pathological and degenerative defects in whichever way such elimination will be carried out; it is necessary first of all to take care of the healthy elements of the race."[32] Not surprisingly, in 1914, Sergi declared the social uselessness of "education of deficients": "The danger is not imaginary; because deficients contain the seeds from which criminals, prostitutes, the mentally unbalanced, madmen, vagabonds and beggars grow."[33]

It was a drastic position, which soon attracted accusations of cruelty from Paolo Mantegazza,[34] and from another eugenicist with Sergi in London: the noted psychiatrist, Enrico Morselli. Morselli, founder of the *Rivista di filosofia scientifica* [Review of scientific philosophy] and illustrious exponent of Italian anthropological psychiatry, offered an original interpretation of eugenics. This was based substantially on two elements: the methodological and epistemological centrality of psychiatry to the new discipline founded by Galton, and its intrinsic links with the "doctrine of race." At the London Congress, Morselli emphasized, first of all, the determinant role of psychology in eugenics, together with biology and sociology.[35] In fact,

30 Giuseppe Sergi, "L'eugenica. Dalla biologia alla sociologia," *Rivista italiana di Sociologia* 18, no. 5–6 (September–December 1914): 630.

31 Sergi, "L'eugenica. Dalla biologia alla sociologia," 632.

32 Sergi, "L'eugenica. Dalla biologia alla sociologia," 632–33.

33 Sergi, "L'eugenica. Dalla biologia alla sociologia," 632–33.

34 Claudio Pogliano, "Eugenisti, ma con giudizio," in Alberto Burgio, ed., *Nel nome della razza. Il razzismo nella storia d'Italia, 1870–1945* (Bologna: il Mulino, 1999), 426–27.

35 Enrico Morselli, "La psicologia etnica e la scienza eugenistica," *Rivista di psicologia* 8, no. 4 (July–August 1912): 290.

it was the work of psychiatry to analyze and explain the principle problem of eugenics, that is, that of "pathological heredity in families."[36] Morselli's nationalist outlook viewed Mendelism as pervaded with a "Germanic mentality affected by metaphysics"[37] and unable to explain the hereditary roots of the most relevant mental pathologies. Instead of Mendel's laws, Morselli preferred Bénédict-Auguste Morel's "theory of degeneration," as he claimed explicitly in 1915: "In substance, eugenics derives from the Morelian doctrine. [...] The exogenesis of illnesses is not only individual: it is becoming, through hereditary transmission, endogenesis, which is collective."[38] The entire "essence of eugenics" can be found in Morel's laws, not only in their scientific aspects, but also in the political and social ones. In fact, since Morel believed in "a well coordinated plan of prophylactic measures for physical and moral hygiene," Morselli felt that "if society does not want to adopt energetic means, such as the sterilization of degenerates, to arrest the physical decadence of the race and the perversion of its intellectual and moral qualities," then the "most competent eugenicists" should at least provide for education.[39]

Having identified the connection between psychiatry and eugenics, Morselli came directly to a differentialist "psychology of races." If the "destiny" of every race was marked out by the stage to which it had attained in the "psycho-physical hierarchy of man," and if the aim of each race could be identified as the "preservation of its own ethnic type," then eugenics must not only aim at the "realization of a uniform type of man," but instead must "vary its practical efforts according to the natural differentiation of work among races and nations during the bio-historical period."[40] In this way, eugenics became a "doctrine and practice of prophylaxis of the race,"[41] becoming a central mechanism in evolutionary anthropology and positivist racism. The "protomorphic races," that is, those that were "enormously inferior in morphological, physiological, psychological and sociological

36 Enrico Morselli, "L'eugenica e le previsioni sull'eredità neuro-psicopatologica," *Quaderni di Psichiatria* 2, no. 7–8 (July–August 1915): 322.

37 Morselli, "L'eugenica e le previsioni sull'eredità neuro-psicopatologica," 323.

38 Morselli, "L'eugenica e le previsioni sull'eredità neuro-psicopatologica," 324.

39 Enrico Morselli, "La rivendicazione delle leggi di Morel," *Quaderni di Psichiatria* 3, no. 11–12 (November–December 1916): 278.

40 Morselli, "La psicologia etnica e la scienza eugenistica," 292.

41 Morselli, "L'eugenica e le previsioni sull'eredità neuro-psicopatologica," 321.

aspects" were distinct from the "archimorphic"[42] races (black, white and yellow); the "fight for ethnarchy," that is, for racial superiority, would necessarily lead to the disappearance of the first group, and the assertion, within the second group, of the "leucodermic" groups. Morselli's "sociological optimism" even theorized a eugenic utopia of the "future man" or *Metanthropos*: "a perfect being in terms of anthropinic specifications, eurhythmic in the proportions of the body, with an advantageous stature, the head always erect, in possession of complete verticality without his current damage."[43] Endowed with "superior intelligence," the *Metanthropos,* thanks to technical-scientific progress, would dominate nature, but with a substantial harmony between the different ethnic groups.[44] If therefore, the course of history realized the perfection of humanity, eugenics would be called to support evolution, forcing the race to follow its destiny, until it reached the utopia of *Metanthropos.*[45]

From the point of view of eugenic policies, Morselli, although stressing the scientific weakness of eugenics, nevertheless proposed the introduction of an obligatory premarital examination, and maintained the importance of educating individuals to have a sense of responsibility towards the collective.[46] Morselli supported, although with some reserves, the education of the insane, and he insisted that it was important to prevent an approximate eugenics from cancelling the therapeutic work of psychiatry, by judging it "useless." This was one of the reasons why he strongly opposed Sergi's affirmations. Education of the insane, according to Morselli, if limited to those few "educable" individuals, who with hard work would be able to reach some awareness of self and the coordination necessary to carry out simple manual work, could not be considered, as Sergi suggested, as an open sore through which degenerative infection would penetrate the social body. There were very few re-educated feeble-minded who were able to re-enter the social circuit, and they usually regressed and ended up in asylums. In

42 Enrico Morselli, "La lotta per l'etnarchia," *Nuova Antologia* 151, no. 938 (1911): 232

43 Enrico Morselli, *Antropologia generale. L'uomo secondo le teorie dell'evoluzione* (Turin: Un. Tip. Ed., 1911), 1335.

44 Enrico Morselli, "Progresso sociale ed evoluzione," *Rivista italiana di Sociologia* 15, no. 5 (September–October 1911): 528.

45 Morselli, *Antropologia generale,* 1336.

46 Morselli, "L'eugenica e le previsioni sull'eredità neuro-psicopatologica," 331.

general, the number of "mediocre, deficient, retarded or insufficient feeble-minded people," who, "enhanced by orthophrenia," would be able to reach the threshold of marriage, had been greatly overestimated. In Morselli's opinion therefore, no orthophrenic "veneer" should prevent eugenics from keeping feeble-minded people at a discreet distance from marriage and procreation.[47] Instead, the problem was the economic and social cost of orthophrenia compared to eugenics. Wouldn't it simply be healthier and more economically advantageous to sterilize the "defectives"? To his friend, the physiologist Charles Richet, vice-president of the French Eugenics Society and 1913 Nobel laureate, who in 1919 had stressed the importance of a radical *sélection humaine*,[48] and to other European and North American supporters of "elimination by death," Morselli responded in 1923, in an essay in defense of a eugenics based not on authorized euthanasia, but on a wise program of social medicine.[49]

2. Pareto's Way: the Problem of the Elite

Between 1896 and 1906—that is in the chronological framework in which *Cours d'économie politique* [Course of political economy],[50] *Les systèmes socialistes* [Socialist systems][51] and the *Manuale di economia politica* [Manual of political economy][52] were published—Vilfredo Pareto developed an anthropological conception of social stratification, which constituted a significant connecting element between his economic and statistical analysis of the distribution of wealth (the well-known "income curve") and the political and sociological theory of circulation of the elite.[53]

47 Enrico Morselli, "Problemi di psicopatologia applicata. È socialmente utile l'educazione dei frenastenici?," *Quaderni di Psichiatria* 2, no. 5 (May 1915): 223–31.

48 Charles Richet, *La sélection humaine* (Paris: Alcan, 1919).

49 Enrico Morselli, *L'uccisione pietosa (l'eutanasia) in rapporto alla medicina, alla morale e all'eugenica* (Turin: Bocca, 1923).

50 Vilfredo Pareto, *Cours d'économie politique*, 1–2 (Lausanne: F. Rouge Lausanne, 1896–97) [ed. used: Turin: Bollati Boringhieri, 1961].

51 Vilfredo Pareto, *Les systèmes socialistes*, 1–2 (Paris: Giard & Brière, 1901–02) [ed. used, Turin: UTET, 1974].

52 Vilfredo Pareto, *Manuale di economia politica* (Milan: Società Editrice Libraria, 1906) [ed. used, Milan: EGEA-Università Bocconi, 2006].

53 On Pareto's social anthropology, see in particular Terenzio Maccabelli, "Social Anthropology in Economic Literature at the End of the 19th Century. Eugenic and Racial Explanations of Inequality," *American Journal of Economics and Sociology* 67, no. 3 (July 2008): 481–527.

The starting point of Pareto's anthropology can be identified in the concept of social heterogeneity, adopted with the intent of providing an explanation for the invariability and universality of the "income curve." The unequal division of wealth did not depend, Pareto argued, on chance or social organization as much as on the unequal distribution of "psychical and physiological qualities" of individuals: society, Pareto declared, was composed "of elements that are more or less different, not only in their evident characteristics, such as sex, age, physical force, health, etc. but also in their less easily observable, but not less important, characteristics, such as intellectual and moral qualities, activity, courage, etc."[54] In 1896, Pareto explicitly declared that he had largely adopted the "doctrine of social heterogeneity" from the writings of Otto Ammon and Georges Vacher de Lapouge,[55] important social darwinists of the late nineteenth century. However, although acknowledging his intellectual debt to *anthroposociology*, the economist rejected Ammon and Lapouge's racial typology and hierarchy, maintaining that the concept of race lacked an adequate level of scientific validity. When one talked about the Latin race, or the Germanic race, etc.—Pareto declared in *Cours*—one was adopting an ethno-linguistic meaning of the term "race," which had no meaning from a zoological point of view. Not surprisingly, in these same years, Pareto was involved in a controversy with Cesare Lombroso, namely about the problem of the scientific value of the concept of "race." Although acknowledging Lombroso's "genius," Pareto reproached him for his lack of "scientific rigor," in particular as regarded his use of the concept of race.[56]

For Pareto, saying "that there exist in society men who possess certain qualities in higher measures than others and saying that there exists a class of men absolutely *better* than the rest of the population is not the same thing."[57] Social hetereogeneity did not imply a racial hierarchy, but instead fed a complex mechanism of "social selection." Yet, as far as social selection was concerned, Pareto was still in debt to Ammon and Lapouge. In Pareto's discourse, selection was a necessary condition for the preservation of

54 Pareto, *Manuale di economia politica,* 94–95.

55 Vilfredo Pareto, "La curva delle entrate e le osservazioni del prof. Edgeworth," *Giornale degli Economisti* 13, no. 10 (1896): 443.

56 See Vilfredo Pareto, "L'uomo delinquente di Cesare Lombroso e Polemica col Prof. Lombroso," in Giovanni Busino, ed., *Écrits sociologiques mineurs* (Geneva: Droz, 1980), 111–25.

57 Pareto, *Cours d'économie politique,* 1–2, 392.

vital organisms. Every society contained "elements unfit to the conditions of life"[58] and if the activity of these elements was not contained within certain limits, then society would be "annihilated."[59] There were three possible measures, of decreasing effectiveness, that could help to avoid this danger: first, "destroy the unfit elements"; second, "prevent the harm they might do, either by instilling fear of the consequences of their actions, by taking away their liberty to act, or by placing them outside of society temporarily or indefinitely"; lastly, "amend them and modify their nature."[60]

The destruction of inferior elements, "widely used by breeders and farmers,"[61] was "incontestably effective"; but, in Pareto's opinion, inapplicable in human society. This was not only because of the "frightening abuses" that would result from its adoption, but above all because it contradicted that "sentiment of altruism and pity that is indispensable for a society to subsist and prosper." Therefore, it was necessary to substitute direct selection with "indirect" selection: according to Pareto, there were "many means, unfortunately very imperfect, with which inferior elements can be eliminated." Regarding the selective effectiveness of penal legislation (death penalty, exile, slavery for criminals) and of war, Pareto kept his distance from Ammon and Lapouge, expressing several reserves. Instead, the "most important selection" would be accomplished by the differential reproductiveness of different social classes. From a "qualitative" point of view—Pareto confirmed—a "higher death rate, particularly for infants, eliminated the weak and deformed in great numbers."[62] In addition, in the human species, the "death rate of adults eliminated many individuals who do not have enough self-control to resist depraved inclinations, at least when pushed to certain excesses." A man of weak character more easily became an alcoholic, accelerating "his degeneration and that of his descendents."

On a quantitative level, demographic selection had the additional advantage of acting on a much higher number of individuals, and according to Pareto, its effectiveness was clearly demonstrated by the immunizing effect of some diseases:

58 Pareto, *Les systèmes socialistes*, 1–2, 541.
59 Pareto, *Les systèmes socialistes*, 1–2, 541.
60 Pareto, *Les systèmes socialistes*, 1–2, 541.
61 Pareto, *Les systèmes socialistes*, 1–2, 542.
62 Pareto, *Les systèmes socialistes*, 1–2, 545.

> The races which were exposed to certain influences, to certain illnesses, ended up resisting them victoriously, precisely because the elements that did not resist were eliminated from selection. A race that is removed from these influences for a long time and is then suddenly exposed could be destroyed, because, not having operated the selection, this race will not have any resistance to the danger that threatens.[63]

Pareto's discourse was an attempt to reconcile selective action with "humanitarian" sentiment:

> The problem to resolve is the following: first of all, are there some means to diminish, reduce to a minimum, the number of birth of individuals unfit to the conditions of social life? Following from this, if it is not possible to decrease these births, if the increase of the number of these individuals becomes a danger for society, how can we eliminate them, with a minimum of error in their choice and in the suffering inflicted on them, and without overly upsetting the humanitarian sentiments, which it is useful to develop?[64]

To answer this question, Pareto first turned on the "philanthropists," the "reformers," the "humanitarians," and in general all people who denied the innate inequality of individuals, claiming to resolve eugenic problems with the tools of education, hygiene and social medicine.

Equally firm was his rejection of eugenic utopias based on rigid control of reproduction, carried out by public authority through coercive means. Although the principle of "appropriate choice of reproducers" in order to "improve the race,"[65] had been recognized "in every age" (and here Pareto cited Theognis of Megara, Plato, Plutarch, Campanella, and finally, Lapouge), the difficulty lay in the "means of execution, to apply this principle to the human species." The coercive eugenic means suggested by Lapouge were received by Pareto, in *Cours*, with "repugnance" and stigmatized as the final outcome of state socialism:

> We consider it useful to see where this path ends up, which, beginning with State monopolies and keeping on with obligatory unions, obligatory insur-

63 Pareto, *Les systèmes socialistes*, 1–2, 546.
64 Pareto, *Les systèmes socialistes*, 1–2, 554.
65 Pareto, *Les systèmes socialistes*, 1–2, 559.

> ance, collective organization of production and the constitution of a welfare state, is leading to the destruction of every individual initiative, the annihilation of all human dignity, and the reduction of men to the level of a flock of sheep.[66]

A measure such as an obligatory premarital certificate would become, as Pareto claimed in the *Systèmes socialistes*, the paradigmatic expression of "medical-hygienist madness."[67] Along the same lines, Pareto cited the case of the collectivist community constituted in Oneida, in the state of New York between 1847 and 1879, as an example of the non-viability of negative and coercive eugenics:

> This community voluntarily placed itself under rigorous discipline, and also practiced a community of goods. As was to be expected, this did not endure for long; after 33 years of existence, it had been transformed into a simple holding, and had no appreciable effect on the improvement of the race.[68]

In this context, Pareto's eugenic proposal rested on two fundamental points.

First, he proposed—citing in particular *La viriculture* of the liberal economist Gustave de Molinari[69]—"automatic internal forces," instead of "external coercive forces." Only a radical change in individual morals could contribute to improvement of the species:

> If the foresight of the results of the sexual act could become one of the principles of individual morals, it would be a great step towards the possible improvement of the species. This foresight would encourage the individual to not bring children into this world, if there were reasons to believe that he would transmit to them some illness or defect, and if there were no means to conveniently relieve it. G. de Molinari, with his usual elevated point of view, has dealt with the problem of automatic internal forces and their relationship with the improvement of the human species.[70]

66 Pareto, *Cours d'économie politique*, 1–2, 394.
67 Pareto, *Les systèmes socialistes*, 1–2, 561.
68 Pareto, *Les systèmes socialistes*, 1–2, 559.
69 See Gustave de Molinari, *La viriculture* (Paris: Guillaumin et Cie, 1897).
70 Pareto, *Les systèmes socialistes*, 1–2, 561.

Second, the theory of the circulation of the elite had, in Pareto's vision, a eugenic function. In a letter from December 1903, Pareto acknowledged the influence of Ammon and Lapouge in the formulation of his theory: "From Mosca I have taken nothing. I have however taken much, a great deal, and I have clearly stated so [...] from Ammon, and a little also from Lapouge. The scholars can moreover see how I partly dissent from them, and have added things."[71]

Despite this intellectual debt, Pareto radically rejected the racial typological description of the elite created by Ammon and Lapouge. The "chosen" subjects—he stated in *Cours*—are simply "individuals whose life activity is more intense" and such activity could "be good as much as bad."[72] No empirical evidence led to the identification of "aristocracy" in the dolicocephalic blonds of Ammon and Lapouge:

> Ammon and De Lapouge specify too much when they wish to give us the anthropological characteristics of this elite, these eugenic races, identifying them as dolicocephalic blonds. For now, this point remains obscure, and lengthy studies are still necessary before we will be able to establish whether the psychical qualities of the elite are translated into exterior, anthropometric characteristics, and before we can know precisely what these characteristics are.[73]

Therefore, it was not the morphological and racial differences that fed social selection, as much as the "invisible hand" of the market, the free competition between individuals:

> If, in fact, it were possible to recognize the character and attitudes of people from some exterior signs, such as form of the cranium, hair color, eye color, etc. the problem would be easily resolved. Unfortunately, these theories have uncertain relationships with reality, and for the moment, there are no other means to select men, if not that of testing what they can do, and putting them in competition, one against the other. This has a place, albeit a very imperfect one, in our societies, and history shows us that their progress is intimately linked to the extension of this use.[74]

---

71 Vilfredo Pareto to Giuseppe Prezzolini, 17 December 1903, in Pareto, *Epistolario. 1890–1923* (Rome: Accademia Nazionale dei Lincei, 1973), 1, 507. See also Pareto, *Manuale di economia politica*, 302 (with reference to Ammon and Lapouge) and Pareto, *Les systèmes socialistes*, 1–2, 131 (with reference to Ammon).

72 Pareto, *Cours d'économie politique*, 1–2, 416.

73 Pareto, *Les systèmes socialistes*, 1–2, 133.

74 Pareto, *Les systèmes socialistes*, 1–2, 342.

In particular, the dynamic interaction between economic conditions and movement of the population explained, in Pareto's view, the circulation of the elite on which the process of social selection depended. In one passage of *Cours*, which focused on the opposition between the stability of the income curve and the internal mobility of the defined area of the curve, Pareto compared the social organism to a living organism:

> The social organism in this way resembles a living organism. The external form of a living organism—for example, a horse—is almost always constant, but internally, there are ample and sundry movements. The blood circulation rapidly moves certain molecules; the processes of assimilation and of secretion incessantly modify the molecules of which its tissue is made up.[75]

The circulation of social "molecules" originated from the "influence of the economic conditions on the movement of the population."[76] In the inferior social strata—Pareto declared—"this influence is a powerful agent of zoological selection"; in the superior strata it "acts at times to limit the number of births, and, in this way, further becomes an agent of selection, facilitating the chosen subjects, born in the inferior strata, to access the superior strata."[77] In the introduction to *Systèmes*, Pareto further defined the role of "pressure of subsistence" on the dynamic of circulation of the elite:

> It seems highly probable that the rigorous selection that occurs in the inferior classes, above all for children, has a more important action. The rich classes have few children and almost all survive; the poor classes have many children and lose great numbers of those who are not particularly robust or well endowed. It is the same reason for which the perfected animal and plant races are very delicate, in comparison with the ordinary races.[78]

From Pareto's point of view, those who wished to persuade the higher social classes to have more children (the "ethicists"), and those who wished to reduce the infant mortality rate of the lower social classes (the "humanitarians") were both mistaken. Both solutions ended in altering the perfect eugenic equilibrium of the circulation of the elite:

75 Pareto, *Cours d'économie politique*, 1–2, 397.
76 Pareto, *Cours d'économie politique*, 1–2, 416.
77 Pareto, *Cours d'économie politique*, 1–2, 416.
78 Pareto, *Les systèmes socialistes*, 1–2, 134.

> If the rich classes in our societies were to have many children, it is likely that almost all would survive, even the frailest and least endowed. This would proportionately grow the degenerate elements in the superior classes and retard the access of the inferior classes to the elite. If selection were to no longer exercise its effects on the inferior classes, these would cease to produce elite members, and the average quality of society would be considerably lessened.[79]

Differing from Ammon and Lapouge, Pareto believed that the lower social classes did not represent a threat to the aristocracy, but rather constituted a reservoir for the continuous formation of the elite: the inferior classes, and in particular, the "rural classes," were the "crucible in which, in shadow, the elite of the future are born. These are like the roots of a plant, while the elite is the flower. This flower fades and must fade, but it is immediately replaced by another, if the roots are not damaged."[80] Experience in fact demonstrated that within the inferior classes, individuals existed who were better endowed than those in the superior classes: "Whoever has spent some time among the manual workers knows that one encounters among them individuals who are more intelligent than this or that scientist, laden with academic titles."[81] And it was this—Pareto controversially emphasized—that made Candolle's and Galton's statistics on the genealogy of men of genius unreliable. In an attempt to explain how "first class elements" could come from the rural classes, Pareto introduced a biological hypothesis which was to have a notable afterlife in Italian eugenics: "It could be that the same fact that the rural classes develop their muscles and rest their brains has precisely the effect of producing individuals who are able to rest their muscles and excessively work their brains."[82] Consequently, preventing the circulation of the elite through the introduction of a rigid caste system could not lead to anything but "decadence":

> Modern authors, in the search for something new, have developed a great love for the institution of the Indian caste system. These authors cannot explain how this excellent system has not prevented the Indians from becoming prey

79 Pareto, *Les systèmes socialistes*, 1–2, 135.
80 Pareto, *Les systèmes socialistes*, 1–2, 134.
81 Pareto, *Cours d'économie politique*, 1–2, 396.
82 Pareto, *Les systèmes socialistes*, 1–2, 135.

> to numerous conquerors, lacking all caste, nor how some thousands of British were enough to maintain British dominion over a country that counts around two hundred million inhabitants.[83]

At the First International Eugenics Congress in London, many of the Italian contributions revealed a clear Paretian influence. The most transparent example was undoubtedly the economist Achille Loria, who—reprising his previous criticism of Otto Ammon's anthroposociology[84]—contested the relationship between the economic elite and the biological elite:

> Economic superiority is by no means an index of superior psycho-physical aptitudes, whether because many of those who now possess that position do not acquire it by virtue of the possession of elevated mental capacity, or because all the others who have inherited these positions from preceding possessors are completely devoid of such aptitudes. Thus, economic superiority cannot in any case be assumed to be the measure or reflection of psycho-physical superiority.[85]

According to Loria, only this argument could inspire a "decisive" and "rational" eugenics,[86] which would not nourish classist prejudices, but on the contrary, would lead to "a minute and positive examination of individual characters."[87]

Roberto Michels' contribution also reflected on elite theory. Although an exponent, in those years, of the nascent neo-Malthusian movement, at the London Congress, socialist Michels propounded the general criteria of a eugenics based not so much on birth control as on the organization of the mass party. On this latter topic Michels had focused a few years earlier his most famous essay *The sociology of the political party in modern democracy*, a fundamental contribution, along with Gaetano Mosca's and Vilfredo Pareto's works, to the elite theory of political power. According to Michels,

83 Pareto, *Cours d'économie politique*, 1–2, 416–17.

84 See Achille Loria, "L'antropologia sociale," in Achille Loria, ed., *Verso la giustizia sociale—(Idee, battaglie, apostoli)* (Milan: Società Editrice Libraria, 1908), 562–73.

85 Achille Loria, "The psycho-physical elite and the economic elite," in *Problems in Eugenics*, 181–82.

86 Loria, "The psycho-physical elite and the economic elite," 183.

87 Loria, "The psycho-physical elite and the economic elite," 183. During the Congress, Loria's position garnered the approval of the anarchist philosopher Kropotkin.

the organization of modern parties favored the selection of a new psycho-anthropological type—that of the political leader—characterized by oratory ability and physical good looks, and additionally, by a series of psychological endowments:

> Firstly, energy of will which enables them to dominate weaker characters; secondly, superiority of knowledge, which compels respect; "catonian" depth of conviction, a force of ideas which often borders on fanaticism and which, from its intensity, commands the admiration of followers; self-confidence pushed even to the point of self-conceit, which has the power, however, of being communicated to the mass; in certain rarer cases, finally, goodness of heart and disinterestedness.[88]

Selecting a form of superiority not linked to income, but to physical and psychological gifts, party organization had a double eugenic function: it guaranteed self-made men from the working classes social access to leadership roles in worker movements; and it favored the affirmation of socialist leaders, indirectly feeding the realization of a social policy which would be more effective eugenically, as it would reduce the economic-social inequality and re-establish "the struggle for life on a more healthy and more natural basis, and allow a greater quantity of men to occupy in society the place to which their special and inborn qualities and their cleverness and energy give a kind of moral and logical right."[89]

Not surprisingly, Michels dedicated a collection of articles entitled *Problems in applied sociology* to Pareto, which was published in German in 1914, and then in Italian in 1919. The first chapter of this essay was specifically devoted to eugenics. The proletariat (or better, the "people"), because of its numeric consistency and the "sad biological conditions in which it finds itself," should be, according to Michels, the subject of specific eugenic study and activity. Product of a synthesis between biology and political economics, eugenics had the crucial job of understanding to what point the inferiority of the poor classes derived from an "unyielding anthropological base"[90] or whether it was a product of economic consequences.

88 Roberto Michels, "Eugenics in party organisation," in *Problems in Eugenics*, 234–35.

89 Michels, "Eugenics in party organisation," 237.

90 Michels, *Problemi di sociologia applicata* (Turin: Bocca, 1919), 4.

Eugenics' objective therefore did not consist in the "artificial production of supermen,"[91] but rather in the "biological improvement of the race," pursued through two principle activities. These were, first, negative measures discouraging the "physically unfit or morally inferior elements" from reproducing (for example, the obligatory sterilization of carriers of hereditary illnesses and of sexual criminals), and, secondly, a social reform policy, aimed at "improving the economic and social conditions of mankind." In particular, it was this last aspect that Michels identified as the "pivot of eugenic work."[92] Not surprisingly, this last form of positive eugenics was to mark in Michels' progressive shift from socialism to fascism. No longer a supporter of birth control and sterilization, but of the eugenic and demographic value of Italian emigration, in the 1920s, Michels did not hesitate to protest against E.W. MacBride, vice-president of the British Eugenics Education Society, guilty of having defined the Southern Italians as a "ethnic group close to Negroes."[93]

Along the same lines of Pareto's anthropology, but with a level of scientific originality far superior to that of Loria or Michels, we also find Corrado Gini's eugenics. Gini's eugenic discourse could not be adequately understood, if not within the systematic process of statistical and demographic revision with which, between 1907 and 1912, he treated the problem of the differential birth-rate of the social classes.

Already in his graduating thesis, published in 1908 with the title *Il sesso dal punto di vista statistico* [Sex from a statistical point of view],[94] Gini dealt with the issue of the "circulation of social classes and populations," introducing for the first time the hypothesis that the cause of differential birth rate could be reduced to the environmental influence on "germinal elements." Animals kept in captivity demonstrated, according to Gini, that "the maturation of the germinal elements is obstructed by captivity, as it impedes muscular activity, makes the environment uniform, and greatly reduces the reactions of the organism."[95] In the same way, in the human

91 Michels, *Problemi di sociologia applicata,* 14

92 Michels, *Problemi di sociologia applicata,* 14.

93 Roberto Michels, "Sulla teoria e sulla pratica dell'Eugenica," *Echi e Commenti* 3, no. 27 (1922): 14.

94 In 1907, the thesis was awarded the Vittorio Emanuele Prize for social and political sciences at the University of Bologna.

95 Corrado Gini, *Il sesso dal punto di vista statistico. Le leggi della produzione dei sessi* (Milan: Remo Sandron, 1908), 454.

species, the "development of sex" appeared favored by those conditions—muscular work, "active rural life," sport—that "command in the organism, and through it, in the germinal cells, a lively reaction, which is obstructed on the other, by the opposite conditions of health and tranquility." This physiological reason could explain, therefore, in Gini's view, the lesser prolificacy of the aristocracy compared to the lower social classes and the decreasing birth rate of the "white races":

> If the stimulus to procreation has lost its intensity, that is due above all, I believe, to the diffuse economic well-being, the decreased physical activity, the broadening and accentuating of that complex of characteristics that we call civilization, the final limit of which is a beatific state, in which every desire is sated and every effort suppressed.[96]

In October 1908, just a few months after the publication of *Sesso,* Gini gave a contribution to the Second Meeting of the Italian Society for the Progress of Science, titled *The different growth of the social classes and the concentration of wealth.* This was later published, in 1909, in *Il Giornale degli Economisti* [The economists' journal]. This essay explicitly proposed the objective of providing the "statistical proof" of the different growth of the social classes. In researching the probable causes of this demographic phenomenon, Gini challenged Pareto's *Systèmes Socialistes,* claiming that it had exclusively emphasized the action of natural selection, without giving enough attention to the role of the environment. On the basis of De Vries' mutation theory, Gini again accentuated the importance of environmental influence:

> In a bad environment, a selected race will worsen, in spite of the most active selection; in a good environment, a race improves, even if subjected to reverse selection. This phenomenon has been ascertained for plants, and seems to hold true for all organisms, and, in particular, for man.[97]

As in *Sesso,* the cause of the "lesser reproductive activity" of the rich compared to the poor was here attributed to their "lower force of sexual instinct."

96 Gini, *Il sesso dal punto di vista statistico,* 458–59.

97 Corrado Gini, "Il diverso accrescimento delle classi sociali e la concentrazione della ricchezza," *Giornale degli Economisti* 2, no. 37 (January 1909): 35.

This conclusion, Gini argued, was "in harmony with the facts of biology, zootechnics and medicine, which demonstrate how the sexual functions are favored, in superior species, by a life of physical fatigue, and in inferior species manifest themselves in alternate generations, under the stimulus of unfavorable environmental conditions."[98]

Having delineated the different growth of the social classes as a "biological law valid for all human societies,"[99] Gini listed the possible applicable consequences of this theoretical result. First of all, Pareto's circulation of the elite was substantially confirmed, although Gini preferred to refer to it as "social exchange," because, on a demographic plane, the upward current did not correspond to a parallel downward current. Also from a eugenic point of view, Pareto's ideas were reaffirmed by Gini, in direct opposition to Karl Pearson's eugenic arguments. In contrast to the beliefs of British mainline eugenics, the poor classes did not in fact constitute a biological threat, but rather a necessary resource:

> The great mass of population is constituted by those whom we call the poor classes; from them, as if from an immense breeding ground, the elect originate, in relatively small numbers, either through personal merit or through force of circumstances. They originate, arise, shine and are extinguished, like rockets; only insignificant traces fall to earth.[100]

A further consequence was relevant for the anthropological field: following the mechanisms of social exchange, the physical and psychological characteristics of the lower classes would be extended to the rest of the population, contributing to the change of their anthropological and cultural characteristics.

Finally, in the economic field, Gini proposed an alternative to Pareto's wealth distribution curve (or Pareto's law), according to which the income distribution was constant in space and time. Gini's new index was based on a mathematical method that took into account not only the number of recipients within the various classes of income or fortune, but also the total amount of their income or fortune. Gini's index, δ, described a general ten-

98 Gini, "Il diverso accrescimento delle classi sociali e la concentrazione della ricchezza," 37.
99 Gini, "Il diverso accrescimento delle classi sociali e la concentrazione della ricchezza," 33.
100 Gini, "Il diverso accrescimento delle classi sociali e la concentrazione della ricchezza," 38.

dency to the concentration of income: it was the first outline of the well-known index, still today referred to as the "Gini index."[101]

Describing the differential growth of the social classes as a universal biological law with several eugenic, anthropological and socio-economic implications, Gini distanced himself from Pareto's influence and paved the way towards the vast research program in demography, biology, statistics and eugenics that he would progressively realize in the following years.

Between 1908 and 1912, Gini extended his methodological reflections to the statistical phenomenon of concentration, creating a wider and more abstract statistical theory of distributions and relations. In 1908, as seen above, the index δ was introduced to analyze an empirical problem, linked to the economic consequences of social exchange. In 1910–11, Gini proposed to "find indices of distribution and relation amongst quantitative phenomena, with enough sensitivity and applicability to usual statistic data, without excessively laborious calculations and without hypotheses too distant from reality."[102] The objective was the development of polyvalent statistical instruments, to utilize not just in economic analysis, but also within the plurality of biological and demographic phenomena, such as, for example, matrimonial prolificacy; the relationship between matrimonial productivity, duration of marriage, and age of the spouses at time of marriage and time of death; the relationship between legitimate fertility and the duration of the marriage or age of the mother.

Two years later, in 1912, Gini introduced a new statistical procedure—the mean difference—to study the variability of quantitative characteristics ("variability index") and qualitative ("mutability index"). The intent was to provide appropriate methodological instruments for the application of statistics to "biology, demography, anthropology and economy."[103] The examples listed by Gini regarding some possible uses of the mutability index were, in this sense, quite explicit:

---

101 On this topic, see Jean-Guy Prévost, *A Total Science. Statistics in Liberal and Fascist Italy* (Montréal: McGill-Queen University Press, 2009).

102 Corrado Gini, *Indici di concentrazione e di dipendenza* (Turin: UTET, 1911), 5. The book was part of the prestigious *Biblioteca dell'Economista* series. A synthesis was presented the previous year at the 3rd meeting of SIPS.

103 Corrado Gini, "Variabilità e mutabilità. Contributo alla studio delle distribuzioni e delle relazioni statistiche," *Studi economico-giuridici pubblicati per cura della Facoltà di Giurisprudenza della R. Università di Cagliari* 3, part 2 (Bologna: Tipografia Cuppini, 1912), 17; offprint.

> Having a measurement of homogeneity for eye or hair color of the inhabitants of a region is no less important, to be able to make a judgment on the purity of races, or on the influence of the environment on human characteristics and other anthropological problems, than having a measurement of homogeneity for certain quantitative somatic characteristics, such as thoracic perimeter, stature, weight, etc.
> Similarly, it could be interesting, in many aspects, to have a measurement of homogeneity for religion, for marriage status, for nationality, for profession etc. of the citizens of a nation; it could be interesting to determine the homogeneity of marriages celebrated by day of the week, or month of the year, the homogeneity of births per month of the year and whatnot.[104]

The methodological theory of the variability indexes on one hand, and the statistical and demographic investigation into the dynamics of social exchange on the other, merged in 1911–12, to become a general theory about the cyclical evolution of nations. The pillar of this theory was the differential fertility and birth rate between social classes.[105] Presented for the first time in 1911 at the Minerva Society in Trieste, Gini's theory provided a scientific response to the nationalist and irredentist anxieties of the Trieste Italians, menaced by the "invasion of the Slavs."[106]

Gini opened the conference with a provocative question: why should "a race rich in intelligence, wealthy, nourished on noble traditions, animated by high ideals" (that is, the Italians) not be able to triumph over "another race, intellectually more limited, economically poorer, for whom the glories of the past can not be a prod for glories in the future"[107] (that is, the Slavs)? Gini's reply was contained in the exposition of his cyclical theory of nations, which can be summarized as follows:

104 Gini, "Variabilità e mutabilità," 113–14.
105 Corrado Gini, *I fattori demografici dell'evoluzione delle nazioni* (Turin: Bocca, 1912).
106 Gini, *I fattori demografici dell'evoluzione delle nazioni*, 3.
107 Gini, *I fattori demografici dell'evoluzione delle nazioni*, 3.

Birth of the nation	Evolution of the nation	Death of the nation
Initial low social and economic differentiation; general growth of the population; high prolificacy in the higher classes, but overtaken by higher prolificacy in the lower classes	Reduction of the death rate is more intense than reduction of the birth-rate	Reduction of the birth-rate is greater than the reduction of the death rate; greater social exchange and demographic reduction of the lower social classes
Low national wealth; high social solidarity	Increase of national wealth, well-being and social tranquility	Diminution and concentration of national wealth; crises; social conflict
Spirit of initiative and social energy	Bureaucratization	Corruption of public life
Increase in the countryside population	Depopulation of the countryside, and urbanism	Depopulation of the city
The lower classes spend their demographic energy on war and emigration	Increase of foreign immigration	Social ascent of foreign immigrants; ethnic modification of the nation

From the point of view of the theory, Italy—which was characterized by an excess of emigration—risked a future of decadence: incessant emigration partly reduced and partly transformed the lower classes of the population, from which the nation was renewed, making social exchange progressively difficult and leading to a period of "demographic and military senescence first, then economic senescence, from which it will be extremely difficult to re-emerge."[108]

Gini's contribution at the First International Eugenics Congress further developed this statistical and demographic outlook. Following in the wake of Pareto, Gini's eugenics radically opposed the Anglo-American position. Instead of artificial selection, he proposed the return to the state of nature; instead of biological protection of the elite, the necessity of social exchange; rather than neo-Malthusianism, a pronatalist policy. In Gini's view, the task of eugenics did not consist in selecting the perfect race, but rather in re-introducing to "civilized society" those "primitive customs" regulating, in the most natural conditions possible, the procreation and raising of children.[109]

108 Gini, *I fattori demografici dell'evoluzione delle nazioni*, 105.

109 Corrado Gini, "Contributi statistici ai problemi dell'Eugenica," *Rivista italiana di Sociologia* 16, no. 3 (May–August 1912): 385.

Having identified "prolificacy" as the primary biological value of the species, Gini listed several factors of counter-selection: the reduced distance between births; the recourse to artificial breast milk; the advanced age of marriage; "the systematic defense of the weak and degenerate." However, the "low reproduction of the higher classes," the nightmare of Anglo-American eugenics, could not be considered as a degenerative factor. In fact:

> Until it is shown that the children of the lower classes—if they were brought up from conception in the same surroundings as the children of the higher classes—would turn out inferior to these, it is not proved that, by stimulating the reproductiveness of the higher classes, one would improve the race more than by leaving their place to be occupied by the children of the working class.[110]

The elite were not degenerate in themselves, but in the fact that their germ plasm was more evolved, and therefore would be the first to decay. Taking the theory of decadence of the germ plasm from Nägeli, and in part from Lamarck, Gini implicitly criticized Mendelian-Weismannian hereditary determinism. On this theoretical basis, he positively welcomed the rapid crisis of aristocratic blood:

> Artificially to stimulate the reproduction in the higher classes, and check that of the lower ones would be equivalent to trying to improve society by increasing the duration of the life of the old and preventing new generations from taking their place.[111]

The renewal of the higher classes by the action of the lowest social classes constituted a biologically justified "normal phenomenon of human societies."[112] While it might generate social conflict, it would not have negative results for the "physical and intellectual characters of the race." The best means for improving the race in Gini's eugenics can be easily summarized: greater intervals between births, natural breastfeeding, earlier marriages, and obstacles to the reproduction of the weak and degenerate.

Seen in the light of the cyclical theory of nations, Gini's eugenics presented two possible interpretations. A first *particular* aspect reconnected

110 Gini, "Contributi statistici ai problemi dell'Eugenica," 381.
111 Gini, "Contributi statistici ai problemi dell'Eugenica," 383.
112 Gini, "Contributi statistici ai problemi dell'Eugenica," 384.

the rise of eugenics to the last stage of society—that of senility—as an extreme attempt to slow down decadence:

> Nations would produce, at the beginning of their civilization, stronger, more intelligent, and happier children; but these advantages would slowly be lost with the progress of the nation and with the rise of marriageable age. Progress in medicine and hygiene, greater care at home, a higher and more intensive and rational education would be more than sufficient to compensate for such physiological impoverishment of the race: but the latter would show itself when progress of this kind came to a standstill, and would contribute towards the decadence of the nation. It is a common custom to speak of young populations and of old populations; and we all feel that in such a phrase there is more than a simple metaphor.[113]

A second *general* meaning led to the interpretation of the life cycle of the population as a sort of natural eugenic process. The biological metaphor adopted by Gini in 1912 to describe the eugenic role of emigration was, from this point of view, enlightening. Like the germinal cells of the organism, emigrants also constituted the "least differentiated" and "most reproductive" elements of the population they were a part of.[114] Although emigration determined the demographic and economic decadence of a nation, nevertheless it was at the same time the "cause of its regeneration in the future nations."[115] Even if the European nations were destined to "extinguish themselves on the shores of Europe," they would—thanks to emigration—be revived "in blood, in language, in thoughts, in the sentiments of the populations of entire new continents."[116]

On his return to London, Gini had the opportunity to further develop his reflections in the eugenic field. In fact, during the prestigious occasion of the inaugural lecture for his course of statistics at the University of Padua in 1913, Gini identified Quételet's *homme moyen* as the "general type of the race," intended as a statistical generalization and aesthetic ideal.[117]

113 Gini, "Contributi statistici ai problemi dell'Eugenica," 370.
114 Gini, *I fattori demografici dell'evoluzione delle nazioni*, 107.
115 Gini, *I fattori demografici dell'evoluzione delle nazioni*, 107.
116 Gini, *I fattori demografici dell'evoluzione delle nazioni*, 139.
117 Corrado Gini, "L'uomo medio," *Giornale degli Economisti e Rivista di Statistica* 48, no. 1 (January 1914); offprint.

Regarding Quételet's model, Gini believed it could not be considered valid as a "type of physical equilibrium,"[118] since this would contradict Darwinian evolutionism. Nor could it be considered the "moral ideal," because this would negate every "stimulus to progress."[119] At any rate, it represented yet another point of reference as a logical construction:

> The average man, and the average soldier, and the average child, and the average newborn, as they respond to the needs of the systematic average, also respond to the facts: respond to the facts, meaning, as all generalizations based on a statistical analysis can and must respond to the facts, that is, not in single cases, but in mass cases.[120]

But beyond a logical-mathematical point of view, the "average man" also constituted an effective "aesthetic ideal."[121] In highlighting the difference between the "average" man and the "handsome" man, considering different races, Gini's cultural relativism revealed a precise racial hierarchy:

> What is there more repugnant for us than the long, pug nose of the Negroes or the Australians, and more distant from the long, straight nose of the Anglo-Saxons? Therefore, when the English disembarked in Australia, the indigenous there derided them for their sparrow-hawk noses. And what is uglier than their swollen lips? [...] And what is said can be repeated for the eyes: eyes which to us seem swine-like appear wonderful to oriental populations, and their natural length industriously lengthened still more with paint is disgusting for us.[122]

According to Gini, the tendency to imitate the aesthetics of the superior social classes and races influenced the formation of the aesthetic ideal:

> The fact that all the populations who have come into contact with European civilization have, sooner or later, more or less completely abandoned their national costume, to adopt our monotonous clothing, is further proof of the influence that the imitation of a superior race exercises on the formation of the aesthetic ideal.[123]

118 Gini, "L'uomo medio," 13.
119 Gini, "L'uomo medio," 23.
120 Gini, "L'uomo medio," 10.
121 Gini, "L'uomo medio," 22.
122 Gini, "L'uomo medio," 22–23.
123 Gini, "L'uomo medio," 24.

Conceived as a kind of universal aesthetic ideal, in Gini's approach, the "average man" became a sort of "pendulum," swinging, driven by the ethnic tendency to stylize racial characteristics on the one hand, and the imitation of superior races, on the other: "In the formation of our aesthetic ideal, the average man acts as a centripetal force, while the tendency to stylize race and sex or to imitate superior models acts as a centrifugal force in many ways."[124]

3. The Italian Committee of Eugenic Studies

The Italian participation at the International Eugenics Congress in London had an immediate corollary, the next year, in the constitution of the first Italian Committee of Eugenic Studies (*Comitato Italiano per gli studi di Eugenica*).[125]

Giuseppe Sergi and Alfredo Niceforo promoted the new committee, during the sitting of 21 March 1913 of the Roman Society of Anthropology (*Società Romana di Antropologia*). The scope of the committee was that of

> studying the factors that could determine the progress or the decadence of the race, both in terms of physical aspect and psychical aspect, carrying out, for example, research on the normal or pathological heredity of characteristics, on environmental influence and the life regime of parents on the characteristics of the children, on the importance of the momentary conditions of the organism in the act of reproduction, or the environment in which the new organism develops.[126]

At the beginning of April, the Board of Directors of the Roman Society of Anthropology nominated an internal Commission to create a program and gather support. The Commission consisted of Giuseppe Sergi (president), Umberto Saffiotti (secretary), Antonio Marro, Alfredo Niceforo, Corrado Gini and Giovanni Mingazzini. The first general meeting of the Italian Committee of Eugenic Studies took place on 17 November 1913, with 16 members present, as indicated in the first—and only—issue of the minutes, published in the *Rivista di Antropologia,* which had

124 Gini, "L'uomo medio," 21.

125 On the Comitato Italiano di Studi Eugenici, see Mantovani, *Rigenerare la società,* 75–85.

126 "Atti del Comitato Italiano per gli Studi di Eugenica," *Rivista di antropologia* 18 (1913): 543–44.

become the organ of the Committee.[127] On this occasion, the statute was approved, nominating the Committee Board of Directors for the 1914–15 term (president, Giuseppe Sergi; vice-president Sante De Sanctis; secretary, Umberto Saffiotti) and promoting (particularly by Corrado Gini) the constitution of an Italian section in the International Catalogue of Eugenic Studies, planned in London in August 1912.[128] On the 17 November 1913, the committee counted 83 members. Compared to the London Congress, the members included experts in physical anthropology (and related disciplines, such as psychiatry, legal medicine and military medicine), demography and statistics; the most interesting novelty was represented by the many members coming from clinical medicine, particularly gynaecologists and hygienists.[129]

Although a part of the Roman Society of Anthropology, headed by Giuseppe Sergi, the Italian Committee of Eugenic Studies was quickly influenced by the figure of Corrado Gini. It was Gini who kept in contact with the Permanent International Eugenics Committee in London, and participated at its first meeting in Paris in August 1913. It was again Gini who promoted, in 1914, the first—and only—scientific initiative of the young Italian committee.

The project consisted of a statistical survey of the members of the Italian academic system (people "who excelled due to physical or psychical characteristics"), in order to evaluate the relationship between order of birth, bio-

127 See "Atti del Comitato Italiano per gli Studi di Eugenica," 543–46.

128 See "Atti del Comitato Italiano per gli Studi di Eugenica," 550–52. In particular, every member was sent a circular that requested them to insert their own publications into a predefined bibliographical scheme, subdivided into "theoretical eugenics" and "applied eugenics," and to send a copy to Corrado Gini's Padua university address.

129 See "Atti del Comitato Italiano per gli Studi di Eugenica," 546–49. The list comprised the following categories:

- anthropologists: Giuseppe Sergi, Sergio Sergi, Fabio Frassetto, Vincenzo Giuffrida-Ruggieri, Enrico Tedeschi;
- legal physicians: Lorenzo Borri, Mario Carrara, Antonio Cevidalli, Salvatore Ottolenghi;
- military physicians: Placido Consiglio, Ridolfo Livi,;
- psychiatrists: Paolo Amaldi, Carlo Ceni, Ugo Cerletti, Ettore Fornasari di Verce, Augusto Giannelli, Giovanni Marro, Giovanni Mingazzini, Giuseppe Ferruccio Montesano, Gian Battista Pellizzi, Augusto Tamburini;
- psychologists: Giulio Cesare Ferrari, Sante De Sanctis, Federico Kiesow;
- clinical physicians (particularly gynaecologists): Mariano Carruccio, Giacomo Cattaneo, Achille De Giovanni, Stefano delle Chiaje, Luigi Mangiagalli, Ernesto Pestalozza, Gaetano Pieraccini, Luigi Pagliani, Tullio Rossi-Doria, Pasquale Sfameni, Pietro Sirena, Pasquale Sorgente, Giuseppe Vicarelli, Giacinto Viola;
- physiologists/zoologists/anatomists: Cesare Artom, Silvestro Baglioni, Paolo Enriques, Carlo Foà, Luigi Luciani, Mariano Patrizi, Achille Russo, Guglielmo Romiti;
- jurists: Guido Cavaglieri, Raffaele Garofano, Raffaele Majetti;
- statisticians: Corrado Gini, Alfredo Niceforo, Franco Savorgnan;
- economists: Achille Loria, Roberto Michels.

logical value of offspring and prolificacy of families. Gini's statistical inquiry was based on 445 responses given to a questionnaire sent to all the professors in Italian universities, assumed to be samples of eugenic value. The results appeared to only partially confirm Gini's theories: in fact the effective number of university professors who were first-borns was greater than predicted in the theory, but for those who were higher in the order of birth, the number of professors was much lower.[130] It was Gini's intention that the Committee extend the survey to other categories, "in literary, artistic, military, bureaucratic, commercial, banking, political and all sporting fields."[131] But the initiative, judging by current archival evidence, was never carried out.

Gini's unique approach to eugenics influenced the studies of several of his students,[132] but was strongly criticized by the gynecologist Felice La Torre, who contested the statistical methodology, claiming instead the eugenic role of prenatal care and assistance of pregnant women.[133] Gini did not hesitate to reply: it was not gynecology, but genetics and statistics that must be the pillars of eugenics.[134]

The conflicting positions of Gini and La Torre, published in the pages of *Rivista italiana di Sociologia* [Italian review of sociology] in 1915, were nevertheless the last, brief flame of activity of the Committee of Eugenic Studies. Its dissolution coincided with the entrance of Italy in the First World War.

---

130 Corrado Gini, "Nuove osservazioni sui problemi dell'eugenica. La distribuzione dei professori delle Università secondo l'ordine di nascita," *Rivista italiana di Sociologia* 18, no. 2 (March–April 1914): 214.

131 Gini, "Nuove osservazioni sui problemi dell'eugenica," 215.

132 See Marcello Boldrini, "Sulle famiglie con pazzi e sulla variabilità del primo nato – ricerche statistiche," *Rivista di Antropologia* 19, no. 1–2 (1914): 411–31; Giovanni Dettori, "Di alcuni caratteri dei neonati secondo l'ordine di generazione e l'età della madre," *Rivista di Antropologia* 19, no. 1–2 (1914): 443–572.

133 Felice La Torre, "I fondamenti dell'eugenica," *Rivista italiana di Sociologia* 19, no. 2 (March–April 1915): 196–218.

134 Corrado Gini, "Genetica e statistica rispetto all'eugenica," *Rivista italiana di Sociologia* 19, no. 2 (March–April 1915): 218–22.

CHAPTER II

EUGENICS AND DYSGENICS OF WAR

At the start of the nineteenth century, the dream of a Greater Italy, with a leading role in the construction of modern civilization, was resumed by political movements that rebelled against Giolitti's liberal so-called *Italietta* [a petty Italy]. The imperialistic nationalism, the intellectual group of *La Voce*, futurism, and revolutionary syndicalism all shared the myth of a national regeneration, and transformed it into a project of total, spiritual, cultural and political revolution, to demolish the liberal regime.

Many interventionists conceived Italy's participation in the First World War as a decisive stage for the regeneration of the Italians through the test of war. Interventionism became a factor of fusion between the myth of revolution and the myth of the nation, producing the conversion of many revolutionary left-wing syndicalists or socialists to nationalism, as in the case of Benito Mussolini. From this, a new revolutionary nationalism sprang out, that conceived the war and revolution as a national palingenesis, which would radically renew not just the political, economic and social order, but also the culture, mentality and character of the Italians.[1] In this context, the Great War also contributed to a notable development in Italy of the eugenic debate.[2]

---

1 Emilio Gentile, "The Myth of National Regeneration in Italy: From Modernist Avant-Garde to Fascism," in Matthew Affron and Mark Antliff, eds., *Art and Ideology in France and Italy* (Princeton: Princeton University Press, 1997), 25–45. See also Emilio Gentile, *La Grande Italia: The Rise and Fall of the Myth of the Nation in the Twentieth Century* (Madison: University of Wisconsin Press, 2009).

2 On the relationships between the First World War and eugenics, see also Weindling, *Health, Race and German Politics*; Soloway, *Demography and Degeneration*; Schneider, *Quality and Quantity*; Broberg and Roll-Hansen, eds., *Eugenics and the Welfare State*; Marius Turda, "The Biology of War: Eugenics in Hungary, 1914–1918," *Austrian History Yearbook* 40 (2009): 1–27 and *Modernism and Eugenics*, 40–63.

Eugenics was involved in debates concerning not only the health of the nation and the protection of society, but also, ultimately, racial supremacy and survival. The wartime effort was accompanied by the demonization of the external enemy, supported by anthropological and biological arguments to explain the differences between "Latin" and German civilization. The nationalist rhetoric described the Germans as barbaric butchers, brutal Huns, a sub-human race. At the same time, the Germans were also presented as a metaphor of modernity, characterized by an inclination to abstract thought, a morbid attraction to material riches, a lack of interior harmony and moral scruples, and by a dramatic scission between spirit and body, rendering them incapable of rising above animal sensuality. This German materialism, egoistic hedonism, and individualism was contrasted with the "Latin genius," expression of racial superiority, power of the spiritual element, sense of limit, and virility.[3] Reinforced by the First World War, in the following decades, the "Latin" myth became one of the most distinctive traits of Italian eugenics.

1. War as Counter-selection

Most of the important figures of Italian eugenics saw the First World War as apocalyptic, considering it an irreversible factor in racial decadence. Franco Savorgnan, professor of statistics at the University of Cagliari, was among the first to speak out about the dysgenic danger of the conflict, in an essay titled *La Guerra e la popolazione* [War and population], in 1917. From the "beginning of humankind," war—Savorgnan declared—was a rigorous determining factor of selection, eliminating the weakest: "war formed those selected races of warriors, conquerors and dominators that founded the first nations and, with this, the first civilized institutions."[4]

3 See Angelo Ventrone, *La seduzione totalitaria. Guerra, modernità, violenza politica 1914–1918* (Rome: Donzelli, 2003), 99–192.

4 Franco Savorgnan, *La guerra e la popolazione* (Bologna: Zanichelli, 1917), 85. On the role of Savorgnan in the Italian reception of Gumplowicz's sociology, see Bernd Weiler, "Ludwig Gumplowicz (1838–1909) e il suo allievo triestino Franco Savorgnan (1879–1963). Analisi del rapporto fra la sociologia austriaca e quella italiana," *Sociologia* 1 (2003): 9–41; Raimondo Strassoldo, "La sociologia austriaca e la sua ricezione in Italia. La mediazione di Franco Savorgnan," in Carlo Marletti and Emanuele Bruzzone, eds., *Teoria, società e storia. Scritti in onore di Filippo Barbano* (Milan: Franco Angeli, 2000), 403–21.

With the enlargement of nations and their populations, however, war's selective power was notably reduced, as only part of the population risked death—the "most chosen," the "best": "In this way, exhausted by continuous wars, many old aristocracies are slowly extinguished, which had dominated through the centuries with wisdom and a strong hand."[5] With the advent of modern warfare and firearms, the war definitively lost its selective power, becoming a "factor of anti-selection": bullets were blind, hitting heroes and cowards, the strong and the weak alike.

The question "is success in war the perfect example of whether the quality of the winning population is higher than that of the conquered?" was, according to Savorgnan, "otiose and scientifically unsolvable,"[6] insofar as it was subjective judgment. Nevertheless, even assuming the winning population were better, modern wars could no longer exercise the same positive influence shown in primitive ages on the "racial development" of humanity. This was due to numerous reasons: it did not result in complete destruction of those conquered; losses were often heavier for the winners; "proliferation" following losses was entrusted, after the war, to "physically and morally inferior reproducers"; destruction of material riches lowered the standard of living and increased poverty, reducing resistance to illness.

To all this could be added the consequences of war on the "physiological wealth of the generations that came to light during and after the conflict."[7] Beyond the inferiority of those fathering the next generation—with physical defects, in the case of those ineligible for service, or, in the case of discharged soldiers, damaged by overexertion and tainted by venereal disease—the integrity of newborns in the time of war was gravely compromised by complicated, difficult pregnancies, "attributable to both nutritional deficiencies that damage the organism of the mother, and to distress and anxieties that upset the nervous system."[8] Even those born in the early days of peace could not expect a better outcome, given the qualitative and quantitative reduction of the "racial type of the possible fathers":

5 Savorgnan, *La guerra e la popolazione*, 86.
6 Savorgnan, *La guerra e la popolazione*, 89.
7 Savorgnan, *La guerra e la popolazione*, 90.
8 Savorgnan, *La guerra e la popolazione*, 92.

> The great majority will be, without doubt, undermined by privations, venereal diseases and tuberculosis, or, in the best hypothesis, will have brought home from the war a nervous system strongly prejudiced by the ceaseless fire of the artillery.[9]

The average father would demonstrate improvement some years after the conflict, but would once again sink when the "sons of the war" reached puberty. Savorgnan's conclusion had an apocalyptic tone: "The dysgenic consequences of the war will have distant repercussions, which will weigh as a curse on the children of our children."[10]

According to Savorgnan, the post-war period would require an intense demographic campaign, concentrated on eugenically selected groups; that is, those who survived the war:

> The category of individual most richly endowed with physical robustness, with courage and energy, will be that of the soldiers who survive the war, even if wounded and mutilated, provided that they are not decaying with tuberculosis and syphilis. Promoting marriage and childbirth among these classes of citizens [...], giving them financial aid, enabling them to found a family and give life to a new generation in which their characteristics appear, will be [...] the sole way to bridge the void left by the war: with people vigorous of body and energetic in character, keeping the best aspects of the race intact.[11]

If the war was characterized by its overshadowing dysgenic effect, then the post-war period needed the imposition of a eugenic policy aimed at favoring national rebirth, not only in terms of number but also of race:

> The future rests with those nations that resolve the problem of population, not with the animal brutality of undisciplined sexual instinct that procreates blindly, but with eugenic criteria, which can be suggested by intelligence, rational thought and science.[12]

Savorgnan was certainly not alone in his denunciation of the dysgenic impact of the conflict. In May 1916, in the pages of *Nuova Antologia* [New

9 Savorgnan, *La guerra e la popolazione*, 93.

10 Savorgnan, *La guerra e la popolazione*, 93.

11 Savorgnan, *La guerra e la popolazione*, 141.

12 Savorgnan, *La guerra e la popolazione*, 141.

anthology], Giuseppe Sergi, commenting on the French demographic decline and summarizing several points from Georges Vacher de Lapouge, offered a bio-sociological interpretation. The "decadence of nations" was not only caused by voluntary birth control, but above all by the war, and not only due to the destruction of the younger generation, but particularly to the tension suffered by society:

> Biological disturbances do not derive only from the destruction of young lives—those most adapted to fertility—but also from those unfavorable conditions in which the nation is suddenly placed. From these come mental and sentimental disequilibrium, psychical and nervous traumas, anxieties, and pain of every sort, intensified by the grave economic conditions that derive from the state of war: all this strikes again in the general organic economy of the population.[13]

The increase in mortality was not the war's only dysgenic effect. Since the nervous system was "the regulator of life and human vitality," its disequilibrium was "a cause of partial or total dysgenia and therefore of relative sterility." Not only the combatants, but also the civilian population far from the front manifested nervous traumas, which "could not help but influence the general state of vitality and genetic development." Such conditions would then be aggravated "by poverty, the difficulties of achieving normal nutrition, the inferior quality of the foodstuffs, and by the terrible uncertainty of tomorrow."[14]

The article closed with a call for eugenic intervention by the state:

> It is therefore incumbent on the state, on the managers, on all those who have power, mind and heart, to support the population in the grave and difficult trials in which we find ourselves. The normal activities of the nation and of daily life must be altered as little as possible; sufficient food must be stored for every class in the city and the country; comfort must be given, not verbal, but an effective assistance of a varied and manifold nature [...] and not only to maintain the high spirits of the nation and the power of resistance to the harsh con-

13 Giuseppe Sergi, "L'eugenica e la guerra," *Nuova Antologia* 51, no. 1064 (1916): 135. See also Giuseppe Sergi, "L'eugenica e la decadenza delle nazioni," in Vincenzo Reina, ed., *Atti della SIPS. VIII riunione (Roma, 1–6 marzo 1916)* (Rome: SIPS, 1917), 181–99.

14 Sergi, "L'eugenica e la guerra," 137.

> ditions of the war, but also to maintain healthy and vigorous bodies for the present and the future.[15]

At the height of the conflict, Sergi proposed a program of eugenic defense against the war, for all those currently aged from birth to twenty years of age. It would be necessary to care above all for the population that was still healthy, "to preserve its integrity as those who will in the future constitute the active population of the nation, which may descend into decadence if the post-war generation is weak and sickly."[16] The first problem to solve was that of nourishment:

> The problem of nutrition must therefore be resolved rationally, might I say scientifically, especially for the lower classes both in the city and the country, in order that the new generations that comprise the first twenty years of life do not experience a decline due to insufficient nourishment. The adults could easily bear a reduction, but not the population in a period of growth, unless we want to see, after the war or in the successive years, a population that is not vigorous and has little resistance to the dangers of various pathogenic ailments, especially tuberculosis, and the diminution of eugenic potential, which would have a serious final effect.[17]

In addition, eugenics would have to satisfy the "vital needs" of "air, light and movement." "Thus, no difficulty should be found in this, save perhaps that of possessing a large zone of ground free from trees and not far from the city, which all the youth, including the children, could access at their ease."[18] As for education, Sergi proposed, first of all, a new type of "technical school":

> Technical work is more educational than the teaching of education with words; it develops the sense of order and discipline, fosters natural logic and invention, while it prepares the future of the man who is formed throughout schooling to be ready for life. The work is also hygienic when it is distributed according to age and gender and on the basis of the physical conditions of the learners. It distracts from vices, easy to develop in the younger years,

15 Sergi, "L'eugenica e la guerra," 139.
16 Giuseppe Sergi, "La guerra e la preservazione della nostra stirpe," *Nuova Antologia* 52, no. 1099 (1917): 11.
17 Sergi, "La guerra e la preservazione della nostra stirpe," 12.
18 Sergi, "La guerra e la preservazione della nostra stirpe," 13.

which increase the possibility of physical decadence. In these aspects, technical schools have a double aim; one is eugenic because the adolescents develop in an ordered way in body and mind, and the other is to prepare them for those industries which will emancipate us from servitude to foreigners, useful for them and for the nation.[19]

The second proposal focused on educational reform, reducing the duration of studies to allow young people to follow the paths indicated by biological development:

> We would impose a scholastic reform on every grade, principally to shorten the length of the various scholastic periods. In this way, the youth are free sooner to take those directions that are more appealing to their nature and tendencies. It is still urgent that all the grades are pruned of what is not necessary to teaching, and in this way the time that is occupied by school is shortened and students left free for more hours.[20]

For Sergi, post-war eugenics assumed the shape of a discipline, a cousin to nipiology and pediatrics. Likewise, Serafino Patellani, student of gynecologist Luigi Maria Bossi and first professor of eugenics in an Italian university,[21] believed that the explosion of the First World War signaled nightfall for the optimistic eugenics of the age of positivism:

> Marriages will be celebrated in houses closed because of recent mourning, and real intimacy will be carried out preceded, accompanied and followed by stories of atrocious human violence, with visions of blood, with the constant echo of the screams of wounded brothers, of the groans of the dying; with the recollection, almost like glory, of a moment of collective madness. And the women, already prepared by anxiety, by the agony of waiting, and notices and false news received, despite their desire to procreate and for love in their homes

19 Sergi, "La guerra e la preservazione della nostra stirpe," 14.

20 Sergi, "La guerra e la preservazione della nostra stirpe," 16–17.

21 Patellani translated into Italian Mendel's *Versuche über Pflanzen-Hybriden* (1865), rediscovered in 1900, simultaneously, but independently, by Correns, De Vries and von Tschermak. See Serafino Patellani, "Gregorio Mendel e l'opera sua," *Il Morgagni*, 56 (1914), 148–54, 161–76, 201–33. Professor of a free course in "social eugenics" at the University of Genoa from 1912, in 1924 he was awarded the first professorship instituted in eugenics in Milan. Patellani's eugenics can be summarized as the defense of the "naturalness" of the human reproductive instinct, which implies, in practice, the refusal of birth control; premarital chastity; condemnation of bachelorhood; state intervention to support the birthrate; protection of motherhood and infancy. See Serafino Patellani, *Prolegomeni di eugenetica sociale* (Milan: Cogliati, 1925).

> being deadened, will be overpowered by a sentiment of maternal humanity. This sentiment, intensified and expanded by the collective poverty, in the joy of seeing the fiancé or husband feared lost, will lead them, though tired and ill, to submit once more for the pleasure of man, reviving life in others.[22]

In the face of the war, the possibility of eugenics was compromised not only by the physical damage to bodies, but by the profound moral and spiritual pollution that menaced the "ethics of procreation." According to Patellani, the life of the barracks, far from being a means of natural selection of the best, had always been a source of dysgenics and immorality:

> In fact a man enters into the army because he is in the fullness of his sexual potential: young, strong, robust, honest and healthy. This situation may become dangerous later on [...]. The idle life of the barracks, the friendship of eventually corrupt companions, life in the large and small cities, the assembly of many men, the distance from relatives, the abandonment of habitual occupations, the ease of sexual rapport with women in the brothels, or worse, on the streets, or with occasional prostitutes, create special conditions that intensify the damage of urbanism, heightening vices and the diffusion of sexual illnesses due to the ease of sexual contact that is offered to him. The advantages of physical education and military exercises are in this way destroyed by dysgenic causes, against which, experience tells us, all the efforts of mankind are not sufficient.[23]

Following the war therefore, dysgenics would manifest above all in the form of moral degeneration, which would then be translated into biological ruin. In dark and melodramatic terms, Patellani went so far as to announce, in 1915, the "death of eugenics":

> The death of eugenics, which, when it was just born offered portents of a bright path, with glimpses of grand benefits for humankind, is not the smallest damage done to science by war. The death of eugenics is the march of infamy that distinguishes our civilization, so atrociously offended in the early years of the 20th century [...]. There will come a day, unfortunately still far off, in

22 Serafino Patellani, "Eugenetica e guerra," *La ginecologia moderna* 8, no. 5–8 (May–August 1915): 225 (Lessons on social eugenics held in the Obstetric-Gynaecological Clinic in Genoa, 6 and 13 March 1915).

23 Patellani, "Eugenetica e guerra," 230.

> which our descendents pronounce a judgment on the events of today and on the arrest of the progress of eugenics, which should have represented a new social religion. On that day, perhaps they will remember that in the period of war, amidst the violence and slaughter, there arose in Italy a free voice of protest and faith.[24]

Together with the sociologists and gynecologists, it was above all the psychiatrists who read an irreversible racial degeneration into the symptoms of war-related trauma.

This was the case for socialist Ferdinando Cazzamalli, a psychiatrist in a mental hospital in Como who published an article in 1916, in *Quaderni di psichiatria* [Psychiatric notebooks], which included several preliminary remarks on the concept of degeneration. In particular, "degenerates do not develop, but are born; however, one becomes a carrier of degeneration [...] when morbid causes modify the body and become fixed in the germ plasm."[25] Degeneration did not come from outside, therefore, but rather in the guise of illness: the environment, "sum factor of all biological phenomena," produced the degeneration of the species through the illness of the individual. To environmental influence, the main source of "morbid causes," could be added the nervous system, as a means of transmission of degeneracy from within the human body. Degeneration could principally be defined, therefore, as an "abnormal state of the nervous system": "organic or functional damage existing in the progenitors, having repercussions in the form of absence of, or congenital defects in, offspring."[26] War had always—continued Cazzamalli—modified the environment, which became a "forge of traumatized, fatigued, or malnourished" people. In particular, the conflict underway did not exercise a direct "psycho-pathogenic effect," but instead equaled an "adjuvant factor."[27] There was, therefore, no "psychosis of war." Instead there existed a "predisposition" to psychosis:

24 Patellani, "Eugenetica e guerra," 182.

25 Ferdinando Cazzamalli, "Problemi eugenetici del domani. Guerra e degenerazione etnica," *Quaderni di psichiatria* 3, no. 7–8 (July–August 1916): 166–67. See also Ferdinando Cazzamalli, "La guerra e le malattie nervose e mentali," in Giulio Casalini, ed., *Almanacco igienico popolare* (Rome: n. p. 1920), 197–209. The book was a supplement to the journal *L'igiene e la vita*.

26 Cazzamalli, "Problemi eugenetici del domani," 166.

27 Cazzamalli, "Problemi eugenetici del domani," 167–68.

> War, by creating neuropsychoses in healthy subjects, even if purely transitory, sows the seeds of more or less latent types of psychopathogens, worsening morbid states that had been overcome, or buried. This will considerably aggravate the static condition of that supreme regulator of human life that is the nervous system [...], with dynamic repercussions for future offspring, certainly badly counterbalanced by the maternal influence, inasmuch as this is affected by the emotional and hyperasthenic (depressive) disorders of these anxious times upon the female organism.[28]

Cazzamalli detailed his synthesis of principal "wartime neuropsychoses," which could all be traced to forms and manifestations of epilepsy: "shellshock," "battle hypnosis," "neurasthenia," "hysteria," and "epilepsy." In general, recalling the definitions developed by Arturo Morselli, son of Enrico and head of neurological and psychiatric consultancy services for the First Army,[29] Cazzamalli defined the psychoses of war as a "pathogenic condition," summed up by the term "asthenia," created by an "organic fatigue" or intense emotion. Healthy people were as likely to suffer from this as were those who were "predisposed." Those "neuropsychoses" which particularly affected those individuals potentially exposed (family pedigrees of alcoholism, psychopathic tendencies and epilepsy), must therefore be added to the list of "wartime psychoses."[30]

In such a pathogenic framework eugenics represented an urgent political and social issue. War had always been of a "degenerative" nature, but the conflict underway represented a biological menace for European civilization:

> The youngest and most vigorous, who promised the irreplaceable generative continuity of the stock, are mown down, cut off. And the survivors? The majority, weakened by physical ailments and by serious emotional depression of the nervous system, will undoubtedly see the lessening of that hereditary biological patrimony that will be transmitted to their offspring.[31]

28 Cazzamalli, "Problemi eugenetici del domani," 168.

29 See Arturo Morselli, "Psichiatria di guerra," *Quaderni di psichiatria* 3, no. 3–4 (March–April 1916): 67–68.

30 Cazzamalli, "Problemi eugenetici del domani," 171.

31 Cazzamalli, "Problemi eugenetici del domani," 173.

The war had transformed "potential" and "latent" sick people into "actual" sick people, turning sane people "neurotic" and "psychopathic." The number of "carriers of degeneration" would increase, as illnesses of the nervous system were transmitted from generation to generation. Epilepsy, through the psychological traumas of the war, would broaden its reach. The survivors of the war, that is, "male procreators" of the future, would be "psychologically traumatized," "neuro-psychoasthenics," "hysterical epileptics," "epileptics" and "carriers of epilepsy." Women, reduced to conditions of "minor physical resistance (depression) and psychological resistance (emotional trauma)," would find, for marriage, only a "damaged male youth," and the resultant offspring would be "scarce, with elevated mortality, definitely neurotic or at least strongly predisposed to psychological disorders." What was to be done, therefore, in the face of the apocalypse that was the First World War? "The pharmaceutical armory of eugenics ranged from castration of those individuals ascertained as degenerate, [...] to perpetual segregation; from marriage limitations (Galton) to interdiction"; but Cazzamalli, in agreement with Enrico Morselli, refused "violent means," and in the end preferred the development of social medicine and the encouragement of an "education of the masses as regards the effects of sexual union."[32]

The theme of "ethnic degeneration" was also taken up in the pages of the catholic journal *Vita e Pensiero* [Life and thought], in an important essay by the psychologist (and future founder of the Catholic University in Milan) Agostino Gemelli. In the article, titled *Eugenica e Guerra* [Eugenics and war],[33] Gemelli adhered substantially to Sergi's hypothesis, but did not deem the war the "exclusive" or "predominant" cause of the decreasing birthrate;[34] negative factors could also be identified in "race-crossing," "illnesses of the female sexual organs" and "criminal neo-Malthusianism."[35] Instead, the war would impact more on future generations, who would inherit psychological traumas from their fathers, worn down by serving on the front, and from their mothers, shaken by poverty, work and violence:

32 Cazzamalli, "Problemi eugenetici del domani," 175–76.

33 Agostino Gemelli, "Eugenica e guerra," *Vita e Pensiero* 4, no. 3 (September 1916): 133–45. During the conflict, Gemelli worked on the development of psycho-aptitudunal tests for the selection of pilots. His interest in eugenics dated from 1915, when he criticised the Galtonian theory of heredity of psychical qualities. See Agostino Gemelli, "Si ereditano le qualità psichiche?," *Vita e Pensiero* 1, no. 3 (1915): 273–83.

34 Gemelli, "Eugenica e guerra," 138.

35 Gemelli, "Eugenica e guerra," 136.

> War reveals, so to speak, psychologically ill people; people predisposed to mental or nervous system illnesses, either hereditarily, or from previous illness, who, due to an emotional effect, experience episodes of nervous and mental illness that were previously concealed. At the end of the war, equilibrium will be re-established, and these ill people, apparently healed, will return to their social life and to their family, and have children, to whom they will transmit this disposition toward illness, or the illness itself.[36]

The sons of war, therefore, could not help but be "neurotic and psychopathic," destined to "carry traces of the terrible event in which their fathers took part, for all their lives, in their nervous systems and psychological structure." According to Gemelli, the war not only "diminished birthrate, but deteriorated the race."[37] In the face of this racial degeneration, negative eugenic remedies could be useful, to "impede or limit marriage between those who could not help but transmit illness or evil dispositions to their children."[38] Instead, the precautions of positive eugenics aimed at raising the birthrate would not be so effective:

> Working fatally against the facts of biological order, we find facts of social order and of economic order, determining among their good and evil effects also this effect of decreasing the birthrate. To raise the birth rate, we must neutralize these factors, that is, we must mutate the current social order.[39]

Gemelli's eugenic project envisioned a fight against neo-Malthusianism, conducted by rehabilitating the Catholic family model, founded on its "natural bases," that is, on "moral Christian laws." Restoring the "natural function" to the family—expressed by "love between the parents," by "mutual respect" and a vision of life not as a pleasure, but as a "test for higher purposes"—would mean, in Gemelli's view, the development of "the most effective and fertile eugenic activities."[40]

36 Gemelli, "Eugenica e guerra," 140.
37 Gemelli, "Eugenica e guerra," 141.
38 Gemelli, "Eugenica e guerra," 142.
39 Gemelli, "Eugenica e guerra," 144.
40 Gemelli, "Eugenica e guerra," 145.

2. War as Gymnasium

In his effort to develop a scientific paradigm that could justify war as a product of the demographic expansion of young nations and as an instrument of modernization of the social organism, nationalist Corrado Gini expounded a series of arguments between 1915 and 1921, aimed at putting the dysgenics of the conflict into perspective.[41]

To begin with, military conscription: is it positive or negative from a eugenic point of view? The people subjected to military service—Gini claimed—married later than peers fit for the army who did not serve. However, they married more frequently, almost as if military service constituted a preference in matrimonial selection. From twenty-five until forty years of age, those who completed military service had fewer living children than their peers, in keeping with the shorter duration of their marriage. But above forty years of age, the number of children clearly increased. Although they married later, those fit for the army had more prolific marriages, as if that same preference found in matrimonial selection allowed them to "marry women who were younger or, independent of age, healthier and more robust, and therefore more fertile."[42] In addition, the greater prolificacy of the military put into perspective, in Gini's eyes, the problem of venereal diseases, which according to many eugenicists was widely diffused among the military lines: in fact, "these would undoubtedly manifest themselves in an unusual frequency of sterile marriages and a high infant death rate, and therefore a lower number of living children, whereas the facts clearly verify the opposite."[43]

---

41 For an analysis of Gini's wartime theories, see, in particular, the articles collected in Corrado Gini, *Problemi sociologici della guerra* (Bologna: Zanichelli, 1921). On the theme of the relationship between eugenics and war, see also Gini's reports at the 2nd International Congress of Eugenics (New York, 22–28 September 1921), and Corrado Gini, "The War from the Eugenic Point of View," in Charles B. Davenport, et al., eds., *Scientific Papers of the Second International Congress of Eugenics* (Held at the American Museum of Natural History, New York, September 22–28, 1921), vol. 2., *Eugenics in Race and State* (Baltimore: Williams & Wilkins, 1923), 430–31. In September 1927, the IFEO nominated Gini president of a Commission for the study of the eugenic or dysgenic effects of war. The first results, preceded by a long report by Gini, were presented at the 3rd International Congress of Eugenics in New York, in August 1932. See Corrado Gini, "Gli effetti eugenici o disgenici della guerra," *Genus* 1–2 (1934): 29–42.

42 Corrado Gini, "La coscrizione militare dal punto di vista eugenico," *Metron* 1, no. 1 (1920), then in Gini, *Problemi sociologici della guerra*, 153.

43 Corrado Gini, "La guerra dal punto di vista dell'eugenica," in Roberto Almagià, ed., *Atti della SIPS. XI riunione (Trieste 9–13 settembre 1921)* (Rome: SIPS, 1922), 45.

Another critical point concerned the purported weak constitution of war babies. Data gathered on still births and infant death rates, particularly "for reasons of weakness or congenital vice" relative to the war years in combatant nations, did not actually show any traces of such a weakness. In fact, statistics of newborns showed in the first years after the war even higher birth weights than before the war. The dysgenic factor, represented by the absence of the best "reproducers" busy at the front, was substantially compensated for by opposite elements, such as selection by social class or number of children:

> In any case, the higher social classes, and families with smaller numbers of children, present, on average, superior physical characteristics, yet give a contribution to the military which is not proportional to their numerical importance.[44]

In addition, the economic disadvantage and the brevity of conjugal contact in times of war favored reproduction by "the people endowed with the most intensive reproductive faculties, able to create better products,"[45] while the long intervals between births contributed to vitality in their offspring.

The third question addressed the death rate. While deaths in combat and due to war injuries had an inevitable dysgenic effect, the excess of deaths due to illness among the military and among the civilians seemed to exercise, according to Gini, a favorable influence on the constitutions of future generations. It was impossible to predict which of these elements would be predominant. Certainly, modern war represented, in Gini's view, a higher dysgenic factor than traditional war:

> Compared to the wars of the past, modern war is more likely to have a dysgenic effect, insofar as deaths in combat or due to injuries are concerned, which now, as has been said, have overtaken death by illness among the military. Moreover, the greater economic prosperity and the better preparation for war have meant that the civilians feel the privations and disadvantages less strongly, and are therefore more subject to a less severe surplus of mortality. On the other

44 Gini, "La coscrizione militare dal punto di vista eugenico," 121.

45 Gini, "La coscrizione militare dal punto di vista eugenico," 121. See also Corrado Gini, *Sulla mortalità infantile durante la guerra,* in Gini, *Problemi sociologici della guerra,* 104–22.

> hand however, the larger recruitment of combatants necessarily carries with it a rigorous selection of the military, which must naturally correspond to a less unfavorable influence of mortality, in combat or due to injuries, and of the difference of mortality due to illness between the soldiers and the remaining population.[46]

As for the impact of the war on the intellectual endowments of the nation, research specifically carried out by Gini on primary school teachers, based on a report by the Deposit and Loan Bank (*Cassa Depositi e Prestiti*) demonstrated that those who died in war did not present a "social value" superior to that of the survivors. Despite being limited to a single profession, the research aimed to put the conflict's dysgenic effect into perspective:

> Other longer investigations will be necessary to judge with precision the selective influence of mortality directly caused by the war; but the investigations carried out in the interim serve only to bear out the suspicion, which had already been considered *a priori*, that the higher death rate during war time does not have the profound dysgenic effects that are generally attributed to it.[47]

The growing rate of deaths and of births immediately following the end of the conflict resulted in favorable selection effects. Death eliminated the weakest, whereas newborn children of the selected military classes were enhanced by the long rest forced on the "reproductive organs of the mother,"[48] and demonstrated a superior constitution. According to Gini, growing birth weights and the frequency of multiple births represented, as much as anything, proof of a favorable eugenic event.[49]

---

46 Gini, "La guerra dal punto di vista dell'eugenica," 49.

47 Gini, "La guerra dal punto di vista dell'eugenica," 62.

48 Gini, "La guerra dal punto di vista dell'eugenica," 63.

49 Gini's students and collaborators attempted to provide further confirmation of this interpretive line: see Marcello Boldrini and Aldo Crosara, "Sull'azione selettiva della guerra tra gli studenti universitari italiani," *Metron* 2, no. 3 (1923), 554–67; Raffaele D'Addario, "L'azione selettiva della guerra in un gruppo di studenti universitari italiani," *Archivio scientifico del R. Istituto Superiore di Scienze economiche e commerciali di Bari* (1926–27 and 1927–28); Giovanni L'Eltore, "Contributo allo studio degli effetti selettivi della guerra dal punto di vista dell'eugenica," *Genesis* 11, no. 1–2 (January–June 1932): 49–62.

3. War as Laboratory

In 1905 the Russo-Japanese war provided a new perception of the psychological impact of modern conflicts, announcing a previously unforeseen role for several sectors of military medicine and psychiatry. In Italy, the issue of the relationships between war and mental illness fed an intense discussion at the 14th Congress of the Italian Phreniatric Society in May 1911, and a psychiatric service was established during the colonial war in Libya. The specter of deviance, particularly in regard to deserters, soldiers suffering "homesickness," or hypersensitive or traumatized people, was nevertheless amplified by the proportion and duration of the First World War.

For the neuropsychiatric body, the war was above all an immense laboratory, a field of clinical experimentation, where it was possible to observe large-scale "trauma, emotion, commotion, privation, mutilation and deviation of every kind, known and unknown, already codified and new."[50] In addition to scientific knowledge, the front enhanced both the organizational and ideological powers of psychiatrists. In August 1915, the Military Supreme Command, on the recommendation of the Military Health Committee (*Ispettorato Medico Generale*), decreed the institution of a special neurological and psychiatric service in each of the four armies' military health systems. The four specialists appointed for the occasion (Arturo Morselli, Vincenzo Bianchi, Angelo Alberti and Giacomo Pighini) organized departments of neuropsychiatry in the first and second lines, with mental illness ward annexes behind the front.[51] In the national exhibition of

---

50 Antonio Gibelli, "La guerra laboratorio: eserciti e igiene sociale verso la guerra totale," *Movimento operaio e socialista* 5 (1982): 346; also fundamental Antonio Gibelli, "Guerra e follia. Potere psichiatrico e patologia del rifiuto nella Grande Guerra," *Movimento operaio e socialista* 4 (1980): 441–64. For a comprehensive reanalysis, see Antonio Gibelli, *L'officina della guerra. La grande guerra e la trasformazione del mondo mentale* (Turin: Bollati Boringhieri, 1991).

51 See "Organizzazione di servizi neurologico-psichiatrici per i Belligeranti," *Quaderni di psichiatria* 2, no. 9–10 (September–October 1915): 396–97; Arturo Morselli, "La neuropsichiatria castrense in Francia," *Quaderni di psichiatria* 3, no. 5–6 (May–June 1916): 131; Francesco Petrò, "Un reparto psichiatrico avanzato d'Ospedale da campo nel suo primo anno di funzionamento," *Quaderni di psichiatria* 4, no. 3–4 (March–April 1917): 71–78. See also "Per il servizio psichiatrico di guerra," *Rivista sperimentale di freniatria* 41 (June 1915): 412–13; "Sul servizio psichiatrico di guerra," *Rivista sperimentale di freniatria* 41 (November 1915): 509–11; Gustavo Modena, "L'organizzazione dei Centri neurologici in Francia," *Rivista sperimentale di freniatria* 43 (August 1917): 344–55; E. Riva, "Il Centro psichiatrico militare di I raccolta," *Rivista sperimentale di freniatria* 45 (May 1919): 308–24 and "Un anno di servizio presso il centro Psichiatrico Militare della Zona di guerra," *Rivista sperimentale di freniatria* 45 (May 1919): 443–59.

works of assistance for the war effort, held in Rome during June–July 1918, psychiatry was well represented, with reconstructions of medical wards, photographs and "products of the sick."[52]

Together with these organizational efforts, the so-called "military neuropsychiatry" (*neuropsichiatria castrense*) looked to eugenics to face the problem of biological selection of the soldiers, aimed at rationalizing and intensifying wartime efforts. In their attempts to guarantee the maximum efficiency of the available biological resources—through the diagnoses of different psychological "anomalies," the identification of "simulations" and the segregation of elements dangerous to military discipline—the physicians soon faced the dilemma of "abnormality": what to do with the defective elements? Keep them well away from the war effort, or utilize them until the end?[53]

Psychiatrists such as Edmondo Trombetta, director of the *Giornale di medicina militare* [Journal of military medicine], and Giacomo Pighini, consulting neuropsychiatrist of the Grappa and Altipiani army, were convinced of the necessity of eliminating the defectives from the army lines, to eventually relegate them to a "special company for deportation to the colonies."[54] However a majority of the physicians at the front favored an approach of extreme Tayloristic re-utilization of "abnormals." For example, Enrico Morselli agreed with the use of the mentally ill and "waste material" as workers:

> It could be that the mildly insane, who are obedient and physically strong, could be used advantageously in military service, even in the active units, if they were surrounded by numerous psychologically healthy people, from whom they would receive some useful influence that would help them to work together, for discipline, perhaps even for courage.[55]

52 Antonio Mendicini, "I centri neurologici nella mostra nazionale delle opere d'assistenza nell'Esercito," *Quaderni di psichiatria* 5, no. 9–10 (September–October 1918): 229–34.

53 For more informations on this issue, see Mantovani, *Rigenerare la società*, 159–65, and also Andrea Scartabellati, *Intellettuali nel conflitto. Alienisti e patologie attraverso la Grande Guerra (1909–1921)* (Bagnaria Arsa: Edizioni Goliardiche, 2003): 100–21.

54 Edmondo Trombetta, "Gli epilettici in zona di guerra (nota critica)," *Giornale di medicina militare* 1 (1918): 54–58; Giacomo Pighini, "Per la eliminazione dei degenerati psichici dall'esercito combattente," *Giornale di medicina militare* 1 (1918): 978–96.

55 La Direzione, "Il lavoro degli anormali psichici e la Guerra," *Quaderni di psichiatria* 4, no. 3–4 (March–April 1917): 79–80.

Every type of illness corresponded with a form of economic use:

> There is much work to be done in the war for which participation without thought is enough: the work of digging and excavating the trenches, transporting munitions, various restocking and repairs, etc. In which case, given that calm and obedient insane people remain among the troops in active service, we do not have to hastily renounce the utilization of their brute strength.[56]

While the automatist comportment of the "pure imbeciles" might be useful, some epileptics might be destined for "custody by the military depots for harmless objects" (*Depositi militari di oggetti innocui*) or for "porterage work."[57] Along the same lines, Cesare Agostini, director of the Perugia military neurological section and neuropsychiatric consultant for the Carnica army, suggested the establishment of centers within war zones specially charged with distinguishing the genuine cases of epilepsy from possible, and frequent, simulations. Serious epileptics would then be sent home "to be secluded in a curative institute or in a criminal mental hospital," while those "affected by rare episodes" could be put to work in special division of troops, "naturally unarmed" and "used precisely behind the front line solely for the work of digging, opening roads, building boardwalks through the trenches, arranging aviation camps and perhaps cultivating the terrain in the zone of operations."[58] Such a solution would prevent, in Agostini's view, the absurd "salvage of social waste" and that form of "counter-selection" that consisted in sacrificing the "physically strongest part of the nation" to the front and repatriating the "physically defective and morally degraded," ready to "multiply the candidates for insanity and criminality."[59]

However, it was medical captain Placido Consiglio who carried the logic of eugenic selection of soldiers to its extremes. Specialist for the War Zone Central Health Commission (*Commissione Sanitaria Centrale della Zona di Guerra*) and director of the military psychiatric diagnosis center in Reggio Emilia, instituted in 1917 as a concentration camp for neuropsychotics identified by the Army Consultancy Board (*Consulenze d'Armata*),

56 La Direzione, "Il lavoro degli anormali psichici e la Guerra," 80–81.
57 La Direzione, "Il lavoro degli anormali psichici e la Guerra," 80–81.
58 Cesare Agostini, "Sulla utilizzazione degli epilettici in zona di guerra," *Giornale di medicina militare* 1 (1918): 31.
59 Agostini, "Sulla utilizzazione degli epilettici in zona di guerra," 32.

Consiglio regarded the conflict as a laboratory of applied eugenics. He saw the military as a highly selected and medical social microsystem:

> The battle against every form of degeneration and abnormality, fought with direct and indirect methods together, can be better realized in the special community, more restricted, more intimate in structure and more homogenized, that is the army [...]. I have always believed that this particular environment must be considered as an instructive form of social experimentalism.[60]

Consiglio's utopia quickly assumed the shape of a eugenically militarized society:

> If every human group could impede the penetration of deviates and psychological degenerates from outside or from internal generation, and eliminate those already born or penetrated, distancing them in such a way as to impede the actual or potential damage to others that comes from their pernicious fermenting actions [...]; well then, the grave problem would without doubt be resolved, and the constitution of that group greatly bettered, in an always progressive mode.[61]

Eugenics, extrapolated from the military microcosm to the social macrocosm, had to be understood as a "function of the State" and managed in first place by physicians:

> It lies with military physicians to undertake the great physical and mental healing process of the military community and the great decrease that we wish to see in the various forms of sickness that inflict humankind. And the same thing must occur in society: in schools, through the work of pedagogical physicians, in social life through the medical sociologists in parallel with active, extended prophylaxis against intoxication and epidemic infections, and moreover, positive moral education, above all in the working classes.[62]

With his strong case-history of hundreds of military psychiatric analyses and studies from the Libyan war,[63] Consiglio did not hesitate to contest

60 Placido Consiglio, "Problemi di eugenica," *Rivista italiana di sociologia* 18, no. 3–4 (May–August 1914): 458.
61 Consiglio, "Problemi di eugenica," 459.
62 Consiglio, "Problemi di eugenica," 460.
63 See, for example, Placido Consiglio, "Studii di Psichiatria Militare; parte I," *Rivista sperimentale di freniatria* 38 (August 1912): 370–407; Placido Consiglio, "Studii di Psichiatria Militare; parte II," *Rivista sperimentale di freniatria* 39 (December 1913): 792–840; Placido Consiglio, "Studii di Psichiatria Militare; parte III," *Rivista sperimentale di freniatria* 40 (December 1914): 881–97; Placido Consiglio, "Studii di Psichiatria Militare;

the more popular Lombrosian arguments: neither genius nor warrior heroism could spring, according to the military physician, from degeneration. "Abnormals" would always be of "no social value, often damaging, always dangerous, and wasters of bio-psychical energy."[64] Moreover, at the request of the War Ministry, Consiglio conducted research on a sample of 772 military prisoners, concluding that resistance to re-education and discipline came principally from families that were carriers of hereditary defects. The Zar family was a notable example—a singular Italian version of the celebrated American eugenic family-case, the Jukes[65]—in which Consiglio counted 44 individuals "in whom neuropsychological degeneration was identifiable, assuming a variety of forms, from psychoses to criminality, epilepsy and madness for four generations and in five families."[66]

This rigid hereditary determinism was clearly the theoretical base for a radical eugenic solution. If anthropological defects were fatally transmitted to generation after generation, then policies centered on education or environmental betterment would be worthless. The sole remedy was selection and isolation:

> Delinquents do not choose to be so, but are constituted in that way in their most intimate cerebral matter: if criminal actions are prevalently determined by constitutional anomalies of the psychophysical make-up, then the human group in which these occur have no work more effective and positive—albeit complex and difficult—than to prevent this evil, combating the impure origins of the parental toxicity, of hereditary morbidity, and of degeneration of offspring. This can be done by isolating and curing criminals such as the insane and neurotic, and so, without any false sentimentalism, supporting human eugenics by impeding reproduction on the part of the many sufferers of tuberculosis, syphi-

parte IV," *Rivista sperimentale di freniatria* 41 (March 1915): 35–73; Placido Consiglio, "Le anomalie del carattere dei militari in guerra," *Rivista sperimentale di freniatria* 42 (October 1916): 131–72; Placido Consiglio, "Nuovi studi sulle anomalie del carattere dei militari in guerra," *Rivista sperimentale di freniatria* 42 (December 1917): 529–44.

64 Consiglio, "Problemi di eugenica," 465.

65 The celebrated study of the Juke family (seven generations of criminals, prostitutes and various degenerates produced by a single couple in the state of New York) was published in 1877 by Richard L. Dugdale, a member of the executive committee of the Prison Association of New York. In 1916, Arthur Estabrook, a field researcher and collaborator of Davenport at the Carnegie Institution, updated and reanalyzed the Juke family data: see Kevles, *In the Name of Eugenics*, 71; Paul, *Controlling Human Heredity*, 43–49.

66 Consiglio, "Problemi di eugenica," 452.

> lis, alcoholism, epilepsy, and degeneration that pollute the font of human life, in such a way that we arrive at a progressive selection of the race.[67]

"For now," Consiglio emphasized, the "traditional human instincts" impeded the practice of sterilization in the "Latin world," but in the meantime, much could be achieved with the "isolation of anomalies from society, for cure, and for re-education of an indeterminate length of time":

> Various widespread methods in Italy and Libya of obligatory insurance against illness, of reformatories in agricultural colonies and similar, could help us to obtain the goal of distancing the dangerous elements from society, and therefore also from reproduction, with enormous moral and social advantages.[68]

At the beginning of the 1920s, in the pages of *Difesa sociale* [Social defense], Consiglio systematized his almost ten years of eugenic reflections, distinguishing between the wartime emergency and the period of peace. The war imposed an "accurate selection" of "degenerates": the major part should be utilized in the "numerous auxiliary services, armed or not, in war zones or domestic territory, according to profession, and to the attitudes and year people were drafted." The "serious degenerates" (in Consiglio's terms, the constitutionally immoral, alcoholics with epileptic tendencies, perverts and those with incorrigible vices, habitual offenders, or those regularly imprisoned), "for special security measures and the defense of the race," would be "segregated and used in work colonies in national territorial zones or overseas, giving them tools, seeds and plots of land." "The most seriously insane and true psychopaths" had to be imprisoned in institutions for rehabilitation, asylums or special colonies; the "degenerate minors" could be, in the end, utilized "in special squads behind the front lines, working within or outside of the war zones, without arms and under strict discipline, to their great re-educative benefit."[69]

Once the war was finished, the prophylaxis at work in the military environment would indicate the "best path" for defending society from "abnor-

67 Placido Consiglio, "La pretesa rieducabilità dei pregiudicati militari in guerra," *Rivista di psicologia*," 9, no. 4 (July–August 1913): 351.

68 Consiglio, "Problemi di eugenica," 461–62.

69 Placido Consiglio, "Come difenderci dagli anormali e dai degenerati nell'ambiente militare," *Difesa sociale* 2, no. 10 (October 1923): 8.

mals and degenerates," based on two fundamental precepts: their "elimination from the civil environment and the reproductive function," and their "symbiotic utilization in diverse work." The two eugenic strategies—elimination/segregation and economic re-utilization—were, however, rooted in only one concept: the "complete knowledge regarding degenerates and abnormals."[70] This could be realized, according to Consiglio, through a vast biographical-clinical survey of degenerates. The project that had matured in nineteenth-century positivist criminal anthropology was destined to have a notably favorable reception among the Italian eugenicists.

4. Eugenics and the "Sons of the Enemy"

Between 1915 and 1917, the violence of the First World War fuelled the diffusion of a specific eugenic "case" throughout the Italian medical context. The "serious problem of eugenics and justice" was provoked by the news, released by the French parliamentary commission and reported in Italian daily news and journals,[71] of "ethnic rapes" being currently committed by German soldiers in occupied Belgium and France.

A discussion on similar acts in Italy was opened in medical circles by well-known gynecologist Luigi Maria Bossi (1859–1919),[72] director of the monthly review *La ginecologia moderna* [Modern gynecology], which deliberately assumed the new subtitle *Review of obstetrics, gynecology, and psychological, eugenical and gynecological sociology* in 1914. Bossi explicitly confronted the question in March 1915, in a discourse addressed to the Genoa Royal Medical Academy (*Reale Accademia Medica*). Sexual violence, with an aim of "Germanizing" occupied France and Belgium, presented a problem, in the case of eventual pregnancy, that was as much ethical

---

70 Consiglio, "Come difenderci dagli anormali e dai degenerati nell'ambiente militare," 9.

71 D. Angeli, "I non desiderati," *Giornale d'Italia* (23 February 1915); A. Polastri, "I piccoli tedeschi," *Giornale di Sicilia* (21–22 February 1915); P. Croci, "Angosciosi problemi della guerra. L'innocente," *Corriere della Sera* (10 March 1915). For a reconstruction of the French debate, see Stéphane Audoin-Rouzeau, *L'enfant de l'ennemi 1914–1918* (Paris: Aubier, 1995). On the Italian situation, see Mantovani, *Rigenerare la società*, 194–97.

72 The first Italian university professor of gynaecology (1887), socialist, with interventionist and Mussolinian sympathies, Bossi was a proponent of a pervasive vision of gynaecology, based on a sociobiological interpretation of uterine pathology, which he considered a "supreme social pathological phenomenon." See Mantovani, *Rigenerare la società*, 101–03.

as it was eugenic. From the first point of view, the birth would further aggravate the suffering of the women:

> Frankly, we must ask ourselves if we have the right to impose further torture, both physical and psychological, on women already heavily tried by human infamy, in homage to a principal of conservation that today is violated everywhere, solely for the egoism of the increasingly widespread, and, what is worse, spread with impunity, curse of illegal abortion.[73]

On the eugenic front, concerns were perhaps even more urgent. The psychological traumas experienced by the mothers; the state of alcoholism or of "morbid, insane, bestial excitement" of the fathers; and the "continuing physical traumas" of the pregnancy would result in children who were "developmentally deficient, destined to be a burden on public charity, or future insane or delinquents." Beyond the danger for families and society, the "children of barbarity" could politically damage the nation in the future, "because it is impossible to eliminate the possibility that the enemy paternal seed, impregnated in a moment of hate, might not be carried by the child in a sad reflection of the same hate."[74]

In the face of such a dramatic situation, Bossi, who had in earlier years led the battle against abortion and neo-Malthusianism,[75] proposed a medical justification for the French and Belgian women who had been victims of sexual violence:

> Now I would not hesitate to confirm, as impudent as such a confirmation may seem, that for exactly those reasons highlighted above, the pregnancies of the Belgian and French women resulting from the barbaric violence of the Germans must be terminated [...].[76]

The initiative not only aimed to eliminate "degenerates," but also to protect the mothers, who, giving birth in conditions of strong psychological trauma, could be risking their lives.

---

73 Luigi Maria Bossi, "In difesa delle donne belghe e francesi violentate dai soldati tedeschi. Una grave questione d'eugenetica e di giustizia," *La ginecologia moderna* 8, no. 1–4 (January–April 1915): 94.

74 Bossi, "In difesa delle donne belghe e francesi," 95.

75 See Mantovani, *Rigenerare la società*, 134–38.

76 Bossi, "In difesa delle donne belghe e francesi," 96.

Bossi returned to this argument several months later, identifying in "ethnic rape" nothing less than the ultimate consequence of the German medical culture, which he deemed a promoter of neo-Malthusianism, gynecological errors and scientific usurpation: "We, in our field, feel we must conscientiously demonstrate that the German culture is, in certain parts, either a dangerous scientific error, a theft of others' genius, or a hypocritical attempt to exploit humanity."[77]

The eugenic appeal of the Genoan gynecologist provoked a limited, but not irrelevant debate: the pages of *Policlinico* [General hospital] (9 May 1915), *Pensiero sanitario* [Sanitary thinking] (10 April 1915), and *Avanti!* [Forward!] (23 November 1915) carried strongly polemical articles, while in *Corriere mercantile* [Trade journal] (21 May 1915) the contrary positions of Enrico Morselli and jurist Pietro Cogliolo stood out. But it was Enrico Ferri's review *La Scuola Positiva* [Positivist school] that confronted the question in the most articulate way, analyzing the legal issues in a series of articles published between April and June 1915. In the first, Salvatore Messina contested Bossi's ideas. He argued that Italian laws punished abortion for reasons that were independent from the circumstances of conception; the absolution that in the past had been given to violated women guilty of abortion did not imply negation of guilt, but was dictated by pity for a moral expiation that overrode the guilt and the respective judicial evaluation. In conclusion:

> Nothing can legitimize the political opportuneness and juridical convenience of temporarily suspending the effectiveness of the normal penal code against abortion and infanticide: that is, to turn a diffused state of deep social compassion for the guilty women into extenuating circumstances for a crime, when there is no need for their wretchedness to unravel the thread of the law in order for them to be treated with justice.[78]

Different beliefs, however, were found in the second article, which justified the right to abortion in the name of a "state of necessity," defined in the penal code:

---

77 Luigi Maria Bossi, "I pericoli e le vittime della cultura tedesca nel campo ginecologico," *La ginecologia moderna* 8, no. 5–8 (May–August 1915): 148.

78 Salvatore Messina, "Le donne violentate in guerra e il diritto all'aborto," *La Scuola Positiva* 6, no. 4 (April 1915): 294.

> Perhaps it is a question of two rights that find themselves in conflict here. The woman who has not contributed voluntarily to this conflict finds her rights concerning her own person in imminent danger; if she cannot otherwise avoid it, she must be able to resolve this by sacrificing, without being legally responsible, the rights that clash with hers. How can we doubt that there is a clash between the rights of the unborn child and the State as regards the physiological development of an embryonic life, and the right of the woman to prevent this seed, forcefully implanted in her, which, should it develop, would see the contrast between the two rights grow ever greater?[79]

The article concluded by reaffirming this last position, citing some tendencies of the Catholic church to favor abortion in the case of rape.[80]

On 25 August 1916, in Benito Mussolini's interventionist newspaper *Popolo d'Italia* [Italian people], Bossi's referendum was published, addressed to "women, physicians, sociologists, jurists and literati," publicly denouncing the German violence and declaring the right to abortion for the women violated. Several responses submitted by readers appeared in what could be considered the final act in Bossi's eugenic debate: the publication in 1917 of an entire issue of *Ginecologia moderna* dedicated to "the defense of women and of the race." Here, Bossi equated the right to abortion for violated women with the political fight against neo-Malthusianism and criminal abortion:

> The defense, therefore, of women and the race, in relation with neo-Malthusianism, criminal abortion and the right to abortion of women systematically violated by the Germans, constitutes a large, complex problem that must be resolved through three indivisible relationships: social, juridical and medical. And it is above all pertinent to gynecologists, because they are responsible, as is obvious, for the basal concept of conservation of the species, that is, the present life and health of the mother; and subordinately, the life and health of the product of conception. The social and juridical sides must naturally be subordinate to the gynecological side.[81]

---

79 Silvio Longhi, "Le donne violentate in guerra e lo 'stato di necessità'" *La Scuola Positiva* 6, no. 6 (June 1915): 485.

80 Bernardino Alimena, "Concludendo sulla violenza carnale e il 'diritto all'aborto,'" *La Scuola Positiva* 6, no. 8 (August 1915): 673–75.

81 Luigi Maria Bossi, "Per la difesa della donna e della razza," *La ginecologia moderna* 10 (1917): 128.

In the face of sexual violence, a "moral war against the perfidiousness of the German culture," in the name of the "defense of women and the race," had to parallel the war raging at the front.[82]

After a brief spark of interest, the debate surrounding the "sons of the enemy" was quickly extinguished in France, suffocated by the growing populationist concerns. In Italy, just a few neo-Malthusian activists kept on supporting Bossi's position in the defense of eugenic quality as opposed to dysgenic quantity. In fact, in 1920, the pamphlets, which Secondo Giorni and Felice Marta—isolated champions of "practical" and "medical neo-Malthusianism"—had published in 1916 and 1915, were republished. The new edition included Giorno's polemic against French pro-natalism and its attempt "to take advantage of the barbaric enemy seed and in this way to procure a greater number of soldiers for the future,"[83] and Marta's concerns regarding race-crossing between the French women and the Senegalese troops:

> But who does not feel that it is grotesque; who can not see the damage and the insult of those wild stallions, next to whom those poor French males must figure as parade horses? [...] Now, if Europe, to remake her race, needs Senegalese crossings and those with syphilitic inheritance, then it seems to us that it is better to choose the lesser of two evils. It is better, after all, to die of listlessness than gangrene.[84]

After the war however, such neo-Malthusian issues as these seemed far from the Italian political and scientific context, which was increasingly eager to listen to the "regenerating" promises of natalism and fascism.

82 Bossi, "Per la difesa della donna e della razza," 130.

83 Secondo Giorni, *Come si prepara la classe del 1916. Il Neo-Malthusianismo e la guerra tra le nazioni* (1916; repr., Florence: Soc. Ed. Neomalthusiana, 1920), 6.

84 Felice Marta, *Neomalthusianesimo medico. Quando e come non bisogna aver figli* (1915; repr., Milan: Società Anonima Editoriale, 1920), VIII.

CHAPTER III

REGENERATING ITALY (1919–1924)

The First World War was a catalyzing event for Italian eugenics. The anxiety over biological regeneration that accompanied the end of the conflict, together with the new dimension assumed by the State as manager of collective biological resources and protector of the health integrity of the social body[1] initiated a new season of growth and development in the eugenic debate. The protagonists of this debate were above all physicians of different political backgrounds, but ready to offer their technical competencies to sustain the economic-productive efficiency of the "human factor."[2] It is no coincidence that the turbulent years of the governments prior to fascism gave rise to the principal institutions that diffused eugenic themes: the Institute of Public Welfare and Assistance (*Istituto di Previdenza e Assistenza Sociale*, known as IPAS) began in 1922 thanks to the organizational effort of Ettore Levi. The Italian Society for the Study of Sexual Questions (*Società Italiana per lo studio delle Questioni Sessuali*, known as SISQS) was created in 1921[3] on the initiative of the historian of science

1 On the expansion of the State functions in public health and welfare policies, after 1918, see Michele Pietravalle, "Per un Ministero della Sanità ed Assistenza Pubblica in Italia," *Nuova Antologia* 1131 (1919): 111; Pietro Bertolini, "Assicurazioni operaie e provvidenze sociali," *Nuova Antologia* 1107–08 (1918): 3–30; 149–76; Pietro Capasso, *L'assistenza di oggi e l'assistenza di domani* (Napoli: Stab. Tipografico Morano, 1920). For a comprehensive framework of the issue, Mantovani, *Rigenerare la società*, 203–23.

2 In February 1914, the Parliamentary Medical Fascio (*Fascio medico parlamentare*), piloted by hygienist Giuseppe Sanarelli in the first decade of the century, became the Parliamentary Medical Committee (*Comitato Medico Parlamentare*) crossing political battle-lines; see Tommaso Detti, "Stato, guerra e tubercolosi (1915–1922)," in Franco Della Peruta, ed., *Storia d'Italia. Annali*, vol. 7, *Malattia e medicina* (Turin: Einaudi, 1984): 880.

3 For the statute of SISQS, see "Società italiana per lo studio delle questioni sessuali," *Rassegna di studi sessuali* 1, no. 4 (July–August 1921): 272–74.

and pioneer of Italian sexology Aldo Mieli,[4] soon to be a protagonist in the debate over pre-marital medical certificates. Finally, the Italian Society for Genetics and Eugenics (*Società Italiana di Genetica e Eugenica,* known as SIGE) was founded in 1919 by Corrado Gini, Cesare Artom and Ernesto Pestalozza.[5]

The debut of this last society in the international eugenic movement was singularly distinguished, in August 1919, by a letter from Gini to Leonard Darwin, in which he proposed the introduction of a racist legislation that would impede matrimonial unions with the "African races" throughout all of Europe:

> At the victorious end of the world war, the allied powers find themselves in increased contact with the African world. It would therefore be opportune if the various eugenic societies aimed to gain legislative orders from the governments of the various nations, where such laws do not already exist, banning marriages between Europeans and the African races, allowing only those with Mediterraneans (Berbers, Egyptians) and with non-colored Arabs. Such bans must be extended to marriages with all those population groups of mixed blood scattered throughout the African continent. The scope of the proposal is to impede the growth of a European–African mixed-blood race, which, from various points of view, is undesirable.[6]

The document was prepared by the anthropologist Vincenzo Giuffrida-Ruggeri and approved several days earlier—on 27 July 1919—by the directive committee of SIGE.[7] The proposal was re-voiced by the engineer Buo-

---

4 For an intellectual profile of Aldo Mieli, see Claudio Pogliano, "Aldo Mieli, storico della scienza," *Belfagor* 5 (1983): 537–57.

5 SIGE was established on 15 March 1919: the president was Pestalozza, vice-president Gini, secretary Artom and vice-secretary Boldrini. SIGE's steering committee included representatives of different disciplines: Vincenzo Giuffrida-Ruggeri (anthropology), Cesare Artom (general biology), Romualdo Pirotta (botany), Giulio Fano (physiology), Alessandro Ghigi (zoology), Bartolomeo Moreschi (zootechnics), Francesco Radaeli (Dermo-syphilopathic Clinic), Vittorio Ascoli (clinical physician), Giuseppe Sanarelli (social hygiene), Ettore Marchiafava (general pathology), Giovanni Mingazzini (psychiatry), Ernesto Pestalozza (obstetrics and gynaecology), Silvio Longhi (juridical science), Achille Loria (social science), Corrado Gini (statistics), Giovanni Marchesini (moral science), Enrico Modigliani (paediatrics). See *Atti della Società Italiana di Genetica ed Eugenica* (Rome: Tipografia del Senato di G. Bardi, 1920), 6–7 and 9. See "Società italiana di genetica ed eugenica," *Rassegna di studi sessuali* 1, no. 1 (January–February 1921): 53. On the modification of the statute, see *Rassegna di studi sessuali e di eugenica* 6, no. 3 (September 1926): 292–93.

6 Corrado Gini to Leonard Darwin (1 August 1919), Wellcome Institute, SA, EUG, c. 123.

7 *Atti della Società Italiana di Genetica ed Eugenica,* 8–9.

nomo, at the general meeting of the African Society of Italy, on 28 August, 1919:

> In the bulletin of this worthy Society (September–October 1919, issue V, year XXXVIII), an article by the engineer Buonomo appeared, based on an important document from the Apostolic Curacy of Eritrea, in which he considered the serious troubles that derive from the union of white men with black women, since, among other things, half-castes seem in general to display a very weak physical constitution and consequently are endowed with very little pro-active energy.[8]

Rejected by the British Eugenics Education Society in May 1920 for being substantially premature from a political point of view,[9] such racist proposals remained the principal initiative of SIGE in its first five years of activity, that is, until the organization of the first Italian Congress of Social Eugenics, in 1924.

The links between IPAS, SISQS and SIGE appeared very close from the beginning, and were further reinforced by their common interests in the eugenic field. From January 1922, IPAS supplied SIGE with valuable technical-organizational support, putting the institute premises at their disposal, printing the proceedings of the Society meetings, somewhat irregular, in the pages of *Difesa sociale*, and allowing the members access to the library, "rich in Italian and international booklets and journals, with many of direct interest for the students of genetics and eugenics."[10] In 1924 also SISQS—which had, in the meantime, seen a notable increase in members and regional groups[11]—attempted to strengthen SIGE, offering a series of special terms for the members who were part of both the societies, and industriously publishing the SIGE's minutes in the *Rassegna di studi sessuali e di eugenica* [Review of sexual studies and eugenics].[12]

8 *Atti della Società Italiana di Genetica ed Eugenica*, 9.

9 *Atti della Società Italiana di Genetica ed Eugenica*, 9 (reply not signed by the Eugenics Education Society, 7 May 1920).

10 See *Difesa sociale* 1 (1922): 18.

11 See Mantovani, *Rigenerare la società*, 242–43. At the start of 1924, the number of members was over one hundred: see "Società italiana per lo studio delle questioni sessuali," *Rassegna di studi sessuali* 4, no. 1 (January–February 1924): 42.

12 See *Rassegna di studi sessuali e di eugenica* 4, no. 3 (May–June 1924): 215–16.

In fact, the sources indicate the existence of a single issue of SIGE's minutes, dated July 1920, comprising several statements written by representatives of the different views of the association: biologist Cesare Artom, psychologist Giovanni Marchesini, anthropologist Vincenzo Giuffrida-Ruggeri and economist Achille Loria.

Artom's contribution summarised the most recent scientific literature regarding Mendelian laws and the chromosomal theory of heredity:[13] for the biologist, eugenics was in fact considered a "subdivision of the study of genetics," since "it resolved to definitively deepen for the human species that which from a complex of data it is already possible to presuppose, which is that (Mendelism apart) the same hereditary laws must hold true for all living organisms, excluding none."[14] As for the "practical scope," Artom argued, eugenics had to follow "completely different directions" from those of genetics, "as for mankind it is not possible to fall back upon artificial selection; and the same genetic isolation of individuals unsuitable for marriage is, for obvious reasons, very difficult to achieve."[15] More than eugenics, it was necessary therefore, to speak of "euthenics" (from the Greek *ευτηνάω*, "to flourish"), that is, of that "special branch of studies that directs all its attention to the influence that the environment has on the occurrence of a number of hereditary factors."[16]

Giovanni Marchesini, on the other hand, insisted on the necessity of investigating the "biological basis of the life of the spirit."[17] In polemics with Benjamin Kidd's position, which was critical toward Galtonian determinism, Marchesini confirmed the relevance of the "bio-psychical predisposition" in defining the "soul of the people":

> The exterior conditions variously influence the life of humanity. Prosperity and poverty, for example, have very different actions, as do liberty and servitude; and the faith in the effectiveness of the reform of institutions is legitimate, as an armor (if I may be allowed the phrase) of the social soul. But we

13 Cesare Artom, "Indicazioni sommarie sugli studi di genetica," in *Atti della Società Italiana di Genetica ed Eugenica*, 15–20.

14 Cesare Artom, "Per gli studi di genetica ed eugenica," in *Atti della Società Italiana di Genetica ed Eugenica*, 13.

15 Artom, "Per gli studi di genetica ed eugenica," 13–14.

16 Artom, "Per gli studi di genetica ed eugenica," 14.

17 Giovanni Marchesini, "Il fattore psicologico nel dominio dell'eugenica," in *Atti della Società Italiana di Genetica ed Eugenica*, 24.

> will act productively on social life, from without, only when we know how to penetrate the biological substrate of the individual psyche.[18]

This approach to the problem of heredity did not however convince Marchesini to share the negative eugenics theorized by the French physiologist, Charles Richet. In his 1919 influential book *La sélection humaine* [Human selection], Richet advocated drastic measures such as sterilization, segregation of defectives and marriage prohibition.[19] "Negative coercion," according to Marchesini, was indeed only a "single and partial aspect of the practical problems of eugenics":

> For improvement of the human species, negative means adopted against the most commonly manifested specific degenerations are not enough. Mental defectives do not fully respond to persuasive action, as an element of their deficiency is their inability to inhibit cruder instincts; but we cannot assert that positive action, psychological, might not anyway be exercised on a large scale, in various aspects and in different ways.[20]

In contrast to Richet's crude prescriptions, Marchesini proposed a "positive eugenics,"[21] based on the radical renovation of educational methods. In particular, he suggested the introduction of a "scientific education" into the scholastic environment, which would promote a "realistic intelligence" among adolescents, a sort of anti-romantic approach to sexual hygiene.[22] Because the "eugenic ideal"[23] took on the "value of a religion," coercive measures would in fact be less effective that those "constrictions that came to the subject from his knowledge and from the intimate persuasion of his spirit."

Anthropologist Vincenzo Giuffrida-Ruggeri agreed that the environment was "not omnipotent": "Where antisocial hereditary factors enter the game, they are stronger than the environment, and effectively antisocial beings exist in any environment."[24] In this sense, the genetic research on the

18 Marchesini, "Il fattore psicologico nel dominio dell'eugenica," 24.
19 On Charles Richet's *La sélection humaine*, see, in particular, Schneider, *Quality and Quantity*, 109–13.
20 Marchesini, "Il fattore psicologico nel dominio dell'eugenica," 26.
21 Marchesini, "Il fattore psicologico nel dominio dell'eugenica," 28.
22 Marchesini, "Il fattore psicologico nel dominio dell'eugenica," 28–29.
23 Marchesini, "Il fattore psicologico nel dominio dell'eugenica," 24.
24 Vincenzo Giuffrida-Ruggeri, "Il problema fondamentale dell'eugenica," in *Atti della Società Italiana di Genetica ed Eugenica*, 31.

existing links between "hereditary factors" and "mental habits" could even furnish "a zoological basis for criminal Lombrosian anthropology."[25] "Acting on an organism" did not signify resorting to Richet's coercive measures. Giuffrida-Ruggeri believed that a "State control of marriages, which offers certain health guarantees,"[26] was indispensable. In contrast to a "barbaric system of castration, propagated by selectionists,"[27] he suggested it was preferable to take direct action aimed at chemically modifying the "germ plasm," and, at the same time, promote genealogical research designed to better define the relationship between the morphological aspects and the best mental and behavioral attitudes.[28]

Achille Loria focused his contribution on the influence of environmental factors, and in particular on socioeconomic conditions. Paraphrasing Rousseau, the economist declared: "Man issues forth from the hands of the Creator healthy and immaculate, but it is the social institutions that corrupt and deprave him."[29] It was not biological heredity but rather the "working class background" that was the "great factory of so-called born delinquents, of prostitutes, of all the degenerations of body and soul, and all the vile pains for which mankind blames nature."[30]

And the recent worldwide conflict worsened a situation that was already dramatic: the cost of provisions, Loria argued, forced "painful and harmful privations" upon the workers; the frequent strikes produced an "ill-omened see-saw of employment and unemployment," strengthening the "moral anxiety"; the wartime fortunes caused "the immediate rise to opulence of the most vulgar and despicable people," creating an "aristocracy devoid of every moral and aesthetic quality"; the worsening of living conditions consigned couples to marry "within the orbit of their own class, stunting the crossbreeding between stocks that are so biologically providential."[31] A final "anti-eugenic influence" was connected not to the poverty of the working class, but to its growing prosperity. With the increase of income,

25 Giuffrida-Ruggeri, "Il problema fondamentale dell'eugenica," 34.
26 Giuffrida-Ruggeri, "Il problema fondamentale dell'eugenica," 33.
27 Giuffrida-Ruggeri, "Il problema fondamentale dell'eugenica," 35.
28 Giuffrida-Ruggeri, "Il problema fondamentale dell'eugenica," 35.
29 Achille Loria, "I confluenti economici dell'eugenismo," in *Atti della Società Italiana di Genetica ed Eugenica,* 37–38.
30 Loria, "I confluenti economici dell'eugenismo," 38.
31 Loria, "I confluenti economici dell'eugenismo," 39.

in fact, they would reduce the "reproduction coefficient," as a result of the diffusion "among the most numerous classes" of the birth control practices that had previously been seen exclusively in the "bourgeoisie and capitalist classes." The worry about undergoing the same sort of depopulation seen in France was oppressive:

> Now it is a whole fertile fount of life, emerging from the youngest and most vigorous spring, which will in this way be exhausted. And from this will come disastrous consequences, already seen in France, where, as a consequence of voluntary sterility, the hearths are empty, and there is female alcoholism, general depravation.[32]

While, at the end of the 19th century, the "health issue" was connected with the social issue, now, according to Loria, the "economic factor" was the core of the "eugenics issue."[33] Although these proceedings of SIGE represent a good example of the theoretical and practical orientations of Italian eugenicists, nevertheless the occasional character of the publication constitutes a clear sign of the organizational difficulties of the association. Notwithstanding this, the eugenic debate that developed in Italy after the war maintained its richness and articulated itself along thematic lines that will be briefly dealt with here: in particular, birth control, premarital certificates, sterilization and mental hygiene.

1. Ettore Levi and the IPAS Campaign for Birth Control

The project for the "creation of an Italian Institute for social hygiene and assistance" was detailed in a pamphlet written by the neuropathologist Ettore Levi in 1921; 6000 copies of which were distributed in cooperative banks, at mutual savings banks, to industrialists, proprietary limited companies, agricultural entities, Colonial institute divisions and to all the provincial physicians.[34]

---

32 Loria, "I confluenti economici dell'eugenismo," 39.

33 Loria, "I confluenti economici dell'eugenismo," 39–40.

34 Ettore Levi, *La medicina sociale in difesa della vita e del lavoro* (Rome: La Voce, 1921).

For Levi—already vice president of the National Society for the Protection and Assistance of War Invalids and member of the Council of the National Board of Health[35] (*Consiglio Superiore di Sanità*)—the war had fully revealed the urgent need for centralized organizational structures with the aim of fighting against social illnesses:

> The war has acted like a revealing photographic wash, abruptly evidencing and multiplying the infinite misery latent in every single individual, constituting the social masses: and so, due to the war, tuberculosis sufferers, psychopaths, cripples, mutilated, blind etc. have become a burden to the State. In this way the State has suddenly seen the social importance, both morally and economically, of the great problems of assistance in times of peace, for which they are totally unprepared, but which they must forcefully shape.[36]

Beyond acting as a "heroic remedy" and revelator, the war had also signaled the definitive transformation of the concept of charity and beneficence in civil assistance. Levi declared:

> The times demand that ancient, insufficient, often hypocritical charitable works be substituted with a vast, enlightened and sincere organization of civil assistance, conceived not as a test of generosity, but as a fundamental duty of the most cultured and fortunate classes to those most ignorant and miserable.[37]

Therefore, while the conflict had demonstrated the impact of social illnesses in all its seriousness and affirmed the need for a secular model of social assistance, the "wartime bleeding" had also taught much, showing the extreme importance of a "unity of command":

> In the fight against social illnesses, the unity of command is no less essential that in facing the wartime enemy: some European states have sought to realize such unity, with the recent institution of ministries of hygiene and social assistance, which however have not yet had the methods, nor the time, to demonstrate their proactive possibilities.[38]

---

35 For the curriculum vitae of Ettore Levi, see the documentation sent by Levi to Mussolini's secretary on 12 March 1930, in ACS, SPD, CO, b. 109005/2, "Levi Ettore." For more informations on him as a eugenicist, see Mantovani, *Rigenerare la società*, 215–25, and Roberto Maiocchi, *Scienza italiana e razzismo fascista* (Florence: La Nuova Italia, 1999), 14–22.

36 Levi, *La medicina sociale in difesa della vita e del lavoro*, 10.

37 Levi, *La medicina sociale in difesa della vita e del lavoro*, 10.

38 Levi, *La medicina sociale in difesa della vita e del lavoro*, 14.

In the expectation that Italy would also create a Ministry for social hygiene, Levi proposed the institution of a centralized bureaucratic organ that would function as an "agent of stimulation and liaison" between the governmental and semi-governmental organizations active in the social assistance sector. In Levi's project, this central organ would be instrumental in managing the problem of social illnesses more efficiently and profitably. It was not only humanitarian intentions, but also, and above all, the exigencies of saving and economy that imposed the organization of "prophylactic social health"[39] against the ills of alcoholism, tuberculosis, syphilis, and mental illnesses.[40]

Levi's economist and productivist logic was complemented by the image of the alliance between capital and work in the face of the common enemy:

> Why shouldn't the fight against social ills be, once and for all, established by men of organizational genius, both industrialist and workers, and conducted with the methods and means that have given rise and caused the prosperity of the great companies that characterize our current civilization?[41]

In fact, Levi's project seemed to quickly arouse sympathy among the liberal right and also among the socialists of the review *Critica sociale* [Social critique].[42]

In less than a year, the inter-classist and technocratic dream, contained in the pamphlet of 1921, was realized. In 1922, thanks to the patronage of illustrious personalities[43] and the financing, among others, of the financial institutions *Credito Italiano* and *Banca Commerciale*, IPAS was born: a "group of study and social action," that was immediately characterized by intense activity in the hygienic education of the popular masses,[44] by the

39 Levi, *La medicina sociale in difesa della vita e del lavoro*, 19.

40 Levi, *La medicina sociale in difesa della vita e del lavoro*, 19.

41 Levi, *La medicina sociale in difesa della vita e del lavoro*, 9.

42 For the debate on *Critica sociale*, see Ettore Levi, *I partiti e la salute della stirpe* (Rome: IPAS 1921).

43 Among others, Luigi Luzzatti, Benedetto Croce, Camillo Golgi (Nobel prize winner for medicine and president of the Superior Council of Public Health), Bonaldo Stringher (director of the Bank of Italy and president of the National Insurance Institute), Gino Olivetti (secretary of the General Industrial Confederation), Pio Foà (president of the Italian Anti-Tubercular Federation), Giuseppe De Michelis (Commissioner General of Emigration), Ettore Marchiafava (malariologist and vice-president of the Italian Red Cross).

44 In particular, propaganda posters for schools and workplaces (*Direttissimo della salute* [Health express], *Alfabeto della salute* [Alphabet of health], *Medusa*, *Conquista della salute* [To conquer health]); the reprint of the volumes dedicated to social hygiene; the establishment in 1924 of the first *Filmoteca Nazionale di Educazione Sociale* [National Film Archive of Social Education]. See Mantovani, *Rigenerare la società*, 218–19.

training of health personnel, and by the broad strategy of organizational connection (national and international) between the numerous associations active in the field of assistance.[45] From 1922 IPAS also published a review, significantly titled *Difesa sociale,* which became, under the direction of Ettore Levi, one of the most authoritative voices of Italian eugenics. While the first editorials insisted above all on the "economic value of human life" and the "struggle of the parties" as "precious instruments of civil progress,"[46] the January 1923 number, with tones of hope, welcomed the rise of fascism, anticipating that the "new man" guiding the country would fully assume the urgent work of biological renewal of the stock, neglected by the preceding liberal government:

> Will the new government impose the realization of this effort separate from every concept of class or party, for the civil greatness and economic power of our country, in a superior vision of defense and reconstruction of the potential individual and collective physical and intellectual energies?[47]

Displaying some ideological analogies with the *Menschenökonomie* of contemporary Weimar eugenics,[48] the eugenic paradigm promoted by Ettore Levi had essentially two characteristics: first, the rejection of coercive eugenics, and, second, the centrality of birth control as a principal selective measure. Well informed on the European[49] and American eugenic legislations,[50] Levi was not, however, disposed to underwrite policies of sterilization and marriage bans. The first barrier to using such tools was, in Levi's opinion, the scanty scientific knowledge of human heredity:

45 In particular, Levi, in his role in the League of Red Cross Societies, supported the foundation of a Central International Committee for the Coordination of the International Federations of Preventive Medicine and Social Relief; see Ettore Levi, *Central International Committee for the Coordination of the International Federations of Preventive Medicine and Social Relief* (Rome: IPAS 1924), cited in Mantovani, *Rigenerare la società,* 220.

46 Ettore Levi, "Per l'avvenire della razza," *Difesa sociale* 1, no. 1 (January 1922): 7.

47 Ettore Levi, "Alle radici dei mali sociali: il fascismo alla prova," *Difesa sociale* 2, no. 1 (January 1923): 3.

48 See, in particular, Weindling, *Health, Race and German Politics,* 399–440.

49 See, for example, "La visita prematrimoniale in Danimarca e in Austria," *Difesa sociale* 2, no. 11 (November 1923): 12–13; "Austria. Visita medica prematrimoniale," *Difesa sociale* 4, no. 10 (October 1925): 23–24; "Belgio. L'esame medico prematrimoniale," *Difesa sociale* 5, no. 4 (April 1926): 18.

50 See, for example, "Stati Uniti. Il certificato medico prematrimoniale," *Difesa sociale* 4, no. 7 (July 1925): 23; "Cenni storici e critici sulla sterilizzazione eugenica," *Difesa sociale* 5, no. 5 (May 1926): 10–11.

> The current knowledge of the laws of heredity is not such as to permit us to stabilize exact rules that indicate who can dedicate themselves to the reproduction of the species and who, causing hereditary defects, should abstain.[51]

In second place, man was not only the product of the determinism of Mendelian laws. On the contrary, Levi claimed, environmental factors also existed, which could not be ignored:

> Man comes into this world with a certain number of tendencies that can then be modified through contact with civilization and the environment, which helps to form the mature man. Evidently, to obtain the best results, the best innate qualities are therefore as necessary as the best environment. Our children need to have the best blood and the best education. Hereditary factors merit great attention; at the same time we must not ignore social reform that concerns the environment. Thinking about those who will be born is a moral duty that must be imposed as a duty on people such as ourselves; it is to these ethical and social ends that the doctrines and suggestions of modern eugenics aims.[52]

Although conscious of the fact that "the danger of physical and intellectual degeneration of the race exists [...] undeniably, and it is connected to the problem of multiplication of physically and psychically defective elements in society,"[53] Levi maintained nevertheless the uselessness of the adoption of "draconian laws": in fact, "who can say where abnormality begins? Who could say when abnormality becomes genius? Therefore preventing the birth of an abnormal does not deprive society of one of its greatest sons?"[54] Neither would social action aimed at favoring the fertility of the so-called "normals" be worthwhile, because the "major prolificacy is always found where poverty, mental deficiency and vice slacken the spirit of prudence and the desire for economic wellbeing visible in balanced individuals."[55] Against "coercive" eugenics, based on the Anglo-Saxon model, Levi pro-

---

51 [Ettore Levi], "Contenuto etico e sociale dell'Eugenica," *Difesa sociale* 4, no. 11 (November 1925): 14. For a discussion of Mendel laws, see R. Righetti, "Le basi scientifiche del movimento eugenico," *Difesa sociale* 4, no. 12 (December 1925): 10–14.

52 [Levi], "Contenuto etico e sociale dell'Eugenica," 15.

53 [Ettore Levi], "La fecondità dei deficienti come problema di Eugenica," *Difesa sociale* 5, no. 1 (January 1926): 15.

54 [Levi], "La fecondità dei deficienti come problema di Eugenica," 16.

55 [Levi], "La fecondità dei deficienti come problema di Eugenica," 16.

posed instead "negative" eugenics, which he interpreted as social and individual hygiene:

> The only other solution to the problem is the negative side of eugenics; that is, that which highlights the causes of the progressive increase of dead weight, hanging always more threateningly over society, pushing individuals and authority to give a stronger importance to individual and social hygiene [...].[56]

The exchange of words between hereditarian Angelo Zuccarelli and environmentalist Levi, in the columns of Pietro Capasso's *Pensiero sanitario* [Sanitary thinking], are highly illuminating in defining the eugenic discourse of *Difesa sociale*. Zuccarelli denounced the curious absence in the journal of references to the essential priorities of eugenics—the discipline of marriage and the prevention of reproduction for degenerates. To this criticism, *Difesa sociale*'s editors responded by claiming that all hygiene and health activities, described and supported by the journal, were "essentially eugenic."[57]

Levi's eugenics did include the broadest and most differentiated medical perspectives, ranging from the prevention of social illnesses to mental hygiene; from the scientific organization of work to medical assistance for maternity and infancy. Nevertheless, in such a vast conceptual and disciplinary articulation, one theme seems to emerge with particular clarity, synthesizing Levi's eugenic positions: that of birth control. In the pre-war period, the ephemeral battle in favor of neo-Malthusianism had been conducted by the radical anarchic left, in particular by the Neo-Malthusian League (*Lega Neomalthusiana*) and the review *L'educazione sessuale* [Sexual education] (not surprisingly subtitled *Review of neo-Malthusianism and eugenics*), which was directed by the Turin physician Giuseppe Berta.[58] In the first post-war years, the eugenic ambitions of gynecologists and puericultors were not directed at birth control, but rather at a program of protection of maternity and infancy. The eugenic paradigm that justified such an

56 [Levi], "La fecondità dei deficienti come problema di Eugenica," 16.

57 See Angelo Zuccarelli, "Al professor Ettore Levi, membro del Consiglio superiore di sanità," *Il pensiero sanitario* 18 (1922): 3–4; Ettore Levi, "Risposta al professore A. Zuccarelli, in tema di eugenica," *Il pensiero sanitario* 19 (1922): 3–4.

58 See Mantovani, *Rigenerare la società*, 128–31.

assistance-oriented approach quickly became famous in the Italian scientific community, such as the "law of Pieraccini," named for its author:

> Women [...] accomplish the task [...] of carrying the organism and the correlating functions (in male and female children) on the physiological-median line. This natural function of women [...] we believe can be expressed in the formula: It is the work of the woman, through reproduction and heredity, to carry the accentuated organic-functional "fluctuations" and the same physiological deviations (degenerative or hereditary; pathological or acquired; male or female) to the respective biological center of the human species.[59]

Preserving the "average man" had been attributed to the female element, thus enabling the insistence of the "eugenic" role of maternity and legitimizing, on a scientific basis, the return of women to their traditional social rank, after the period of exceptional participation and emancipation created by the war. It also fuelled the development of a "social obstetrics" that aimed to further extend the power of the State—through the mediation of physicians as "regenerators" of the stock—to the management of the national biological patrimony.[60]

The first Congress of Social Obstetrics was held in Rome on 6–8 January 1919, to discuss the "problems of eugenics which could be vital for the events of the nation and the race."[61] On this occasion, abortion and therapeutic sterilization were harshly condemned, and a resolution was approved, proposed by the gynecologist Tullio Rossi-Doria—socialist[62] and early eugenicist[63]—which rationalized maternal assistance through the creation of an Institute of Maternal and Infantile Assistance (*Istituto di*

59 Gaetano Pieraccini, *La stirpe dei Medici di Cafaggiolo. Saggio di ricerche sulla trasmissione ereditaria dei caratteri biologici* (Florence: Vallecchi, 1924), 461–62.

60 For more information on these aspects, see Francesco Campione's book, *Per i germi della specie* (Bari: Laterza, 1920) and the articles in the journal *L'igiene e la vita* by physician and socialist member of Parliament Giulio Casalini. See Mantovani, *Rigenerare la società*, 187–90.

61 Tullio Rossi-Doria, "Congresso di ostetricia sociale," *Il Policlinico - Sezione pratica* 3 (19 January 1919): 79.

62 For an analysis of the clash between the activities of Rossi-Doria in the field of "social medicine" to protect the weaker classes and the maximalist left-wing of the socialist party, see Tullio Rossi-Doria, *Medicina sociale e socialismo. Scritti per l'educazione politica e igienica dei lavoratori* (Rome: Mongini, 1904).

63 Tullio Rossi-Doria supported, at the end of the 1800s, the Lamarckian theory of the heredity of acquired characteristics, to reinforce the importance of "preventive medicine" in the rational and hygienic management of the reproductive process: see Tullio Rossi-Doria, *L'eredità delle malattie* (Milan: Vallardi, 1893). In 1913, he became a member of the Italian Committee of Eugenic Studies.

Assistenza Materna e Infantile),[64] in which the foundations of the future fascist ONMI could easily be seen.

From the beginning, faced with the "regenerative" and "quantitative" eugenics of Italian gynecologists and "pediatricians," the assiduous work of Levi's IPAS in support of birth control assumed the form of an arduous and isolated intellectual undertaking. A clear testimony to this was the debate that took place at the conference held by Levi in January 1924, at a meeting of the Roman section of SISQS, on the theme of *Birthrate and eugenics.*

The central nucleus of Levi's presentation was the reaffirmation of the eugenic value of birth control: a "rule of special conduct, to be observed in married life, so that healthy and physically and mentally normal offspring could be had at the most opportune and desired moment, with the noble objective of allowing people to raise and educate children in the best way, with the superior aim of giving families and society intelligent and proactive elements."[65] The list of advantages of birth control was long: individual (economic safety, improvement of women's health, balanced growth of children), collective (reduction of social tensions and conflicts), medical-eugenic (less reproduction of defective individuals, reduction of social illnesses), moral-religious (rational discipline of the sexual impulse, fight against abortion and infanticide).

Supported by a broad display of data relative to the international context, particularly Anglo-Saxon, Levi proposed basing the legitimacy of birth control on eugenic responsibility and efficiency. This, Levi argued, would aim at reinforcing, rather than damaging, moral tradition:

> The supporters of birth control aim to introduce to the masses, especially the inferior classes, a sense of responsibility, which has until now been lacking, since in such classes, more than in the others, they are free to give vent to blind and at times brutal instinct. The aim is moreover, or rather, above all, to rein-

64 Tullio Rossi-Doria, "Congresso di ostetricia sociale," *Il Policlinico - Sezione pratica* 4 (26 January 1919): 113. The Congress was characterised by papers on the "protection of legitimate pregnancy" (E. Truzzi); "illegitimate pregnancy" (E. Alfieri); the "assistance for illegitimate children" (O. Viana); the public diffusion of "obstetrical hygienic norms to advantage the mother and the unborn child" (T. Rossi-Doria); the "tuberculosis in pregnancy and anti-tubercular prophylaxis in infancy" (L. Mangiagalli), on the prophylaxis of syphilis (I. Clivio); "alcoholism and maternity" (E. Ferroni); and on methods to slow "the increasing frequency of criminal abortions and neo-Malthusian practices" (E. Pinzani).

65 Ettore Levi, "Il controllo delle nascite (neomalthusianismo)," *Rassegna di studi sessuali* 4, no. 1 (January–February 1924): 24–25.

> force the institute of marriage, condemning healthy people who choose voluntary celibacy and advising (contrary to Malthus, who preached protraction) marriage at a young age.[66]

However, the debate that followed Levi's presentation was certainly not favorable to his hypotheses. Senator Pestalozza expressed "deep reserve regarding recourse to contraceptive means, underlining the damage that could be done to women's health." Silvestro Baglioni, president of SISQS and director of the Institute of Physiology of the University of Rome, doubted the effectiveness of contraceptive means in achieving eugenic aims, because it wasn't possible "to apply certain laws to men, which hold true for plants and animals." Pietro Capasso however, was moderately favorable. After contesting the connection between prolificacy and national wealth, he declared himself in favor of a eugenic campaign regarding birth control and obligatory premarital certificates.

The most articulate criticism of Levi's position—and the most influential politically—came from Corrado Gini, firm opponent, from 1922 and in the same column of *Difesa sociale*, of neo-Malthusianism and "Anglo-Saxon" eugenics.[67] In particular, it was the essay *Le basi scientifiche della politica della popolazione* [The scientific bases of population policies] in which Gini developed a systematic analysis of what he considered the three principles of the "quantitative" and "qualitative" rationalization of births: the selection of "reproducers," the eugenic control of marriage, and the limitation of births.

Regarding the first aspect, Gini emphasized the difficulty of defining the hereditary mechanism with certainty:

> Our knowledge of heredity is still too uncertain to allow exact prognostications on the hereditary transmission of certain defects, and even less to be able

66 Levi, "Il controllo delle nascite," 29.

67 Gini was the only Italian, together with Ettore Levi, to participate in the Sixth International Malthusian and Birth Control Conference, with a presentation titled "On Birth Control," later published in *Difesa sociale* 4, no. 3–4 (March–April 1925): 83–87. See also Corrado Gini, "Il neomalthusianismo," *Difesa sociale* 1, no. 8 (August 1922); Corrado Gini, "Prime ricerche sulla fecondabilità della donna," *Atti del Regio Istituto Veneto di Scienze, Lettere ed Arti* 83, part 2 (1924): 315–44; Corrado Gini, "Nuove ricerche sulla fecondabilità della donna," *Atti del Regio Istituto Veneto di Scienze, Lettere ed Arti* 84, part 2 (1925): 269–308; Corrado Gini, "Decline in the Birth-Rate and the Fecundability of Woman," *Eugenics Review* 17 (January 1926): 258–74.

> to say with precision if such defects will be transmitted in amounts that would cause serious harm to society.[68]

The phenomena of induction, the transmission of functional diathesis and the evolution of the germ plasm[69] made the identification of effectively hereditary characteristics arduous:

> When nature is left free to exercise selection on a stock, we understand that, finally, the selection of the best will occur through the difference of mortality; but if, on the other hand, we wish to pre-emptively choose the good reproducers, from the eugenic point of view, it is too easy to err, by confusing congenital characteristics with acquired ones, and induced congenital characteristics with those that are truly hereditary.[70]

And, equally, "we do not have the ability to distinguish individuals who are the best due to the truly superior quality of their germline from those individuals who are the best only because their germline is currently in its full bloom."[71] In addition, further complicating the situation, the possibility of an illness considered hereditary being, on the contrary, a "transitory illness of the germ" that had an immunizing effect, contradicted the idea of selection of the best reproducers, because "healthy reproducers could at times be worse than others from the point of view of the next generations, on whom they will not confer any immunization."[72]

As for matrimonial selection, Gini came back to Mendelian determinism, in order to substantiate his condemnation of any kind of eugenic regulation of unions between spouses. The question was very clear: given Mendel's laws, was it more advantageous to favour marriages of "defective" individuals with healthy ones, "in the hope of gradually obtaining, in this way, a decrease in the illness," or alternatively, was it better to favor unions between healthy people, leaving ill people to marry among themselves,

68 Corrado Gini, *Le basi scientifiche della popolazione* (Catania: Studio editoriale moderno, 1931): 103. The book is the fruit of the conference held in 1927 at the Italo-Brazilian Culture Institute (*Istituto di Alta Cultura italo-brasiliano*) in Rio de Janeiro, integrated with the university lessons from the years 1927–28 and 1930–31.

69 See ch. 4, 179–183.

70 Gini, *Le basi scientifiche della popolazione,* 112.

71 Gini, *Le basi scientifiche della popolazione,* 117.

72 Gini, *Le basi scientifiche della popolazione,* 118.

"counting on a more rapid extinction of their stock, due to its lesser organic resistance"?[73] Since the major part of pathological characteristics showed recessive behavior, the question could be reformulated in these terms: was a generation of healthy heterozygotes better, even if it would give rise to a certain percentage of ill ones, or was it better to have two distinct classes of homozygotes: healthy and sick? Gini's response was again direct:

> This second solution appears definitely preferable, at least at first glance, as the ill individuals tend toward extinction; but this would not be the case if it was demonstrated that [...] the carriers of certain diseased factors had superior reproductive powers. If this were the case, the system of isolating and coupling ill people among themselves could lead, instead of to a decrease, to a multiplication of the pathological sources.[74]

This was without also counting the enormous complications of hereditary transmission in cases of crossings between individuals of different stock.

However, as can be easily imagined, while the second point of a "program of reproductive rationalization" was "extremely problematic for practical realization,"[75] Gini's judgment on the third point—birth control—was absolutely negative. In his eyes, the "rearing of man could not constitute an economic act."[76] Rationality would only induce couples to desire one or few children:

> It is indisputable that, in the majority of cases, a family of whichever social class can not, with only work, obtain the means to maintain, at an appropriate social level, eight children. In the working classes, where the costs of raising a child are much lower, I would say that a married couple could, on average, with appropriate work, and maintaining a proper level of education for their children, raise no more than four, and in the middle classes, no more than two.[77]

Consequently, if reasoning was based on the economic advantages of raising children, it would "finish with raising few."[78] And the first damage of birth

73 Gini, *Le basi scientifiche della popolazione,* 125.
74 Gini, *Le basi scientifiche della popolazione,* 125.
75 Gini, *Le basi scientifiche della popolazione,* 130.
76 Gini, *Le basi scientifiche della popolazione,* 137.
77 Gini, *Le basi scientifiche della popolazione,* 136.
78 Gini, *Le basi scientifiche della popolazione,* 137.

control would be of an economic character: the production of men, Gini claimed, followed the laws and rhythms of biology, not those of the market, and therefore could never be rational. To this must be added the negative consequences from a psychological and moral point of view, with the triumph of individualistic egoism and the disintegration of the family unit:

> If we commence reasoning on the question of procreation, we will not finish reasoning only at that certain point of rationalization of the birth rate that the partisans would like. We pass quickly to considering why we should identify personal interest with the interests of the family unit, often concluding that it is not reasonable to sacrifice our individuality to it.[79]

Birth control, for Gini, was an even more dangerous weapon because it threatened to escape from the hands of neo-Malthusians, leading in the end to the political collapse of the nation. In fact,

> when rationalistic practices take strong hold in a country, and the birth rate continues, for a certain period, to diminish, it is very difficult to arrest this descent. Individuals may subsist, but the nation, the race, is condemned: it will disappear, or at least lose its proper place in the world, to the benefit of those nations that, obeying instinct, still have the necessary vitality to maintain themselves and multiply.[80]

Therefore, while Gini developed the theoretical synthesis between "Latin" quantitative eugenics and fascist pronatalist policy, Ettore Levi continued—from an increasingly isolated position—his campaign in favor of birth control.

In 1924, at the first Italian Congress of Social Eugenics, where Gini's influence was strongly seen, Levi's presentation was one of the few to maintain the eugenic value of birth control. And the next year, at the fourteenth meeting of the Italian Society for the Progress of Science in Pavia (May, 1925), Levi once again repeated the necessity of considering the problem of the "quality" of the population, advising the creation in Italy of a Constructive Birth Control and Racial Progress Society, such as that founded in Great Britain by Mary Stopes. Levi declared:

79 Gini, *Le basi scientifiche della popolazione*, 142.
80 Gini, *Le basi scientifiche della popolazione*, 142.

> Control must be understood not in the restricted sense of limitation to a minimum, but in the broader and more logical sense of a regulation based on rigorous scientific criteria.
> Control until now has been applied without any scientific criteria and without eugenic aims, and it must be added that it has been abused, and is abused even now, causing damage, instead of advantage, to the quality and future of the race.
> Such abuse must cease.
> To achieve this goal, the scientific sphere, and particularly the medical class, must assume management of control, saving it from empiricism, and above all, profoundly studying the question.[81]

In reality, in the Italian context, Levi's hopes were evidently lacking any political future: in October 1924, Mussolini—forgetting his youthful positions—declared his hostility to Malthusian ideas.[82] It was not the "quality" but the "quantity" that concerned fascism, and the Ascension Day Speech (in May 1927) would clearly demonstrate this.

In February 1926, Levi, in order to save his eugenics column in *Difesa sociale*, had no choice but to turn to Corrado Gini:

> [The column] for various reasons, which you know well, has not been realized as I wished. Perhaps you, either personally, or through one of your students or friends, could assure me some articles in the next issues, so that I do not need to end the column?[83]

But while the director of *Difesa sociale* withdrew from public debate just a few months after this letter, due to strong nervous exhaustion, Silvestro Baglioni, the new president of SISQS, published—as part of Capasso's series *Piccola Biblioteca di Propaganda Eugenica* [Small library of eugenic propaganda][84]—his *Principii di eugenica* [Principles of eugenics], which sanctioned orthodox fascist eugenics. On the basis of a statistical "validity curve," in which the first children of a couple were assumed to be also

81 Ettore Levi, "Demografia ed eugenica in rapporto al movimento contemporaneo per il razionale controllo delle nascite," in Roberto Almagià, ed., *Atti della SIPS. XIV riunione (Pavia, 24–29 maggio 1925)* (Rome: SIPS 1926), 120.

82 Anna Treves, *Le nascite e la politica nell'Italia del Novecento* (Milan: LED, 2001), 128.

83 Ettore Levi to Corrado Gini, 1 February 1926, ACS, Gini Papers (hereafter AG), b. b5.

84 For a complete list of the titles in the series, see Mantovani, *Rigenerare la società*, 241.

the least biologically desirable, the physiologist attributed the responsibility for a dangerous "anti-social selection" to birth control:

> Evidently, the nation needs the best, the strongest, the most valid, and not the first two of a series of a married couple, who are or could be, compared to successive children, the least strong and the most degenerate. The clash between the egoistic ideal of the individual and the complex ideal of racial improvement could not be more manifest.[85]

"Eugenic activities" did not therefore consist of the "application of a badly understood voluntary limitation of births,"[86] but of "all the general works that lead to the betterment of the physical and psychical conditions of the parents, aiming above all to combat serious social diseases, such as syphilis and tuberculosis, and those grave poisons of civilization: alcoholism and toxic drugs, abuse of food, and general intemperance."[87] Since "individual cure" would lead to "cure of offspring," eugenics could be seen, in Baglioni's view, as the "corollary" and the "implicit conclusion of general hygienic propaganda."

According to Baglioni, more than biological or medical sciences, eugenic principles should derive inspiration from the "spiritual life," and in particular from art and sentiment. The cult of art represented the beginning of an aesthetical education process, which manifested its eugenic effectiveness in the choice of spouse. The beauty of art was transmitted from the artwork to the spectator, and from this, to the spouse and children:

> These [spectators], in the choice of their spouse and lovers, choose that type of beauty that stays in memory and fantasy, lit by works of art. And as the children born to this couple will share similar characteristics with the parents, in this way we will see the perpetuation of special types of beauty, under the positive perennial action that we can therefore say is the true eugenic action of the works of art.[88]

---

85 Silvestro Baglioni, *Principii di eugenica* (Naples: Edizioni del Pensiero sanitario 1926), 44. See also Silvestro Baglioni, "Problemi eugenici e demografici nei riguardi del rafforzamento della razza," in Lucio Silla, ed., *Atti della SIPS. XXVI riunione (Venezia, 12–18 settembre 1937)* (Rome: SIPS, 1938), 1, 363–96.

86 Baglioni, *Principii di eugenica*, 46.

87 Baglioni, *Principii di eugenica*, 47.

88 Baglioni, *Principii di eugenica*, 51.

But if the "cult of art" could be applied only to a cultivated minority, as far as the majority was concerned, eugenicists had to recourse "to sentiment, and to the most intimate and deep-seated instinct," that of "love for children." This must start with education on marriage and birth, which would precociously involve "the youth, from the start of their sexual life":[89]

> We must search, therefore, to always increase the love for children, even before they are born. It is this antenatal love of children that must be the principal motive keeping the young from the dangers of illness and intoxication that, debilitating their organisms, brutally strikes their germinal elements.[90]

In the same year, although by then in an increasingly isolated position, it was the physician and socialist reformist member of Parliament, Pietro Capasso, who denounced the "dangers" hiding behind the "current incitement to procreate and aggravate the population increase."[91] The excess of births, "continuous and inexorable, not balanced by adequate, healthy and intelligent emigration, far from constructing a 'great venture'" represented, for Capasso, a serious risk "to the well-being and tranquility of the nation."[92] The restriction of immigration achieved in the United States with the Johnson-Reed Restriction Act of 1924, from one side, should definitively shatter the illusion of those who still hoped to find an outlet abroad for the growing Italian demographic pressure. From the other side, Capasso claimed, it was a further proof of how "there, the demographic problem and that of eugenics, primitively understood as protection of the race aiming to give it predominance and superiority in the contact and conflict with other races, are deeply and seriously regarded."[93] Italy should also follow, in other ways, the North American example, tapping into eugenic resources to improve the quantitative and qualitative assets of the population. Perfectly aware of the ideological climate of the day, Capasso drew a clean distinction between eugenics and neo-Malthusianism:

---

89 Baglioni, *Principii di eugenica*, 53.

90 Baglioni, *Principii di eugenica*, 54.

91 Pietro Capasso, *Pressione demografica, emigrazione ed eugenica* (Naples: Edizioni del Pensiero sanitario, 1926), 58. For a similar position, see Leonardo Bianchi, "Iperpopolazione ed eugenica," *Il pensiero sanitario* 3 (1928): 12–16.

92 Capasso, *Pressione demografica, emigrazione ed eugenica*, 17.

93 Capasso, *Pressione demografica, emigrazione ed eugenica*, 28.

> Eugenics is not Malthusianism, nor is it neo-Malthusianism. It has the means to improve the psychical-physical qualities of the offspring and therefore its horizons are not limited to the pallid theories of Malthus, overly linked to an economic determinism that does not greatly consider the laws of biology, basis of the current sociological doctrine.[94]

Eugenics must not be, on the other hand, confused with "theories that refer to acts of physical mutilation," considered "dangerous for every psychical-physical function of the human organism":[95] particularly sterilization should be used only for "recidivist criminals," a "social measure, this, of high value and with merit for reflection and study."[96]

To "prepare a healthy generation," Capasso instead suggested a "severe matrimonial prophylaxis":

> Delay, when the candidate is temporarily able to harm procreation; impede in serious, exceptional cases in which marriage would constitute a true crime for the offspring; avoid mindlessly giving life to syphilitics, idiots, rachitic persons, epileptics, persons with hydrocephaly, and abnormals.[97]

Only by confronting the demographic problem in a "eugenic sense," Capasso concluded, could Italy achieve an "enviable ascent and a race of unexpected supremacy, in an atmosphere of serene effort, creator of restoration, comforts, and happiness," that would contribute "without implications and without hypocrisy to world peace."[98] One year after these words, in 1927, the start of the natalist campaign of the fascist regime would shatter Capasso's hopes.

2. A Concrete Proposal: Premarital Certificates

The history of eugenics legislation started in the United States, when Connecticut enacted a statute in 1895 prohibiting any man who was "epileptic, imbecile or feeble-minded" from marrying a woman under 45 years of

94 Capasso, *Pressione demografica, emigrazione ed eugenica*, 44–45.
95 Capasso, *Pressione demografica, emigrazione ed eugenica*, 45.
96 Capasso, *Pressione demografica, emigrazione ed eugenica*, 49.
97 Capasso, *Pressione demografica, emigrazione ed eugenica*, 46.
98 Capasso, *Pressione demografica, emigrazione ed eugenica*, 58.

age, the presumed limit of child-bearing. In Europe, the first eugenic marriage laws were introduced only after the First World War, as a form of prevention of the diffusion of venereal or mental disease: precisely, in Norway (1919), Germany (1920), Sweden (1920), Turkey (1921), and Denmark (1922).[99] In Italy, in the 27 January 1919 meeting, the social hygiene section of the Post-war Commission, accepting the proposal by the syphilographer Ferdinando De Napoli and his colleagues Achille Sclavo and Cesare Ducrey, approved, in principle, the introduction of medical premarital certificates, which "in regards to syphilis will be more easily acceptable, inasmuch as it will be imposed exclusively on the future husband, in almost all cases responsible for conjugal contagion."[100]

For De Napoli, who recalled Tommaso Campanella more than Francis Galton, it was the duty of every citizen to consider marriage not as an individual act but as "a national service," while the State, for its part, using "all means compatible with nature and sacred human liberty," must appeal to the citizens to "impede the decadence of the race." While it was "not human to regulate the reproduction of mankind as we regulate that of animals or vegetables, it is not prudent or moral to leave marriage without any sanitary control, which could avoid at least the dangers of syphilis."[101]

In the summer of 1920, the Italian Society of Dermatologists and Syphilographers developed a proposal for a law in six articles, signed by professors Radaeli, Fiocco and Fontana, on the prophylaxis of marriage: the male candidates had to obligatorily present a certificate, written by a commission composed of a physician chosen by the candidate and an expert syphilographer. In the case of existing infection, the candidate had to present himself to the municipal authority after a period of time congruent with effecting a cure. Presented to the General Direction of Public Health (*Direzione Generale di Sanità*), the proposal did not have any legislative outcome.[102]

The next year, in October 1921, it was female physicians who supported the introduction of a medical premarital certificate, at their first national

---

99 See Marie-Thérèse Nisot, *La Question Eugénique dans les divers pays*, 2 vols., (Brussels: Librairie Falk Fils, 1927 and 1929).

100 Ferdinando De Napoli, "Lue, maternità, eugenica e guerra in rapporto alla Politica Sanitaria," *Il Policlinico-Sezione pratica* 45 (1919): 1323.

101 De Napoli, "Lue, maternità, eugenica e guerra in rapporto alla Politica Sanitaria," 1326.

102 See Mantovani, *Rigenerare la società*, 179.

conference in Salsomaggiore,[103] and again at the Congress for Family Education (*Congresso per l'Educazione in Famiglia*), convened in Rome in May 1923 by the National Council of Italian Women (*Consiglio Nazionale delle Donne Italiane*).

In November 1922, the Parliamentary Medical Group (*fascio*) again confronted the question, approving a more radical resolution in comparison with the "De Napoli proposal," as it considered premarital certificates obligatory for both spouses and with injunctive powers.[104] In the wake of the parliamentary initiative, Aldo Mieli's *Rassegna di studi sessuali* caused an intense debate, discussing an essay by the Berlin social gynaecologist Max Hirsch, *Chi debbo sposare? Consigli di un medico* [Who should I marry? A doctor's advice],[105] and encouraged a referendum that put the fundamental questions on the table:

> Can matrimonial certificates [...] achieve their predicted scope? And, if they can, is it possible, or useful, to limit personal freedom in this way? In the end, could the certificate, even with its hygienic and health advantages, have drawbacks, perhaps more serious than those it is intended to eradicate?[106]

The first contributor to the debate was the syphilographer Vincenzo Montesano, who clearly expressed his doubts on the effectiveness of the certificate, starting with the organizational and bureaucratic difficulties:

> Don't even ask me if this certificate should be issued by a national commission or by whichever doctor under his own responsibility. In the first case, as usual, we will have another bureaucratic organism, cumbersome and slow, which will complicate things rather than facilitate them. In the second case, we can guarantee that the pseudo-specialists will crowd around, ready to offer, for an adequate recompense, all the certificates people could want.[107]

---

103 See "Primo convegno italiano delle dottoresse in medicina," *Rassegna di studi sessuali* 1, no. 5 (September–October 1921): 278–79.

104 A. M., "Il certificato sanitario prematrimoniale," *Rassegna di studi sessuali* 2, no. 6 (November–December 1922): 357. On 10 February 1923, the Parliamentary Medical Group entrusted Pietro Capasso with preparing a draft bill on "Health certificates for marriage contracts" (*Certificato sanitario dei contraenti matrimonio*): see "Il fascio medico parlamentare," *Rassegna di studi sessuali* 3, no. 1 (January–February 1923): 74.

105 Max Hirsch, *Chi debbo sposare? Consigli di un medico* (Rome: Leonardo da Vinci, 1923).

106 A. M., "Il certificato sanitario prematrimoniale," 357–58.

107 Vincenzo Montesano, "Il certificato prematrimoniale e la profilassi sociale della sifilide," *Rassegna di studi sessuali* 2, no. 6 (November–December 1922): 359.

In addition, the introduction of matrimonial prohibitions would inevitably cause an increase in illegitimate unions and births, as well as abortions and "Malthusian practices," causing serious damage to "social interests."[108] According to Montesano therefore, new laws and new bureaucratic organisms were useless. His motto was: "Let us educate, let us cure."[109] Instead of legislative action, he again invoked the eugenic effectiveness of education:

> If instead of creating new laws against which tricks will be sooner or later easily found, we intensified by all means the anti-venereal propaganda for all the social classes, especially those less advanced, workers, farmers, etc., wouldn't we better achieve the aims we are proposing?[110]

For Montesano, the adoption of a premarital certificate would be at best "a complement to a vast prophylactic organization able to help every individual and family understand the dangers of venereal diseases and defend themselves against them in a rational way, and give all diseased people the easiest and most energetic means to take care of themselves."[111]

Domenico Barduzzi, director of the Dermatology and Syphilis Clinic and the venereal diseases ward, at the University of Siena, agreed with Montesano's position, emphasizing the problem of establishing, in regards to syphilis, "without severe or repeated inquiries, the recovery, when the disease reappears after years and years of latency, especially when due to deficiency or lack of treatment." Instead of premarital certificates, according to Barduzzi, "it would be simpler and less odious to have individual health cards from birth, or a health passport, to accustom the population to value the great importance of national health in every contingency of life."[112]

Ferdinando De Napoli however, was in favor of premarital certificates, not released by a commission, but by a single physician, and limited to men:

> For the man, who is almost constantly the one who carries venereal infection to the marriage bed, very frequently contaminating the purity and poetry of

---

108 Montesano, "Il certificato prematrimoniale e la profilassi sociale della sifilide," 359.

109 Montesano, "Il certificato prematrimoniale e la profilassi sociale della sifilide," 360.

110 Montesano, "Il certificato prematrimoniale e la profilassi sociale della sifilide," 360.

111 Montesano, "A proposito di certificato matrimoniale e di abolizionismo," *Rassegna di studi sessuali* 3, no. 1 (January–February 1923): 122.

112 Domenico Barduzzi, "Sul certificato sanitario prematrimoniale," *Rassegna di studi sessuali* 3, no. 1 (January–February 1923): 45.

> the nuptials, I would ask the intervention of a physician, together with the mayor and a priest, to give their assent to an enduring and sacred tie, that must by now represent not an individual act, but a national one.[113]

Even though the introduction of a certificate carried an increase in the possibility of corruption, this did not cancel its importance. In fact, De Napoli asked, "must we proclaim the uselessness of the law in general because some judge (meaning some judge, as we must say some physician; for the dignity of our class!) is dishonest, or because the guilty turn to fraud in order to elude the law?"[114]

For De Napoli, the sexual question needed, in fact, "discipline and brakes" and, thus, the premarital certificate—equipped with an appropriate informative record—could fulfill a role beyond sanitary, prevalently educative:

> This form of propaganda would act positively on everyone, illuminating everything on the nature and seriousness of venereal danger [...]. And I believe that, if nothing else, this egoistic sentiment [...] will induce anybody to voluntarily accept the suggested measures.[115]

De Napoli's positive view was joined by that of Pietro Capasso, who, in reflecting on the problem of eventual fraud related to the certificates, reversed the relationship between sexual morals and prophylactic health, as suggested by Montesano:

> It is truly strange that, while against the certificate we grasp with much preciousness at the weapons of morals and offended modesty within the patriarchal purity of the current family life, it is the little request of the legal and competent guarantee of the health of the spouse that is considered sufficient to upset and crumble the domestic morals [...] pushing the potential spouses into concubinage![116]

According to Capasso, the Italian population, that had supported the sacrifice of the war, would not refuse a new State intervention with sanitary

---

113 Ferdinando De Napoli, "Visita prematrimoniale," *Rassegna di studi sessuali* 3, no. 1 (January–February 1923): 50.
114 De Napoli, "Visita prematrimoniale," 50.
115 De Napoli, "Visita prematrimoniale," 52.
116 Pietro Capasso, "Intorno al certificato prematrimoniale," *Rassegna di studi sessuali* 3, no. 3 (May–June 1923): 188.

aims in the private sphere. The families involved would not abandon themselves to violent or illegal reactions, but on the contrary ask for news and information:

> The ignorant, upon application of the new law, will ask, naturally, the reason for the certificate. This will be the hour for good, timely propaganda, especially from the physician. And when the high concept of defense of collective health that has inspired the new institution is understood, the predicted rebellion and prudish disdain will abate.[117]

For Capasso, if the certificate was the last piece of a vast national prophylactic reorganization—as Montesano believed—it was no longer useful for anything. On the contrary, it would be important to impede the "current crimes of the generation" without waiting for an inevitably slow maturation of the collective hygienic education. Capasso's solution attempted a "gradualist" mediation:

> We should adopt the certificate for now, with an informative scope, not limiting it however to sexual diseases. Tuberculosis, epilepsy and serious alcoholism are equivalents to syphilis [...]. When the certificate is adopted, an educational and informative campaign will become more topical, more requested by those same interested individuals and their families, and this [campaign], together with other national prophylactic means, will give the individual, the family and the race those benefits for which we are fighting this worthy battle.[118]

Another decided supporter of the certificate was Aristide Zippari Garola, who, referring to syphilis, suggested a model that would be obligatory, for men and for women, and would comprise clinical laboratory analysis.[119]

Guido Verrotti, on the other hand, had the opposite view. To the preceding arguments (fraud, diagnostic difficulties, negative reactions of patients) he added a singular refusal of coercive methods: coercion could be justified in war, but was counterproductive "in ordinary regimes."[120]

117 Pietro Capasso, "Intorno al certificato prematrimoniale," *Rassegna di studi sessuali* 3, no. 4 (July–August 1923): 229.
118 Pietro Capasso, "Intorno al certificato prematrimoniale," *Rassegna di studi sessuali* 3, no. 3 (May–June 1923): 189.
119 Aristide Zippari Garola, "Ancora sul certificato matrimoniale,'" *Rassegna di studi sessuali* 3, no. 4 (July–August 1923): 328.
120 Guido Verrotti, "Il certificato medico prematrimoniale," *Rassegna di studi sessuali* 3, no. 3 (May–June 1923): 333.

Far from having a hygienic propaganda function, an eventual adoption of the premarital certificate would lead to opposite results, ending in "distancing the intensification of real prophylactic means, on which we should instead be spending the efforts of all physicians, sociologists and political men, because those that exist leave much to be desired, due to the insufficiency and incompleteness with which they are applied."[121] Capasso's reply was not long in coming: how could the immediate post-war period be considered normal? In reality, it was exactly in such a moment, characterized by intense international clashes, that eugenics was called upon to reinforce the "physical forces of mankind." Thus:

> Defending the race—when it is taken seriously—is neither a small thing nor a small responsibility for a state that, not having other riches, must rely on the muscles of its population. To achieve such an end, every means is good.[122]

And it was not by chance that it was this very same Pietro Capasso, by then leader of the Neapolitan Eugenic Group (*Gruppo Eugenetico Napoletano*), who carried the issue of premarital certificates to the attention of the fascist government, in December 1923. The report of the meeting between Mussolini and Capasso bears witness to the conflicting positions of *il Duce*'s populationist orientation and Capasso's qualitative eugenics:

> He [Capasso] therefore showed the President the immense question of the eugenic defense of marriage and the prevention of bad births, demonstrating the recommended reasons for the first step of adopting a premarital certificate with a purely informational scope.
>
> The hon. Mussolini remembered that several years before he had been interested in this topic, and he had translated a book by Gobineau: he realized therefore, the ideal need of defending matrimony from the hidden dangers of serious social diseases. He, however, did not gloss over the serious difficulties that would be met in the adoption of the means: above all the great and small domestic tragedies destined for young spouses, hypersensitive beings, due to the resultant bans. He added to this, furthermore, that *we must intensely procreate.*

121 Verrotti, "Il certificato medico prematrimoniale," 333–34.

122 Pietro Capasso, "Il certificato prematrimoniale," *Rassegna di studi sessuali* 3, no. 6 (November–December 1923): 380.

> The hon. Capasso objected that it is useless to procreate when that implies the birth of beings who are useless and damaging to society. He then explained how the measures mentioned would be limited only to the obligation of the presentation of the certificate, a simple informative reciprocal gesture, without any powers of prohibition.[123]

Although the debate on premarital certificate continued at least until the end of 1927, Mussolini's political and ideological disagreement was already perfectly clear at the 1923 meeting.

However, the following year—1924—was marked by relevant achievements. In April–May, Pietro Capasso again proposed the introduction of a premarital certificate on the stage of the 2nd National Meeting of the Italian Society for the Study of Sexual Questions. The state—Capasso declared—could not wait for "divulgation and information campaigns to create a public awareness of the need for the voluntary avoidance of bad marriages."[124] The economic and social damage that would derive from the degeneration of the race was, in fact, incalculable and, in the face of this, the public power had the "duty" to intervene: in such a way, the state would "defend families, individuals, the generation; in other words, defending itself." Illnesses such as syphilis, tuberculosis, epilepsy, alcoholism, and blennorrhagia were damaging not just for the individual but also "for the family and the race": "To ensure that in such conditions procreation is impeded or postponed is an act of humanity by biologists, psychologists and sociologists, and the duty of the State, because its validity and richness resides in the validity of the race."[125]

In Capasso's opinion, in the face of the collective usefulness represented by the premarital certificate—even in a moderate version, non-obligatory and limited to only men—the criticisms of the opposition (the possibility of fraud, the uncertainty of diagnosis, the dangers of "concubinage") faded.[126] The essential concept was this: "It is necessary, in the highest inter-

---

123 See "Notizie. Problemi di eugenica e profilassi in un colloquio dell'on. Capasso con S.E. Mussolini," *Rassegna di studi sessuali* 3, no. 6 (November–December 1923): 438; italics added.

124 Pietro Capasso, "Il certificato prematrimoniale," *Rassegna di studi sessuali e di eugenica* 4, no. 3 (May–June 1924): 179.

125 Capasso, "Il certificato prematrimoniale," 183.

126 Capasso, "Il certificato prematrimoniale," 181.

ests of public health, the integrity of the race, the happiness and morals of the family, to impose a premarital certificate."

Not by coincidence, the first session of the SISQS congress concluded with the approval of a resolution, proposed by Ettore Levi, which adopted the "gradualist" interpretation of Capasso:

> The congress, having heard the relation of the hon. Professor Capasso on the eugenic legislative measures realized after the war in the international field, present the government with the opportuneness of inaugurating a premarital certificate with an informative scope, and as an element of the harmonious refusion of Italian legislation for the defense and improvement of future generations.[127]

Three years later, the debate on premarital certificates reached its apex, and also its definitive sunset, in the two inquiries published respectively by the newspaper *Il Resto del Carlino*, in January–February 1927, and by *Difesa sociale* in March–April of the same year.[128] Although the major part of the contributors declared themselves in favor of a form of premarital medical visit, very few of these approved the immediate introduction of an obligatory premarital certificate. While Ferdinand De Napoli and Pietro Capasso once again denounced the paradoxical absurdity of the "right to sexual choice" and the "stupid selfishness" that surrounded marriage,[129] the Parma pathologist Umberto Gabbi—already a supporter, in his speech to the Chamber of Deputies in 1926, of "family criminal records" for every citizen and of a national health record[130]—attempted to demonstrate the substantial existing harmony between "regenerative" fascism and obligatory medical examinations:

> The State, fascistly conceived as a force and as an ethical reality, can not slowly proceed nor be deprived of its dominion over the individual and the collective.

[127] The resolution was reported in *Rassegna di studi sessuali e di eugenica* 4, no. 3 (May–June 1924): 189.

[128] On the two inquiries, see also Massimo Ciceri, *Origini controllate. La prima eugenetica in Italia (1900–1924)* (Rome: Prospettiva Editrice, 2009).

[129] See Ferdinando De Napoli, "Difendiamo la stirpe," *Il Resto del Carlino*, (26 January 1927). De Napoli continued to support the necessity of a "state prophylaxis of marriage" even after the Ascension Day Speech: see Ferdinando De Napoli, *Da Malthus a Mussolini. La guerra che noi preferiamo* (Bologna: Cappelli, 1934).

[130] See Ipsen, *Dictating Demography*, 186–87.

> When a great and enlightened national social interest is at stake, the right of the State to penetrate the family should not find obstacles in the Chinese Wall of sentiments based on selfishness.[131]

In order to maintain the argument of a matrimonial union not subject to eugenic controls, Gabbi therefore claimed that the right of individual liberty would represent, in a fascist State, a sort of non-sense:

> Once, perhaps, when "personal liberty" was synonymous with abuse and moral and political discipline, we could have waved the ghost of individual liberty: now that the Italian population is regimented under the firm laws of Fascism and have been convinced of the great benefits that this new political orientation, imposed by the regime, has given and will give the nation, there is no sense in considering the concept of individual liberty with an ancient mentality.[132]

According to Gesualdo Ciarrusso, of the University of Bologna, the right to individual liberty "can not comprise the liberty of the individual to harm the species by handing down to its offspring characteristics which will make its existence wretched." The sanitary control of marriage, whether it be in the form of premarital certificate or the sterilization of women, had to be imposed:

> Impose it! That is the necessary word. By now the general conscience is mature. The Regime, free from useless and harmful sentimentalism, alien to deleterious compromises, conscious, energetic, decisive, must act fascistly also here. We must request this from the fascist government, radical reformer of the customs of the population through judicious laws. We must request this, especially now that social refinement, degeneration of the senses and the unpardonable lightness with which life is lived have reduced the sexual instinct to an instrument of pleasure, stripping it of its attributes of high aims for the conservation of the species.[133]

For Guglielmo Bilancioni, from the University of Pisa, there could be no doubt that "as the races of the animals and plants are selected, in its greater

131 Umberto Gabbi, "Sentimento e necessità," *Il Resto del Carlino* (28 January 1927).
132 Gabbi, "Sentimento e necessità."
133 Gesualdo Ciarrusso, "Risposta affermativa," *Il Resto del Carlino* (30 January 1927).

capacity, the fascist State [...] has the right to protect the physical and moral integrity of the stock." But for "practical realization," "a climate of superior civil progress" was necessary, which seemed to still be lacking in Italy.[134]

Enrico Ferri, socialist deputy and leader of the Italian school of positivist criminology, did not mince words in his defense of obligatory premarital certificates: "There is no doubt that human breeding in the same way, but more importantly, as the breeding of horses, cattle, sheep, pigs etc. is an urgent necessity, in order to improve the stock, according to Darwinian and Mendelian facts. The fascist regime has the will to realize a rational program, passing from ideas to action. It is therefore a question of practical means and modes."[135] As with criminals therefore, two forms of prevention were also necessary for the "procreation of healthy and strong beings": the first, "direct or enforced," included the premarital certificate, the "prohibition of marriage of certain persons," and the "sterilization of serious abnormals"; the second, "more complex and slow and difficult," involved "propaganda and education in schools and after school," "the training of a hygienic awareness in the population," and the "improvement of household hygienic conditions, nutrition etc."[136]

The reservations of the opponents of a mandatory premarital certificate were mainly concentrated on the difficulties of diagnosis. A syphilitic with evident lesions—confirmed, for example, the hygienist and epidemiologist Arcangelo Ilvento[137]—would never submit himself to a medical examination. Whoever considered marriage either ignored the illness or believed himself to be healed, because he had had no manifestations for some time. In these cases, "the clinical diagnosis is often uncertain" and moreover "the danger of transmission to the wife and children remained for a certain period" that varied from five to ten years.[138] Along the same lines, hard proof to demonstrate the heredity of tuberculosis and alcoholism was lacking:

---

134 Guglielmo Bilancioni, "Questione di civiltà," *Il Resto del Carlino* (2 February 1927).

135 Enrico Ferri, "Visita prematrimoniale obbligatoria?," *Difesa sociale* 6, no. 4 (April 1927): 1–2.

136 Ferri, "Visita prematrimoniale obbligatoria?," 1–2.

137 For a bibliography on the figure of Arcangelo Ilvento, see Maiocchi, *Scienza italiana e razzismo fascista*, 32.

138 Arcangelo Ilvento, "Visita medica prematrimoniale?," *Difesa sociale* 6, no. 3 (March 1927): 3–4.

> At least until now no sure proof has been provided that alcoholism is eventually sufficient to produce of itself permanent hereditary defects. Therefore in these last cases it is sufficient that the measure of social protection stops at the drunken relative, while we lack a solid reason to extend it to the son and discuss if he is permitted to marry.[139]

If some eugenicist had sterilized Beethoven's drunken parents—Aldo Mieli added—the world "would not have had one of its great artists."[140] It was easy to heal from syphilis—Vincenzo Montesano ironically declared in *Resto del Carlino*—but there was no surety of complete healing: the long period of latency of the disease made an occasional medical examination, a few weeks before the marriage, totally useless.[141] The gynecologist Ernesto Pestalozza[142] showed the same scientific prudency, as did the clinician Vittorio Ascoli,[143] the psychiatrist Giovanni Mingazzini[144] and Giuseppe Montesano.[145] Nevertheless, in Montesano's view, a second argument was raised against the obligatoriness of the certificate:

> I see in this campaign for premarital certificates a manifestation of the self-interest of the majority to the detriment of the minority. It is a self-interest that could appear in those States where economic values dominate, but is unthinkable in those others which aim toward the holistic progress of humanity.
> The most certain index of this progress is the development of a sentiment of solidarity with all members of the group, with the weakest more than with the strongest. Illness is combated not by sacrificing its carriers but by seeking to account for all the other causes of its diffusion and energetically eliminating those already known.[146]

The theme of respect for the individual returned significantly with the contribution of endocrinologist Nicola Pende:

139 Ilvento, "Visita medica prematrimoniale?," 6.
140 Aldo Mieli, "Proposte pratiche," *Il Resto del Carlino* (9 February 1927).
141 Vincenzo Montesano, "Risposta negativa," *Il Resto del Carlino* (2 February 1927).
142 Ernesto Pestalozza, "Visita prematrimoniale obbligatoria?," *Difesa sociale* 6, no. 4 (April 1927): 4–5.
143 Vittorio Ascoli, "Visita prematrimoniale obbligatoria?," *Difesa sociale* 6, no. 4 (April 1927): 2–3.
144 Giovanni Mingazzini, "Sul certificato prematrimoniale," *Difesa sociale* 6, no. 3 (March 1927): 2–3.
145 Giuseppe Montesano, "Il certificato prematrimoniale," *Difesa sociale* 6, no. 4 (April 1927): 3.
146 Montesano, "Il certificato prematrimoniale," 3.

> A law that commands the future spouses to present themselves to a civil official of the State with a certificate of physical, intellectual and moral health [...] puts the physician and the State under the moral and juridical obligation to act, creating laws to protect the sexual rights (once the legitimate procreative rights are legally impeded) of the rejects of the matrimonial ordinance. The sexual function cannot be suffocated by a law: and if, in the interests of the family and the State, procreation by defectives or ill people should be avoided, it is not possible, neither theoretically or practically, to inhibit such individuals who, manifestly or secretly, exercise their sexual function. On the contrary, as we frequently see, in certain ill people [...] [the sexual function] is heightened and, every moral brake being removed, often results in immoral, or even criminal, acts.[147]

Also for the psychologist Sante De Sanctis, it would be "offensive for human dignity to deprive the individual of the liberty to sacrifice himself, when it does no harm to the good of the community": humanity, in fact, "is not a herd, nor a stud farm for racehorses."[148] Neither—claimed Lucchetti in *Resto del Carlino*—would it be easy to discipline the "hearts and sentiments—not even by a government that, among its skills, has shown, every time it was necessary, force, and an invincible force."[149] According to Alessandro Stoppato, professor of law at the University of Bologna, a heavy intervention by the State in the lives of the citizens "would bring serious humiliations and upsets to the families, and the investigation would worry them, agitating public opinion, and could cause further prejudices, different and distinct from that of not being able to contract their desired marriage."[150]

The problem of maintaining social order was common to a good part of those against the obligatoriness of the premarital certificate. For the psychiatrist Pellacani, for example, the introduction of coercive measures would

147 Nicola Pende, "Sul certificato prematrimoniale: obbligo legale od obbligo morale?," *Difesa sociale* 6, no. 3 (March 1927): 10.

148 Sante De Sanctis, "Visita prematrimoniale obbligatoria?," *Difesa sociale* 6, no. 3 (March 1927): 2. In his role as president of the Lazio section of the Italian League of Hygiene and Mental Prophylaxis, Sante De Sanctis approved, in January 1927, the adoption of a "campaign scheme," full of "hygienic-prophylactic" rules for engaged couples, which had to be printed on the back of prescriptions written by the neuropsychiatric wards, see "Lega Italiana d'Igiene e Profilassi Mentale. Sezione laziale," *Difesa sociale* 6, no. 1 (January 1927): 13–14.

149 Giuseppe Lucchetti, "Le difficoltà del certificato," *Il Resto del Carlino* (6 February 1927).

150 Alessandro Stoppato, "I vantaggi e i danni," *Il Resto del Carlino* (28 January 1927).

profoundly pollute public morality: "If the subjects, defective males and females, were denied legitimate procreation through marriage, would they therefore be removed from the sexual circulation of society, and therefore from illegitimate procreation? No, certainly not."[151] The criminal news—Lucchetti continued—could even "increase its columns, due to the intensification of homicides and suicides of passion; the tribunals would be increasingly busy, due to the multiplied troubles offered by concubinage and illegitimate children."[152] Nicola Pende had the same warning: "The reasonable doubt arises as to which is the greater evil, the increase of prostitution or more frequent illegitimate sexual rapport, or the increase of illegitimate births, and sicknesses of unmarried men, if not the fact that some epileptics or chronic alcoholics or tuberculosis sufferers or syphilitic or psychically abnormal person might be able to trick another person into marrying him!"[153]

And all this was without counting the reluctance on the part of physicians to be transformed into agents of the State, renouncing professional privacy. Pellacani maintained: "Certain ill people could be prevented from seeking a cure, to avoid the discovery of their illness by a physician who is no longer bound by professional privacy [...]. The transformation of the physician into a possible fiscal agent could present, from this point of view, a serious danger."[154] No colleague—Pende repeated—"wants to deceive themselves that they are inspired by God, or so knowledgeable as to be infallible, like a pope of medicine. Clinical medicine must today honestly declare itself incapable of giving a sure verdict."[155] Also Leone Lattes , professor of legal medicine, believed that the "delicate and essential liberty to found a family" could not be handed over "to the discretional powers of physicians."[156]

According to Pellacani, to be able to adopt a truly effective eugenic legislation, that is, one based on the sterilization of defectives and on obligatory premarital examinations, it was necessary above all to diffuse "eugenic sentiments in society":

151 Giuseppe Pellacani, "Basta la visita prematrimoniale?," *Il Resto del Carlino* (30 January 1927).
152 Lucchetti, "Le difficoltà del certificato."
153 Pende, "Sul certificato prematrimoniale," 10.
154 Pellacani, "Basta la visita prematrimoniale?."
155 Pende, "Sul certificato prematrimoniale," 8–9.
156 Leone Lattes, "Dalla teoria alla pratica," *Il Resto del Carlino* (6 February 1927).

> In the sexual field, even if legislative coercion, mild, such as premarital examinations, or radical, such as obligatory sterilization, can appear—in the case of the first—not totally useful and not without inconveniences, and—in the case of the second—completely premature, it is necessary to spread the knowledge of the fundamental social importance of the germ plasm and its integral protection and conservation over the course of generations.[157]

And while for Antonio Dal Prato the principal objective was to "form the hygienic conscience of the masses,"[158] for Salvatore Ottolenghi, more than a new law, it was necessary to consider the intensification of the "physical and moral hygiene."[159] According to Francesco Bonola, lecturer at the University of Genoa, it was enough to simply trust the eugenic instinct of the intended spouses: "eugenicists seem to me to count too much on the practical spirit of our time. Those who marry, man or woman, search for their other half in the best conditions possible. Also of physical conditions."[160] Aldo Mieli's position was more articulated: it could be opportune to introduce a health passport and premarital examination, impeding the unions "in extreme cases" and adopting, for all the rest, an "indirect" strategy, for example facilitating the life "of all those that the State would desire not to procreate, in such a way that they would not, in desperation [...] contract a marriage that would bear painful fruit."[161]

On the whole, the proposal supported by those against an obligatory certificate was aimed at the introduction of a form of optional premarital prophylaxis, in the wider context of a "complete and integral realization of fascist ideals [...] in the field of social sanitary organization,"[162] principally based on the protection of pregnancy, and on "hygienic propaganda." Within this general framework, there were also specific suggestions, such as those of Arcangelo Ilvento, who proposed the "personal health passport" and the "hereditary record" inspired by Lundborg's Swedish model,[163]

157 Pellacani, "Basta la visita prematrimoniale?."

158 Antonio Dal Prato, "Basta la pratica igienica," *Il Resto del Carlino* (6 February 1927).

159 Salvatore Ottolenghi, "I rimedi legali sono insufficienti," *Il Resto del Carlino* (6 February 1927).

160 Francesco Bonola, "Soluzione negativa," *Il Resto del Carlino* (9 February 1927).

161 Mieli, "Proposte pratiche."

162 Ilvento, "Visita medica prematrimoniale?," 8.

163 Ilvento, "Visita medica prematrimoniale?," 7. See also Arturo Donaggio, "La visita medica prematrimoniale obbligatoria," *Difesa sociale* 6, no. 4 (April 1927): 4.

or those of Pende, who recommended the "constant penetrative work of the physician" for the "somatic and psychical benefit of the individual, from infancy until marriageable age."[164]

The impossibility of reaching a majority consensus between the supporters and negators of obligatory premarital certificates nevertheless is reflected in the conclusions drawn by the editors of *Il Resto del Carlino* and *Difesa sociale*, at the end of the two inquiries. For *Il Resto del Carlino* Cesarini Sforza declared that the fascist state could not avoid limiting individual liberty in the name of the superior interests of the "race": "Must the State that dominates and controls every manifestation, we can say, of social life, in the name of a superior ethical interest, which intervenes with all its force even for the smallest infraction of the social solidarity, then neglect the sometimes serious offences to human and social solidarity brought by those who [...] contribute to the decadence of the race?"[165] Fascist laws for the protection of maternity and infancy were not enough, just as the "moral and physical prophylaxis" was not sufficient: "It is necessary that the State intervenes directly, reawakening with every means the sense of individual responsibility to offspring, which is not equally awakened and energetic in everyone. Moral and physical prophylaxis—education and hygiene—can do something; but it is in vain to hope that civilization will spread over all levels of the population."[166] From this, the necessity arose, according to Cesarini Sforza, of introducing an obligatory premarital examination, but without prohibitive powers, as in Denmark and Norway.

At the opposite end of the scale, Augusto Carelli, new editor of *Difesa sociale*, stressed the importance of solidarity with the weakest and maintained the "necessity of pain" against eugenic utopias:

> Today's men seem a bit drunk on their conquest of so-called mechanical progress, and seem always more inclined to devalue certain moral values that oppose them. Against their ideals of physical power, those of humility, pity and human charity seem to increasingly disgust them: and it seems they have forgotten that pain is not only the unavoidable companion of existence, but

164 Pende, "Sul certificato prematrimoniale," 10.

165 Widar Cesarini Sforza, "Perché approviamo la 'visita prematrimoniale,'" *Il Resto del Carlino* (17 February 1927).

166 Cesarini Sforza, "Perché approviamo la 'visita prematrimoniale.'"

> also has great moral, and therefore social, value. Pain is a true teacher of life; we know how much we have learnt with our truly painful experiences, and we can say that every great human work is the fruit of pain. Any force directed to the improvement of society, a force that is always legitimate, laudable and only right and proper, must remember this necessity of pain, and must consequently carry with it the conviction that it will never be possible for man to eliminate evil from his existence simply by virtue of scientific postulates.[167]

Beyond the substantial lack of agreement between the technicians of public health, it was Mussolini's Ascension Day Speech that in May 1927, suffocated the debate over premarital examinations, considering them a dangerous variant of birth control. The formation of free and optional premarital consultancy clinics, which, starting from 1924, spontaneously developed in Milan,[168] Turin,[169] Genoa,[170] Trieste,[171] and Bologna,[172] was in fact cut off on the precise order of Mussolini, whose political and ideological objective coincided, by now, with the demographic (quantitative and pronatalist) development of the nation. This is demonstrated clearly by a letter the chief officer (*prefetto*) of Bologna sent to Mussolini on 9 April 1928:

> Excellency, I am honored to communicate: in accordance with your esteemed signature of the 15th there is no longer a prenuptial medical consultancy in Bologna, which was spoken about in Resto Del Carlino on the 14th, page six. I am informed that as they are in majority excellent fascists, they were extremely pleased to obey the Chief's nod.
> I remain, Excellency, with profound respects, your devoted servant—
> Giuseppe Guadagnini.[173]

167 Augusto Carelli, "Visita prematrimoniale obbligatoria?," *Difesa sociale* 6, no. 4 (April 1927): 6. After Mussolini Ascension Day Speech (May 1927), Carelli became a firm opponent of neo-Malthusianism and "Nordic" eugenics: see Maiocchi, *Scienza italiana e razzismo fascista*, 33; Mantovani, *Rigenerare la società*, 300–01.

168 The initiative was promoted, in 1924, by gynaecologist Emilio Alfieri and again in 1928 by gynaecologist and president of the Milan Red Cross, Alfonso Cuzzi.

169 For the initiatives of the local group of SISQS and, in particular, of syphilographer Arturo Fontana, see *Rassegna di studi sessuali e di eugenica* 6, no. 4 (December 1926): 326–28 and *Rassegna di studi sessuali, demografia e di eugenica* 8, no. 1 (January–March 1928): 25ff.

170 For the initiatives of the Ligurian group of SISQS and the lecturer of legal medicine Gian Giacomo Perrando; see *Rassegna di studi sessuali e di eugenica* 8, no. 2–3 (April–November 1928): 164ff.

171 For the initiatives of the local Sanitary Group of *Fascio femminile*, see *Difesa sociale* 5, no. 7 (July 1926): 167.

172 Promoted by the *Poliambulanza Felsinea*; see "Il consultorio medico prenuziale," *Il Resto del Carlino* (14 April 1928).

173 ACS, SPD, CO, b. 509.560/III, "Istituto Centrale di Statistica," sf. 1: "I.C.S. – Provvedimenti legislativi nell'Interesse Demografico." See also Mantovani, *Rigenerare la società*, 291, for a deeper analysis of the topic.

In the pages of the journal *Archivio fascista di medicina politica* [Fascist journal of political medicine], the founder and editor Umberto Gabbi, only a few months before among the most solid supporters of an obligatory premarital certificate, publically humbled himself, declaring his error,[174] and promptly dedicating an entire issue to comments on *Numero come forza* [Number as force], the Mussolinian equation formulated in the foreword of statistician Richard Korherr's book *Regresso delle nascite: morte dei popoli* [The decline of births: the death of peoples].[175]

3. Sterilization and Euthanasia

In the context of Italian eugenics, sterilization as an extreme surgical solution to the problem of dysgenic degeneration, constituted a very minor theme. At the end of the nineteenth century,[176] this measure had found a firm supporter in Angelo Zuccarelli, chief physician of the interprovincial mental hospital of Nocera, founder and director from 1893 of the Museum of Criminal Anthropology at the University of Naples, where he taught from 1887, and from 1890 published the review *L'anomalo* [The defective].[177]

174 Umberto Gabbi, "La battaglia per la natalità," *Archivio fascista di medicina politica* 2 (1928): 267–68.

175 See "Politica demografica e crisi di natalità," *Archivio fascista di medicina politica* 2 (1928): 283–359. Korherr's volume was published by the *Libreria del Littorio* in 1928, with a preface by Mussolini. On the figure of Umberto Gabbi, right-wing liberal, interventionist, nationalist in 1919, fascist in 1923, member of Parliament in 1924 and senator for "exceptional scientific merit," see Maiocchi, *Scienza italiana e razzismo fascista*, 33–38; Mantovani, *Rigenerare la società*, 269–70 and 286–87.

176 In United States, in 1907, the state of Indiana (USA) approved the first sterilization law for criminals, idiots, imbeciles and those guilty of sexual violence. By 1924, approximately 3,000 people had been involuntarily sterilised in America; the vast majority in California. By the late 1930s, more than fifteen states of the United States, as well as the parliamentary governments of several European states and Canada, had adopted compulsory and "voluntary" legislation authorizing sterilization, castration, and abortion on eugenic grounds. In Europe, the first example of sterilization legalisation with a eugenic aim was in the Swiss canton of Vaud, in 1928. It was followed by Denmark (1929, 1934, 1935), Germany (1933, 1935), Norway (1934), Sweden (1935, 1941), Finland (1935) and Estonia (1936). Among the most recent studies, see Marius Turda, "'To End the Degeneration of a Nation': Debates on Eugenics Sterilization in Inter-war Romania," *Medical History*, 53 (2009): 77–104; Mark A. Largent, *Breeding Contempt: the History of Coerced Sterilization in the United States* (New Brunswick: Rutgers University Press, 2008); Gisela Bock, "Nationalsozialistische Sterilisationpolitik," in Klaus-Dietmar Henke, ed., *Tödliche Medizin im Nationalsozialismus. Von der Rassenhygiene zum Massenmord* (Cologne: Böhlau, 2008): 85–99; Ian Dowbiggin, *The Sterilization Movement and Global Fertility in the Twentieth Century* (Oxford and New York: Oxford University Press, 2008); Natalia Gerodetti, "From science to social technology: eugenics and politics in twentieth-century Switzerland," *Social Politics: International Studies in Gender, State and Society*, 13, no. 1 (2006): 59–88.

177 On Zuccarelli, see also see Maiocchi, *Scienza italiana e razzismo fascista*, 12; Mantovani, *Rigenerare la società*, 52–53.

Zuccarelli's proposal was explicit since its first formulation in 1894. Convinced of the necessity of artificial selection, which was "readier and more effective than the natural one" in order to combat the "excessive multiplication of defective humanity," the anthropologist maintained that the introduction of "sterilization of extremely degenerate people" was indispensable and he justified it as an extension, on a "social" prophylactic level, of the "individual" prophylactic method already in use for some time, as regarded, for example, women with tuberculosis at risk of death in case of pregnancy. The categories to which sterilization had to be applied were epileptics, the tuberculosis sufferers, lunatics, alcoholics, syphilitics, the mentally ill with "degenerative" pathologies, and delinquents ("instinctive" and "habitual"). Zuccarelli's solution was presented in these terms in 1898, in an essay that was critical of the legislation mandating involuntary sterilization introduced in Michigan in 1897, which he judged overly rigid and discriminating.[178] It was reprised in 1901 in a communication to the Naples Gynaecological Society (*Società Ginecologica di Napoli*) at the 5th Congress of Criminal Anthropology in Amsterdam and the 11th Congress of the Italian Phreniatric Society (*Congresso della Società Italiana di Freniatria*) in Ancona. For Zuccarelli, sterilization represented the sole rational remedy for the menace of physical-psychical degeneration. He maintained:

> We must not overly fear the erosion of the respect due to individual liberty [...] nor must we exaggerate such a sentiment. We are not speaking of healthy life; instead we are dealing with illness, with anomalies of the most serious kind, and [therefore] restrictions, limitations to such a liberty, with the scope of avoiding one of the biggest collective damages—the great damage to human perfectibility—must appear more than right and reasonable, dutiful, necessary, indispensible.[179]

Enrico Ferri brought Zuccarelli's proposal to the attention of Parliament, where it was contested by the Member of Parliament and professor of law, Luigi Lucchini. It was later discussed in 1906 at the International Congress for the Assistance of the Insane in Milan, in front of the great figures of Ital-

178 Angelo Zuccarelli, "Asessualizzazione o sterilizzazione dei degenerati," *L'anomalo*, 8, no. 6 (1898–99), offprint.

179 Angelo Zuccarelli, "Per la sterilizzazione della donna come mezzo per limitare o impedire la riproduzione dei maggiormente degenerati," *Bollettino della Società Ginecologica di Napoli* 1 (February–March–April 1901): 3.

ian psychiatry, such as Lombroso, Morselli, Bianchi, and Tamburini. Zuccarelli emphasized three points that were to be voted upon in the day's agenda:

> 1) to recognize the necessity of effective prophylactic actions, aimed at preventing the procreation of abnormals as much as possible;
> 2) to advise, as the most adapted and secure means to achieve this aim, sterilization of highly degenerated people [...];
> 3) to associate ourselves with the unanimous vote expressed by the "Congress on Work-related Illnesses" so that the fight against tuberculosis becomes a function of the State, advising that every health treatment in favor of the tuberculosis sufferers be preceded or accompanied by their sterilization.[180]

But, demonstrating the lukewarm reception of sterilization projects in the Italian medical culture of that period, the assembly limited itself to approving only the first point, the most generic and moderate, leaving the other two, certainly more radical and operational, to future discussions.

Also in this case, the First World War reignited the debate, and once again Zuccarelli was ready to propose his deterministic hereditarianism, this time by criticizing Ettore Levi's moderate eugenics. For Zuccarelli, sterilization was the "capital problem of eugenics," as he affirmed many times between 1924 and 1925: "real and substantial 'eugenics' can never be achieved, without the 'sterilization' of the excessive number of considerably defective and degenerate individuals already in existence."[181]

With the war just recently finished, in a speech to the 3rd Congress of the Italian Pro Abnormals Society, the psychologist Francesco Umberto Saffiotti radically contested Giuseppe Sergi's position on the eugenic value of educating abnormals:

> There are two profoundly different aspects to the problem: the biological aspect and the social aspect [...] and these two aspects are not reducible to one or the other [...]. In subordinating the interests of the individual to the interests of the race we also feel the huge weight of tradition, and selfish and humanitarian sentiments, and we also feel a timidity and lack of certainty in supporting the necessity of extreme measures, if not for the secure conviction

---

180 Angelo Zuccarelli, "La proposta della 'sterilizzazione' dei più anormali quale misura profilattica sociale contro la degenerazione," *L'anomalo* (1909):16–17, offprint.

181 Angelo Zuccarelli, *Il problema capitale della "eugenica"* (Ferrara: Industrie Grafiche Italiane, 1924), 8.

> of their necessity, then for the opportunistic considerations of the moral and juridical lack of preparedness in which we find ourselves as regards the legitimacy of extreme sanctions. We must overcome this timidity and uncertainty and resolutely affirm that the real and proper solution to the eugenic problem, regarding those who are physically and psychically insufficient, consists only in rendering them unable to procreate.[182]

Stalwart supporter of sterilization using X-rays, Saffiotti was among the few in Italy to oppose the Lombrosian refrain, which underlined the utility of degeneration in producing immortal geniuses such as Leopardi or Manzoni. Saffiotti was instead of the opinion that degenerate geniuses, on the contrary, shone with brilliant light only in comparison to the grand obscurity that surrounded them: "The fact that a genius arises from a family of degenerates does not compensate for the multitude of individuals who damage social progress."[183] For Saffiotti, the education of "lunatics" was a necessary method in the fight against degeneration, but it was not sufficient:

> Measures of assistance, of hygienic improvement, of physical education, of individual and social prophylaxis are all highly useful means to try and contain physical and psychical degeneration, but their effects are uncertain, slow, difficult, and certainly inadequate to compensate for the deleterious effects of the spreading of the causes of degeneration.
>
> And if we do not have the courage to resolutely affirm the necessity of extreme remedies for extreme ills, we will never be prepared for us, and for humanity, to achieve the progress of physical and psychical health.[184]

In the name of the "health of the stock," the State had the right, therefore, to mandate the sterilization of those who were "dangerous to the species" and "to impose artificial selection, both direct and preventive: direct on the individual adults, preventive in the suppression of newborns that present undoubted manifestations of hereditary degeneration."[185] In order to

[182] Umberto Saffiotti, "Eugenica e anormali," *L'infanzia anormale* 5–6 (1920): 1–86; 79–80. On the figure of Saffiotti, see also Maiocchi, *Scienza italiana e razzismo fascista*, 18; Mantovani, *Rigenerare la società*, 176–77.

[183] Saffiotti, "Eugenica e anormali," 81.

[184] Saffiotti, "Eugenica e anormali," 81.

[185] Saffiotti, "Eugenica e anormali," 81.

achieve this ultimate and resolute goal, Saffiotti suggested a "minimal program" for eugenics, consisting of premarital certificates, health passports, and a campaign against syphilis and tuberculosis:

> The path for achieving certain stages of human progress is long and full of difficulties [...]. However, there is a practical eugenic action that is immediate: a minimal program. This minimal program reconciles the supporters and opposers of extreme measures. The first will renounce extreme solutions due to the necessary contingencies of the moment; the others will contribute to this minimal program with all the fervour of their humanitarian sentimentalism.
> In this minimal program, the problem is not a strictly biological problem, but a principally social one.[186]

The theoretical position of Gaetano Pieraccini,[187] socialist physician and long-time supporter of coercive measures in the field of social medicine,[188] is also of particular interest. In the concluding chapter of a long essay in 1924, dedicated to the study of the heredity of biological characteristics in the family pedigree of the Medicis of Florence,[189] Pieraccini, believer in the interaction between heredity and the environment in the transmission of morphological and psychical characteristics of the species, called himself both "eugenicist" and "euthenicist." For Pieraccini, the "genocratic" dream was still far away, but "Society" could contribute to the acceleration of the evolutionary process regulated by natural selection, intervening in "environmental factors"—with social and hygienic medicine—and introducing a form of "matrimonial prophylaxis."[190] Not genius, but the "average man" embodied Pieraccini's eugenic ideal:

---

186 Saffiotti, "Eugenica e anormali," 82–83.

187 Important exponent of Florentine socialism from the start of the nineteenth century, anti-interventionist and anti-fascist, on 10 June 1925 Pieraccini was detained by authorities, while, with Carlo Rosselli and Alessandro Levi, he was laying flowers on the tomb of Garibaldi in memory of Giacomo Matteotti. In 1930, he was arrested for handing out commemorative manifestos about Matteotti. The sentence of a year of imprisonment was commuted to an admonishment. His house was a meeting place for anti-fascists, and the police believed he was in contact with anti-fascist emigrants. He became the first mayor of Florence after Liberation. See ACS, CPC, b. 3954, f. 5944, "Pieraccini Gaetano." For a biographic profile, see Maurizio Degl'Innocenti, *Gaetano Pieraccini. Socialismo, medicina sociale e previdenza obbligatoria* (Manduria: Lacaita, 2003).

188 See Gaetano Pieraccini, *La difesa della società dalle malattie trasmissibili* (Torino: Bocca, 1895).

189 Pieraccini, *La stirpe dei Medici di Cafaggiolo*. For a review of this "brilliant book," see M. Carrara, "Le leggi dell'eredità in una storica famiglia italiana," *Difesa sociale* 5, no. 4 (April 1926): 6–9.

190 Pieraccini, *La stirpe dei Medici di Cafaggiolo*, 445.

> We must consider this crowd of average men as a great force of civil life. The social machine cannot be constituted only of propellers; the most disparate elements are necessary to produce and maintain a regular useful effect. Now the mass of median people [...] if they are conservative, are also regular methodical producers of global wealth.[191]

Instead of the construction of a homogenous elite of superior men, Pieraccini proposed a sort of *eugenic socialism,* that, reducing the negative influence of environmental factors, allowed all individuals to freely develop their true hereditary biological potential:

> A political and social constitution which, with the public ownership of the means of production, makes it possible for all individuals to develop their proper aptitudes: this removes the inequality between those who have too much and those too little, giving everyone the possibility to freely follow the trajectory for which their biological dowry has destined them (both hereditarily and innately). We will not level anything, putting all individuals on the same plane or destroying (as is usually repeated) the single personality. On the contrary, we will favor natural and anthropological differentiation, renewing in this way the fortunes of the human family.[192]

To carry out this objective of a healthy biological *medietas,* Pieraccini did not limit himself to indicating the improvement of the economic and hygienic life conditions of the most disadvantaged social classes, but went so far as to promote the methods of matrimonial prophylaxis and sterilization. The principal aim of the premarital certification was clearly identified in the segregation of "degenerates":

> In this way we should avoid marriages with lepers, consumptives, people with venereal disease, insane people (certain forms of insanity, as with manic-depressives and precocious dementia, present a high rate of heredity), with imbeciles, many epileptics (essential epilepsy), with alcoholics, habitual morphine and cocaine users, and delinquents, as is already happening in several American states in the north, and to a smaller degree, in Europe.[193]

---

[191] Pieraccini, *La stirpe dei Medici di Cafaggiolo,* 445–46.

[192] Pieraccini, *La stirpe dei Medici di Cafaggiolo,* 447–48.

[193] Pieraccini, *La stirpe dei Medici di Cafaggiolo,* 457.

Again the American eugenics model—with its highly celebrated family-studies, such as *The Jukes*—motivated Pieraccini's support to introduce a sterilization law, even if limited to "only the cases of strong organic degeneration and only after the judgment—case by case—by a competent medical tribunal."[194] In particular, according to the socialist physician, discharge from mental hospitals should be conditional upon a preventive sterilization procedure.[195] Against the predictable criticism of those who rejected sterilization as an intolerable offence against individual liberty, even Pieraccini invoked the lessons of the war, the "point of no return" in the definitive consecration of the superiority of the state over the individual:

> When sons are plucked from their parents, the husband from his wife, the father from his children to embrace death; when men are obliged to kill other men for controversial ends; when the citizens are constrained, against their political and philosophical convictions, to slaughter other men, including those who are surely of equal beliefs; when all this can be done with manifest pernicious damage to the human race [...]; well then, if we have a realistic and serious concept of eugenics and don't want simple academic amateurism, and if through artificial sexual selection we want to leave the breeding farms of horses, cows, dogs, pigs, to also benefit humans, then we can not continue to hide behind the classic reserve of respect for individual liberty.[196]

While waiting for science to reach its conclusions regarding "progressively degenerative heredity," the eugenicist could not rest, but had to "direct [people] to good," that is, present to the "conscience of the citizens" an effective documentation that attested to the important social problem represented by "the relationship of biological heredity with the destiny of the races."[197]

In the same year, the position of Pieraccini was reinforced by Paolo Enriques'[198] essay *L'eredità nell'uomo* [Heredity in man]. This brief tract emphasized the incontestable validity of Mendel's laws as the mechanism of heredity transmission of not only morphological and physiological char-

194 Pieraccini, *La stirpe dei Medici di Cafaggiolo*, 459.

195 Pieraccini, *La stirpe dei Medici di Cafaggiolo*, 458.

196 Pieraccini, *La stirpe dei Medici di Cafaggiolo*, 460.

197 Pieraccini, *La stirpe dei Medici di Cafaggiolo*, 461.

198 On the figure of Paolo Enriques, see Claudio Pogliano, "Bachi, polli e grani. Appunti sulla ricezione della genetica in Italia," *Nuncius. Annali di Storia della Scienza* 14, no. 1 (1999): 150–52.

acteristics in the human species, but also of psychical and behavioral ones, such as musical and artistic talent on one hand, and prostitution, criminality and pauperism on the other.

Director of the Institute of Zoology and Comparative Anatomy of the University of Padua and popularizer of American eugenic literature, Enriques did not hesitate to propose, in these pages, the introduction of obligatory sterilization for criminals, and voluntary for the "seriously constitutionally ill," as well as a sanitary passport and premarital certificates.[199]

According to Enriques, the binding necessity of a eugenic legislation appeared to be confirmed by the degeneration of a society increasingly exposed to the "danger of a progressive lowering of the average physical and intellectual level of the population." The development of medicine allowed, in fact, the survival of a "quantity of weak and constitutionally feeble people, who would in other times have died";[200] charitable institutions, "besides protecting the temporarily unfortunate and the elderly," contributed to "raising the weak and the unhappy due to constitutional defects, and to conducting the depraved along the moral path"; the "socialist spirit" tended to "level the masses and protect the unfit," while the "democratic-bourgeois spirit" of the ruling classes limited "procreation of intellectually superior people."[201] What was to be done therefore? For Enriques, the essential point was not the right to life, which must be guaranteed and protected, but the right to produce life:

> To correct the damaging actions of these institutions and habits, we must at least create a series of measures which favor the reproduction of the best and inhibit that of the worst; the "best" and "worst" in a eugenic sense, that is, endowed with physical and psychical assets, or, respectively, weaknesses.[202]

Revisiting an argument already supported immediately after the end of the First World War,[203] Enriques repeated, in conclusion, the necessity to overcome the "dysgenic"[204] concept of justice inspired by the French Revolu-

199 Paolo Enriques, *L'eredità nell'uomo* (Milan: Vallardi, 1924), 380.
200 Enriques, *L'eredità nell'uomo*, 381–82.
201 Enriques, *L'eredità nell'uomo*, 384–85.
202 Enriques, *L'eredità nell'uomo*, 385.
203 Paolo Enriques, "Eugenica e diritto," *Studi sassaresi* 1 (1921).
204 Enriques, *L'eredità nell'uomo*, 386.

tion and followed by Socialism, in the name of a new "eugenic right" in which "the laws and all morals would be orientated toward the improvement of the race." In particular, there was only one principle, according to Enriques, that would lead any process of social change: "Respect those who have been born and help them; but inhibit the reproduction of the worst, and facilitate that of the best."

Despite being positively and lengthily reviewed by Carlo Foà[205]—the Milan physiologist who, in the column *Cronache scientifiche* [Scientific chronicles] of the leading fascist theoretical journal *Gerarchia* [Hierarchy], declared, in the same years, his ambiguous sympathies for sterilization[206]—Paolo Enriques' eugenic theories did not become the official position of the fascist regime.

The proceedings of the First Congress of Social Eugenics in 1924, demonstrate this clearly.[207] Beyond several important, but isolated positions—such as that of Roberto Michels, who believed it was right to "eliminate the physically unsuitable or morally inferior elements from sexual circulation"[208]—the majority of Italian eugenicists seemed, on the contrary, to share the views expressed in 1923 by the elderly psychiatrist Enrico Morselli, in his essay *L'uccisione pietosa (eutanasia) in rapporto alla medicina, alla morale e all'eugenica* [Mercy killings (euthanasia) in relation to medicine, ethics and eugenics]. In these pages Morselli criticized not just the American legislation on sterilization, but also the most radical side of German Weimarian eugenics, favorable to the euthanasia of "incurable feebleminded" individuals, as expressed in the essay of the psychiatrist Alfred

205 Carlo Foà, "L'eredità dei caratteri normali e patologici. 1," *Gerarchia* 9 (1925): 609–13; Carlo Foà, "L'eredità dei caratteri normali e patologici. 2 ," *Gerarchia* 10 (1925): 677–82; Carlo Foà, "L'eredità dei caratteri normali e patologici. 3," *Gerarchia* 11 (1925): 745–50; Carlo Foà, "Conseguenze sociali dell'eredità biologica," *Gerarchia* 12 (1925): 815–19.

206 Professor of human physiology at the University of Milan, collaborator of Pende—see Carlo Foà and Nicola Pende, *La fisiologia e la clinica degli increti* (Milan: Istituto Biochimico Italiano, 1927)—president (from 1929) of the Italian Society of Social Medicine (*Società Italiana di Medicina Sociale*), Carlo Foà became, starting from 1927, one of the most orthodox voices of the pronatalist population policy and of "quantitative" eugenics of the regime: see Carlo Foà, "Eugenica e matrimonio italiano," *Politica sociale* 4 (1932): 191–200. In 1938, he fell victim to the racial laws. For his views on sterilization, see Carlo Foà, "Eugenetica e diritto," *Gerarchia* 1 (1926): 58-61; Foà, "Opere e leggi di medicina sociale," *Gerarchia* 2 (1927): 151–52.

207 See ch. 4, 147–58.

208 Roberto Michels, *Problemi di sociologia applicata* (Turin: Bocca, 1919), 1–14.

Hoche and the jurist Karl Binding, entitled *Die Freigabe der Vernichtung lebensunwerten Lebens* [The authorization for destruction of life unworthy of living].[209] The crude right-wing nationalistic utilitarianism of the two scholars, who identified the value of individual life in its productive efficiency, could not help but seem foreign—Morselli affirmed, repeating his anti-German nationalistic prejudices—to "us, Latins," endowed with a different sense of humanity and proportion.[210] Adopting similar criteria would impose a true and proper decimation of the social body, which risked the disappearance of a Byron, a Leopardi, an Aesop or "other men with similar taints of the body, but excelling in intellect."[211] Morselli therefore roundly denounced the notion of eugenic euthanasia:

> I must say that among the means of human selection examined in all aspects, advised and pushed by the eugenicists, the violent, Spartan suppression of harmful or useless individuals through euthanasia, is only a remote possibility, an extreme measure in case the other means [...] do not achieve the scope of arresting the undeniable current increase of the morbigene and degenerate causes that can be managed with "social control."[212]

Instead of euthanasia, Morselli proposed an "ethnarchic selection," achieved through "the sexual isolation of whites, that is, the absolute prohibition of reproductive unions with the races of low intellectual and social value":

> Racial crossings of individuals of the white race with those of any inferior race must be impeded, not excluding the yellow man; above all we should aim at the conservation and increase of the mental quality that characterizes the superior races, that is, ours: intelligence, the inventive, and at the same time, assimilative spirit, social solidarity, the sense of individual duty, the consciousness of the moral and social importance of work, the formation of an intellectual aristocracy devoted to the development of science, art and religion.

---

209 Karl Binding and Alfred Hoche, *Die Freigabe der Vernichtung lebensunwerten Lebens, ihr Mass und ihre Form* (Leipzig, F. Meiner, 1920). On the Binding-Hoche polemic, see among others: Henry Friedländer, *The Origins of Nazi Genocide: From Euthanasia to the Final Solution* (Chapel Hill: University of North Carolina Press, 1997), 13–16.

210 Enrico Morselli, *L'uccisione pietosa (eutanasia) in rapporto alla medicina, alla morale e all'eugenica*, (Turin: Bocca, 1923), 89.

211 Morselli, *L'uccisione pietosa*, 66.

212 Morselli, *L'uccisione pietosa*, 232.

213 Morselli, *L'uccisione pietosa*, 237.

> All that is lacking or is rudimentary in the Negro races of the colonized territories.[213]

In the context of every race, instead of euthanasia and sterilization—of which Morselli repeated the "relative practical appropriateness," yet postponing any application to a future "evolution of morals and sentiments of the civil population"[214]—he proposed prevention, social medicine, hygiene:

> Here is a program of social medicine that is much more valid for eugenics than authorized euthanasia, insofar as it grasps the causes of painful phenomena and is not content to combat the effects; here, the highest moral principle of respect for life is satisfied, without which civil progress would not exist.
> The collective good must remain the supreme aim of eugenics, but first we must ensure that this collectivity is purged of all that determines and maintains blameless deficiencies, monstrosities and annihilation of the physical-psychical personality of the individual.[215]

The "selectionist doctrine," applied through "extreme" measures (euthanasia) or "mutilating" (sterilization)—although "the most secure and most abiding by the principle of defense of the race"—must, however, withdraw, not only for moral and juridical reasons, but also because of its "current practical unfeasibility," to make way for means that were "softer and perhaps also more effective, insofar as they penetrate the viscera of the social body, and involve the reproductive functions of the organism, its conditions of life, its relation with the natural forces."[216]

The physical and moral "reclamation" of society was, therefore, the premise for a moderate form of eugenics:

> The improvement of the species, the regeneration of a race beset with ills that seem inseparable from the progress of civilization, until now based on the principle of individual liberty, must happen gradually, evolving with the diminution of this liberty. This must be achieved above all in relation to sexual union for reproduction, and in second place by forbidding individuals the false right to squander their patrimony of physical and mental energy, for example,

214 Morselli, *L'uccisione pietosa*, 250.
215 Morselli, *L'uccisione pietosa*, 253.
216 Morselli, *L'uccisione pietosa*, 258.

by poisoning themselves with alcohol. Then we must mention the increasingly energetic fight against the great factors of degeneration, which are independent of the will of individuals and are of an exogenic nature, such as syphilis, tuberculosis, malaria, pellagra, infective fevers, morbid epidemics and regional illnesses, above all tropical.[217]

In conclusion, according to Enrico Morselli, a progressive limitation of the reproductive freedom and a vast program of social medicine and social hygiene constituted the core of "Latin" eugenics and the secret of its superiority compared to the "Nordic" model.

4. The Work of the "Useless": Mental Hygiene in Italy

Due to the intensive organizational activity of Ettore Levi and Giulio Cesare Ferrari,[218] the Italian League of Hygiene and Mental Prophylaxis (*Lega Italiana di Igiene e Profilassi Mentale*, known as LIPIM), was instituted on 19 October 1924, in the hall of the provincial council of Bologna. The presiding board was made up of Ferrari, Levi and Eugenio Medea. The honorary presidency was assigned to Leonardo Bianchi, Eugenio Tanzi and Enrico Morselli, while the central committee was comprised of the presidents of the thirteen regional sections.[219] The first assembly debate culminated in the definition of the aims of the League:

1) Research, gather and assess information, documents etc.; conduct or stimulate inquiries, investigations, research etc. on the causes of mental illnesses, on

---

217 Morselli, *L'uccisione pietosa*, 259.

218 Invited in 1923 to a meeting in Paris in his role as vice-president of the International Commission for the Study and Prophylaxis of Mental Illnesses, Ferrari was urged to institute also in Italy a section of the new International League of Prophylaxis and Mental Hygiene created in New York on the initiative of Clifford W. Beers. In Italy, Ferrari contributed to the creation of a provisory committee for participations in future congresses in New York of the International League and, in view of the formation of a National League, coordinated with Ettore Levi, already supporter since 1921 of a similar project: see Giulio Cesare Ferrari, "La lega italiana per l'igiene mentale," *Difesa sociale* 3, no. 6 (June 1924): 4–6.

219 The list of regions and their relative presidents is as follows: Piedmont (Lugaro), Lombardy (Medea), Veneto (Cappelletti), Liguria (Vidoni), Emilia (Ferrari), Tuscany (Amaldi), Marche (Modena), Lazio (De Sanctis), Abruzzo (Del Greco), Campania (D'Abundo), Apulia and Sicily (the clinical psychiatrists of the Universities of Catania and Bari), Sardinia (De Lisi): see "Costituzione della Lega Italiana di Igiene e Profilassi Mentale. Resoconto ufficiale della seduta inaugurale. Bologna 19 ottobre 1924," *Difesa sociale* 3, no. 11 (November 1924): 8.

the resulting moral and economic damage to the individual and the community, on legislative and preventive medicine measures, enacted to correct such causes and avoid such damage.
2) Carry out extended, energetic and continual works of disseminating information, collected and duly analyzed: propaganda aimed at stimulating the ruling classes and the political sphere, propaganda with an educational aim, wherever the masses can be influenced (schools, factories, barracks, agricultural communities, emigrant centers, etc.).
3) Coordinate the actions of the League with that of the public and private associations, national or regional, that conduct similar campaigns (alcoholism, venereal diseases, intellectual and moral deficiency of adults and youth) with particular regard to the prevention of criminality.
4) Stimulate the cooperation of teachers, scholastic physicians and scholastic wardens for the immediate selection of children predisposed to certain illnesses; and carry out analogous activities in the work and military environments.
5) Cooperate in the preparation of specialized personnel (health and social assistants) for this special form of prophylaxis.
6) Promote the institution by provincial administration and other public and private entities of dispensaries for the early diagnosis and ward care of those predisposed to nervous and mental illnesses, of those at the beginning of the sickness, and those discharged prematurely by hospital psychiatrists.
7) The same institutions should promote the formation of open wards and all the innovations of medical assistance aimed at prophylaxis and cure of mental illnesses.[220]

LIPIM's scientific program and internal composition recalled many issues discussed by Italian psychiatry, in particular immediately after the First World War, with regard to the inadequacy of Italian legislation on the assistance for mental diseases and to the failure of asylums as curative hospitals.

Leonardo Bianchi himself had sounded a cry of alarm in 1918 in front of the National Post-War Commission, and again in the Senate in 1922, appealing to the Prime Minister, Luigi Facta. Called to discuss the problem

220 "Costituzione della Lega Italiana di Igiene e Profilassi Mentale," 8.

of the "social defense against neuroses and psychoses," the National Post-War Commission completely approved Bianchi's proposal:

> 1) Institute the care in sanatoriums of the curable forms of psychotics, removing them from asylums and placing them in the psychiatric university clinics, suitably enlarged.
> 2) Provide more comprehensively for schools for deficients;
> 3) Intensify the fight against alcoholism and all the causes of physical degeneration.[221]

Four years later, Senator Bianchi returned to his theme, stressing the growing number of feeble-minded and the consequent necessity of modifying the ineffective 1904 laws:

> In Italy in 1874 there were roughly 12,000 committed feeble-minded; today there are around 45,000. [...] But I maintain that the number of admitted mentally ill people in asylums represents only a small part of the sick people. When, for example, we consider that in asylums, in accordance with our laws, we can admit only those who are judged dangerous to themselves or others, it is easy to guess at the enormous numbers of infirm, neurasthenic, epileptic and degenerate people in general.[222]

Based on the concept of "public security," the Italian legislation, according to Bianchi, had transformed asylums from places of cure to incurable wards, characterized by a simple custodial function, in which the sick arrived when it was already too late. To escape this vicious circle it was necessary that the mechanism of admittance and discharge was left in the hands of medical staff, and liberated from the ties implicit in the concept of "danger," and that a vast network of prevention finalized at "retarding the degeneration of the race" be placed side by side with asylums.[223] Bianchi believed above all in interventions that would take effect on those "social illnesses" (alcoholism, syphilis, malaria, tuberculosis) deemed to be the origin of the "psychosomatic weakness of men and their offspring." He also valued actions that caught the early symptoms of illness,

221 La Direzione, "Per la psichiatria nel dopo-guerra," *Quaderni di psichiatria* 6, no. 3–4 (March–April 1919): 98.
222 Leonardo Bianchi, "Medicina preventiva e malattie nervose e mentali," *Difesa sociale* 1, no. 6 (June 1922): 3.
223 Bianchi, "Medicina preventiva e malattie nervose e mentali," 4.

in those places where it manifested publicly for the first time, that is, in the schools.[224]

Next to psychiatric dispensaries and "special schools for the feeble-minded," a eugenic legislation was the third remedy suggested by Bianchi. As "the value of a race, in social conflicts, is strictly linked with physical and mental health, and above all with the vigor of the character," the introduction of a eugenic legislation focused on the control of marriage answered a political need even more than a health one:

> It is good to know that for every person admitted to an asylum there are no less than 50, or maybe 100 sick people headed for degeneration; we know that many of these come from marriage between imbeciles, criminals, epileptics, chronic alcoholics and various other forms of degenerates. The time for eugenic legislation will come.[225]

In his address to the Prime Minister, Bianchi described eugenics as an instrument of redemption and optimization of the nation: "The stronger a nation is," he claimed, "the less it produces infirm or incapable people, who disturb the ordinary life and work of the nation; or even when it does produce them, possesses *strong organs of correction and elimination*."[226]

In the same years, Enrico Morselli and his review *Quaderni di psichiatria* supported Bianchi's arguments. A new column of the review, inaugurated in 1919, referred to Bianchi's position in order to delineate the features of a "post-war psychiatry," that is, a psychiatry fully conscious of the new social dimensions implied by the transformations which the worldwide conflict had triggered:

> The War has been won, but with victory we have not satisfied that larger aspiration to a renewal of all the assets of our ancient Civilization, which animate and agitate the European populations today. [...] A confused bureaucracy, hostile to every innovation, fixated on its own passive resistance, makes ren-

224 Bianchi, "Medicina preventiva e malattie nervose e mentali," 4.

225 Bianchi, "Medicina preventiva e malattie nervose e mentali," 7.

226 Bianchi, "Medicina preventiva e malattie nervose e mentali," 7; italics added. Bianchi's eugenics was based on neo-Lamarckian theories that used the "engrams" or "mnemes" of Richard Semon to describe the evolution of the "germ plasm": see Leonardo Bianchi, *Eugenica, igiene mentale e profilassi delle malattie nervose e mentali* (Naples: Idelson, 1925).

> ovation very difficult; but nevertheless, we must prepare and diffuse this program in every field of national activity.
> Psychiatry, which has multiple and strict ties with social life, must first be aware of its own needs, of services that it could render, of its role in the renewal of the nation; and so we are dedicating a special column to the post-war period, and it will deal with or even only indicate those points that are now assuming importance.[227]

In full agreement with Morselli, the socialist Cazzamalli called for a "convention of psychiatrists" in which the "competent people" would directly confront the age-old question of reforming the asylum system. Cazzamalli declared: "For the psychiatric workshop the psychiatrists in the front line must know how to be demolishers of the old, constructors of the new, and wise organizers."[228] In the same issue of *Quaderni di psichiatria*, Morselli, acting as spokesman for a multiplicity of requests coming from all over Italy, presented the platform for a convention dedicated to issues of *Psychiatry in the post-war period*, polemically set against the Congress of the Italian Phreniatric Society, scheduled for 1920:

> Currently, we are aiming to put psychiatry into contact with real life, and to have it accomplish, in its social function and technical organization, those steps that respond to the greatly felt need for a general renewal. Therefore, leaving the "Phreniatric Society" to its program of contents more theoretical than practical, there are many colleagues who believe it necessary to hold a convention of a different nature, before the end of 1919, in which psychiatrists congregate to deal with the most pressing themes of "post-war psychiatry," developing a different program from that mentioned above, which will not hamper the execution of 1920, and which will be a program more in tune with the urgent exigencies of the current historical moment.[229]

Among the diverse issues listed in detail by Morselli—legislation on mentally ill and asylums, reorganization of asylums and psychiatric institutions,

227 La Direzione, "Per la psichiatria nel dopo-guerra," 96.
228 Ferdinando Cazzamalli, "Una riforma della Spedalità psichiatrica," *Quaderni di psichiatria* 6, no. 5–6 (May–June 1919): 138.
229 La Direzione, "Per la psichiatria del dopo-guerra. Proposta di un Congresso Alienistico pel Dopoguerra," *Quaderni di psichiatria* 6, no. 5–6 (May–June 1919): 144.

improvement of the professional psychiatric class—it is worth underlining the points comprised in the field of "social psychiatry":

> 1) An immediate and human definition of psychical illnesses of war and relative measures (special pension, allowance, care, etc.);
> 2) Social prophylaxis against neurosis and psychosis, and eugenic measures (see the report of Leonardo Bianchi);
> 3) Fight against alcoholism, syphilis, tuberculosis and pellagra;
> 4) Fight against criminality, particularly underage;
> 5) Social measures for the mentally ill, abnormal and amoral people, that the schools are uncovering;
> 6) Severe applications of appropriate acts to develop obligatory physical education of children and youth of both sexes [...].[230]

In November 1920, the Congress of Italian Psychiatrists (*Congresso degli Alienisti Italiani*) desired by Morselli was held in Genoa and a resolution was approved that involved a precise scheme of reform of "asylum-prisons," summarizable in three elements: 1) Hospitals for extreme cases, with institutions for prophylaxis and mental hygiene; 2) Hospitals for chronically unable to work and special institutes for feeble-minded children and for criminals; 3) Agricultural colonies and industrial laboratories for chronically ill workers.[231]

This was obviously not a plan for closing asylums, but for restoring them to their presumed curative function, inserting them into an open system, differentiated (into extreme cases, incurably disabled, chronically ill workers) and "prophylactic." This was the position that Morselli defended in 1920, in the pages of *Quaderni di psichiatria,* in an interesting discussion with Enrico Ferri. Asylums were not only to "defend the social body against the disease of insanity," as Ferri maintained, but on the contrary had "a medical function, therapeutic and prophylactic":[232]

230 La Direzione, "Per la psichiatria del dopo-guerra." 144–45.

231 La Direzione, "I nuovi indirizzi della assistenza neuro-psichiatrica," *Quaderni di psichiatria* 14, no. 5–6 (May–June 1927): 108.

232 Enrico Morselli, "La funzione sociale del Manicomio," *Quaderni di psichiatria* 7, no. 5–6 (May–June 1920): 135.

> Asylums are not only houses of custody, where the insane are closed up to take away their means to behave according to their whims, impulses or their deliriums in the midst of society, damaging its interests and disturbing its sentiments. This tutelary function has unfortunately come to prevail in the medical-social aims of asylums, due to the coercive legislative principles of admittance of the mentally infirm; but this prevalent juridical method is absolutely damaging to the patients themselves.[233]

For Morselli, the social defense of the "painful fact of insanity" should not be a main function of asylums, but must rather be relieved through "medical, hygienic and socio-political measures":

> The fight against alcoholism and tuberculosis; the regulation of customs and protection against sexual illnesses; the organization of schools for the feeble-minded; the measures against pellagra and malaria; the general improvement of economic, hygienic-sanitary, etc. conditions. These other social defenses [...], which are more effective than the functions of asylums, are those fore-seen and demanded by eugenics. For example, the limitation of marriages between people hereditarily disposed or certified syphilitic and alcoholic, and perhaps also between tuberculosis sufferers; the restriction of unions between relatives, particularly among defective families; the facilitation of unions with young races [...] etc.[234]

According to Giuseppe Muggia, director of the Sondrio asylum, the creation of wards and dispensaries could transform asylums, making them "suitable for their high social function."[235] But it was above all necessary that psychiatrists lengthened their gaze "beyond the walls of the asylum," not limiting themselves to the brief period of confinement but concentrating their energy on "wise works of prevention," as useful on the sanitary plane as on the economic one.[236]

The fundamental problem—confirmed Giulio Cesare Ferrari in 1923—was that "asylums are not good for social productivity, and are not worth

233 Morselli, "La funzione sociale del Manicomio," 134.

234 Morselli, "La funzione sociale del Manicomio," 136.

235 Giuseppe Muggia, "Per l'avvenire della Psichiatria e dell'assistenza psichiatrica," *Quaderni di psichiatria* 9, no. 9–10 (September–October 1922): 192.

236 Muggia, "Per l'avvenire della Psichiatria e dell'assistenza psichiatrica," 194.

what they cost."[237] The solution indicated in Germany by Hoche and Binding could be shared, but it was not the only viable option. According to Ferrari, it would rather be better to rationalize the asylum system, lightening the weight of the "effectively non-dangerous incurables," who could advantageously be utilized in work colonies, while the "psychiatric hospitals" would be wholly dedicated to the cure of "a few acutely ill."[238] Also Cesare Agostini's psychiatric experience was focused on the organization of a "more rational and economic treatment of the mentally ill." He was the director of the Perugia asylum, which he described as a model to imitate: the reduction of costs and the best assistance for the "acute cases" were the fruit of a vast operation of evacuation of the "tranquil and innocuous incurably demented" to "new departments in pre-existing poor asylums in Rieti, Foligno and Spoleto."[239]

Again in 1923, on the occasion of the National Hygiene Convention in Milan, the physician Ernesto Ciarla called for the institution of a mental prophylaxis service. For assistance to be effective, it was necessary to intervene in favor of the sick "before the mental illness is declared," in the so-called "premonitory period." To this end, the institution of special clinics and dispensaries was needed, and, "to make the prophylactic measures as effective as possible, also preventoria": the early care of subjects in whom mental illness was still in an initial stage would "impede a future incurable illness, and therefore the burden of a long maintenance of the part of public administrations."[240]

Therefore, in the moment of its constitution, LIPIM had behind it at least twenty years of debates from physicians that clearly indicated the eugenic path to follow: prevention of mental illnesses in dispensaries and identification of defectives in schools.

Significantly, in 1935, Giuseppe Pellacani saluted the Italian movement for mental hygiene as the start of a new era for the history of psychiatry. Following the "Latin" phase (from Chiarugi and Pinel to Esquirol) that

237 Giulio Cesare Ferrari, "Il prossimo avvenire dell'Assistenza psichiatrica in Italia," *Quaderni di psichiatria* 10, no. 5–6 (May–June 1923): 112.

238 Ferrari, "Il prossimo avvenire dell'Assistenza psichiatrica in Italia," 114.

239 Cesare Agostini, "Per un trattamento più razionale ed economico degli alienati di mente," *Quaderni di psichiatria* 10, no. 9–10 (September–October 1923): 193.

240 Ernesto Ciarla, "Per l'istituzione di un servizio provinciale di profilassi delle malattie mentali," *Quaderni di psichiatria* 11, no. 9–10 (September–October 1924): 192.

described psychopathological syndromes and constructed asylums, and the "German" phase (Griesinger, Wernicke and Kraepelin), in which the individual anatomic-clinical orientation prevailed, Pellacani believed it was now time for a "social" or "prophylactic-hygienic" phase, concerned above all with analyzing the exogenic causes of psychopathology:

> Today we have gone from individual psychiatry to the social phase, hygienic-prophylactic, of psychiatry, characterized on the theoretical-practical plane by a definite approaching of psychiatry to neurology and to general medicine, and by the necessity to know and combat the group of liminal infirmities, of light and initial infirmities, which appear in the neuropsychiatric practice of dispensaries.
> This dynamic psychiatry (opposed to the static psychiatry of the old asylums) directs all its scientific interest and its practical interventions to the evolving forms [of illness] [...] Boundary forms: showing how many carriers of psychopathic anomalies there are in society, with no consciousness of their infirmity [...].[241]

Pivot of the "diagnostics and practice of hygienic-prophylactic psychiatry,"[242] dispensaries underwent an intense development in the twenties and thirties, due to the strong propulsion of LIPIM. An internal census of the League in 1936 registered the organized presence in 26 provinces: Agrigento (1931), Alessandria (1933), Ancona (1910), Arezzo (1904), Ascoli Piceno (1928), Belluno (1920), Bergamo (1931), Bologna, Catanzaro (1914), Cuneo (1932), Genoa (1928), Gorizia (1932), Mantua (1930), Milan (1924), Novara (1936), Parma (1932), Pesaro (1927), Reggio Calabria (1935), Rome (1929), Siena (1933), Sondrio (1932), Teramo (1928), Treviso, Trieste (1927), Venice (1927) and Verona (1930).[243] However, the hope that the new eugenic-prophylactic apparatus would translate into a consistent possibility of economic savings, reducing the number of admissions in asylums, was quickly revealed as illusory. The new structure, managed directly by psychiatric hospital personnel, received, in fact, prevalently "psychopaths," that is, "individuals between healthy and mentally

[241] Giuseppe Pellacani, "Psichiatria e psicoigiene," *L'igiene mentale* 15, no. 1 (1935): 8.

[242] Pellacani, "Psichiatria e psicoigiene," 9.

[243] Carlo Ferrio, "Nota conclusiva sull'Assistenza Psichiatrica non coattiva in Italia," *L'igiene mentale* 1 (1936): 101. The date of foundation of the Bologna and Treviso dispensaries are missing in the source.

ill," already discharged from asylums. Moreover, the effectiveness of the dispensaries was measured, according to the census cited, more in terms of "general social utility" than in the diminution of assistance costs. These social functions were to be found:

1) In the facilitation of the readmission into social life of those discharged from psychiatric hospitals;
2) in combating relapses;
3) in the study, according to scientific, statistical and medical-social criteria, of all psychical illnesses and abnormalities generally found outside of psychiatric hospitals;
4) in transmitting to the public every sort of knowledge of social hygiene, especially psychical (prophylactic campaign).[244]

As regards the eugenic "selection" in the field of education (differential classes, autonomous schools, medical-pedagogical institutes), a report by Eugenio Medea at the 2nd European Meeting for Mental Hygiene (Rome, 27–28 September 1933) listed, in particular, the differential classes in Rome, Milan and Genoa, hoping they would become "obligatory" in all the principal Italian elementary schools; the Zaccaria Treves Autonomous School in Milan; the kindergarten-school founded by Sante De Sanctis in Rome from 1899 and, also in Rome, the Montesano Orthophrenic School; the Autonomous Schools in Genoa, directed by Giuseppe Vidoni; the Medical-pedagogical Colony of Marocco in Venice, started by Tumiati; and Medical-pedagogical Institutes in Trieste, Florence, Thiene, and Bologna.[245]

In Genoa and Venice, the provincial administrations instituted centralized services of mental prophylaxis, directed respectively by Vidoni and Tumiati, which controlled the neuropsychiatric dispensaries, the charitable institutions for disabled mentally ill, and the consultancies for abnormal infancies. In Genoa, an "extended medical-pedagogical assistance" paralleled the scholastic system, involving Nicola Pende's Biotypological Institute [see chapter 4]:

> For all those enrolled in the Genoa Autonomous Schools, connected to scholastic charities, to several charitable institutes and to the Biotypological Institute of the Genoa Medical Clinic, a bio-psychological diagram has been completed.

244 Ferrio, "Nota conclusiva sull'Assistenza Psichiatrica non coattiva in Italia," 103.

245 Eugenio Medea, "L'igiene mentale e la scuola," *L'igiene mentale* 13, no. 3 (1933): 10–12.

> Such schools include a supplementary section for the eldest, until 14 years old, for instruction and introduction to work. In the elementary school, in industrial and agricultural schools, the psychotechnical work orientation [of the children] is determined.[246]

In Naples, the scholastic problem was at the center of the activities of the regional section of LIPIM, which coincided greatly with the Neapolitan Eugenic Group: both, not coincidentally, presided over by Leonardo Bianchi. The latter—responding to the question "What does 'mentally healthy' mean?" for a questionnaire published by LIPIM's review—discoursed at length on the theme of mental hygiene in classrooms:

> The Italian primary schools are not well organized everywhere, they do not all offer that which is required by hygiene: very seldom are there medical-scholastic workers, prepared for all the exigencies of modern-day schools. Not all schools select students, who, due to organic conditions, or to a certain level of psychical insufficiency, need particular schools and pedagogical methods designed to develop their physical and mental organism.[247]

In the secondary schools, the situation was, if possible, worse: "the programs are confused, the culture is too broad; the choice of books is not guided by sound criteria, mnemonic methods prevail; no consideration is given to the individual disposition and particular attitudes; the number of overtired students is worrying."[248] These same worries were expressed by the Neapolitan Eugenic Group in 1926, concluding a "broad debate on mental hygiene, intimately linked to the healthy psychical-physical evolution of human life and therefore to the prosperous future of the individual and the race." The group hoped:

a) That suitable schools be provided as soon as possible in big and small cities, which are still behind in fulfillment of their fundamental social duties;

b) That school buildings include where possible a free area, or at least wide verandas [...];

[246] Giuseppe Pellacani, "I servizi di profilassi neuro-mentale in Italia," *L'igiene mentale* 14, no. 1–2 (1934): 17.

[247] Leonardo Bianchi, "Che vuol dire 'sano di mente'?," *L'igiene mentale* 6, no. 2 (1926): 3.

[248] Bianchi, "Che vuol dire 'sano di mente'?," 3.

c) That the function of vigilance of health and mental hygiene in childhood and adolescence, through the figure of the school physician, should be strictly disciplined and observed as a function of the State, with a national character;

d) That certain weak or more slowly developing students are selected and instructed in special classrooms and schools, with methods and programs more suitable for the development of their physical and mental conditions.[249]

In Rome, the key figure was undoubtedly that of Sante De Sanctis, president of the Lazio section of LIPIM from 1924 and national president, from 1930.[250] De Sanctis had been a pioneer of Italian experimental psychology, had studied methods of "scholastic selection" from the beginning of the century, was the author, in 1907, of a scale of mental tests for the evaluation of I.Q., and in 1909, of a broad "medical-pedagogical and assistance classification of mentally ill and neuropsychopathic children."[251] De Sanctis contributed to LIPIM's eugenics with his decades of experience in the field that he himself defined as the "scientific organization of mental work."

As in the factories, also in the classrooms Taylorism should be applied. De Sanctis claimed: "There is no doubt that the school is a factory, where students work, and which needs to be productive. School productivity has a value analogous to all the economic values."[252] In order to be "guarantor of the greatness of the nation,"[253] schools had to be "organized in a scientific way, as an industry of the State would be organized." There were three phases in particular that would define the "mental and moral reclamation" of the "school factory": the intellectual and moral "evaluation" of the scholars; the "selection," with the "scope of eliminating those pupils from the student body that, following their evaluation, demonstrated, in their mental level or in their character, variations below normal, and therefore dam-

249 "Sezione campana," *L'igiene mentale* 6, no. 4 (1926): 18.

250 The Lazio section was particularly active in the field of optional premarital examinations, neurological examinations of underage prisoners or those in corrective facilities, and genealogical researches in schools: see "Lega italiana di igiene e profilassi mentale, sezione laziale," *Difesa sociale* 4, no. 12 (December 1925): 19–20; "Lega italiana di igiene e profilassi mentale, sezione laziale," *Difesa sociale* 5, no. 8 (August 1926): 17–19.

251 For some biographical sketches, see "IV Assemblea generale dei soci (Milano 19 marzo 1934): Commemorazione del prof. Sante De Sanctis (Antonini, Medea, Corberi, Albertini)," *L'igiene mentale* 15, no. 2 (1935): 6–18.

252 Sante De Sanctis, *Igiene Mentale* (Turin: Paravia, n. d.), 6–7.

253 De Sanctis, *Igiene Mentale*, 6.

aging to the community";[254] the "differentiation" of the group, aimed at distinguishing, through mental tests, the subgroups of "differentiated inferior" from the "differentiated superior" (that is, the "very intelligent" and "very wise"). After carrying out these three preliminary operations, the hygienist-teacher had to impose an *optimum* working regime on the scholars, following the principle of "temporary maximization of work," with an educational, not productive, scope:

> When the students find themselves, for intrinsic reasons, in the phase of under-achievement and imitate, without meaning to, certain unionized workers, we think it would be useful to whip them to maximum work. If the under-achievement is then involuntary, or is due to extrinsic causes, it is scholastic hygiene that must intervene to re-establish an optimum regime. [...] The experience has made me appreciate the practice of dividing the work as an excellent means of maximizing purely mental work, without increasing the speed and the quantity: that is to say, to teach the students secondary work, whether it is work of memorization or composition, which is subordinate to the principal work.[255]

As well as being one of the most illustrious and active members of LIPIM, De Sanctis, from 1926, was the head of the Roman Provincial Federation of the National Maternity and Infancy League (*Federazione provinciale romana dell'Opera Nazionale Maternità e Infanzia*), known as ONMI. This new institutional role was instrumental in linking his competencies as a pioneer of Italian experimental psychology and his social fascist political ambitions. In 1928, in the pages of Mussolini's *Gerarchia*, De Sanctis welcomed the constitution of ONMI with enthusiasm, but at the same time, invited the regime to listen to the advice of experts: "Social assistance monitored by the State demands expertise; and it demands it in the name of its aim, which is the defense of the stock."[256] The problem of selection between the "recoverable" and "rejected" was once again central:

> To the "rejected" group, we must assign the juvenile deficients, paralytics, serious idiots, deficients, serious epileptics, invalids with complications, such as:

[254] De Sanctis, *Igiene Mentale*, 8.

[255] Sante De Sanctis, "L'organizzazione scientifica del lavoro mentale," *Rivista italiana di sociologia* 20, no. 5–6 (September–October 1916): 520–21.

[256] Sante De Sanctis, "I problemi di rieducazione," *Gerarchia* 12 (December 1928): 962.

> deaf-mutes, or deficient blind people, or blind epileptics, and the mentally ill with tuberculosis.
> To the "recoverable" group, we can easily assign so-called "differentiated" juveniles, and deficients and epileptics with paresis or not, but not the most serious [...].[257]

The grouping into two categories was to be based on "psychological methods," fast and flexible, rather than on the extremely useful, but "long and delicate," "polymorphic investigations" of Pende's Biotypological Institute. According to De Sanctis, the judgment of "educability" had to be made "as technical as possible" and, to such an aim, it was necessary that the selection operate according to criteria of "social and productive adaptation of the mentally ill":[258]

> The technical assistance of the "recoverables" is done, because (to speak in banking, therefore brutal, terms) what is spent represents a species of advance or loan on the part of the community, which will be compensated in time by the future productivity of the assisted.[259]

The same internal ONMI memorandum, on the 20 February 1928, declared, however, that assistance should only be granted to elements *functional* to national interests: "Assistance from ONMI is justified only for those individuals who, in the appropriate conditions, could operate socially as useful and productive elements for the nation."[260] The "training in a profitable job" was therefore, for De Sanctis, "the most serious and most economic instrument of correction, of social redemption for our unhappy youth":

> Now, if the deficients of the kindergarten/school children from 12 years onwards can achieve an "individual economic value" that is from 50–80% of the value of normal children of the same age, this will inevitably have one consequence: social legislation that imposes *obligatory work* on abnormal children and youths, whether feeble-minded or unstable of conduct. In sum: "ruralize" and "industrialize" the "recoverable" deficients.[261]

257 De Sanctis, "I problemi di rieducazione," 963.
258 De Sanctis, "I problemi di rieducazione," 966.
259 De Sanctis, "I problemi di rieducazione," 965.
260 De Sanctis, "I problemi di rieducazione," 965.
261 De Sanctis, "I problemi di rieducazione," 969.

As well as being economically advantageous for the "defense of the stock," the "techniques of work" exercised an obvious therapeutic function:

> Work "recovers," because it develops common sense. [...] Common sense can certainly not be obtained with education; since there are no rudiments and no training. It can only be hoped that it will develop in "recoverables" through the acquisition of the consciousness of their own force to overcome the extrinsic and intrinsic obstacles of life. Consciousness that cannot be acquired if not through work that is visibly and tangibly productive, transformable into easily comprehensible values. Therefore, the ideal education of deficients must be *homo faber*, not *homo sapiens*.[262]

"Of useless men, make productive elements":[263] this was the objective that sanctioned the happy meeting of the fascist social assistance system with the specific current of eugenic thought that was rooted in the development of experimental psychology and mental hygiene in Italy. Reform of asylums, psychiatric dispensaries and "differential classes" were, ultimately, three aspects of a single eugenic project, of which the ultimate goal was the maximum level of economic rationalization of national biological resources.

In the 1920s, this eugenic approach had represented an alternative method of efficient management to the Hoche and Binding proposal, while in the 1930s it was at odds with Nazi eugenic sterilization policies. Borrowed from a precedent Prussian project of 1932, the German sterilization law was approved on 14 July 1933, with the name of "Law on the Prevention of Genetically Deficient Progeny" (*Gesetz zur Verhütung erbkranken Nachwuchses*). It opened the offensive against the disabled, only six months after Hitler came to power, and became the "cornerstone of the regime's eugenic and racial legislation."[264] At the First European Congress for Mental Hygiene (Paris, 30–31 May 1932), the Italian delegates, with Tumiati

---

262 De Sanctis, "I problemi di rieducazione," 970.

263 Luigi Maggiore, "L'assistenza dello Stato agli invalidi, storpi e mutilati," *Politica sociale* 4 (1932): 477–81, cited in Mantovani, *Rigenerare la società*, 317.

264 Friedländer, *The Origins of Nazi Genocide. From Euthanasia to the Final Solution*, 26. The law defined a person "affected by a heredity illness," and therefore a candidate for sterilization, as anyone who was afflicted by the following disorders: congenital lunacy, schizophrenia, circular insanity (manic-depressive disorder), hereditary epilepsy, Huntington's chorea, hereditary blindness, hereditary deafness, serious hereditary physical deformations, serious alcoholism. On the rejection of Nazi negative eugenics by the fascist regime and the Catholic Church, see Giorgio Sale, *Hitler, la Santa Sede e gli ebrei* (Milan: Jaca Book, 2004): 115–24.

and Corberi at their head, impeded the approval of a resolution, based on the report of the Swiss psychiatrist Ernst Rüdin[265]—director from 1931 of the Kaiser Wilhelm Institute for Psychiatry in Munich and from 1933 head of the *Deutsche Gesellschaft für Rassenhygiene*—according to whom "the only effective weapon for mental hygiene [...] is that of impeding the fertility and the development of the sexually defective."[266] A year later in Rome, during the Second European Congress for Mental Hygiene, it was Sante De Sanctis who hurled himself against the theoretic premise of the 14 July 1933 law:

> Rüdin's empirical hereditary prognosis is supported by statistics that are too limited to be the basis of a method of demonstration. In order to be convinced, you must already have... prejudices that not everyone shares. For instance, faith in "Morel's harmful progressive heredity" or in the absolute applicability of Mendel's Laws to human generations; or the certainty of a "progressive cerebration" in the sense of von Economo; or the mystic adoration of Nietzsche's "superman" or the perfect model of a privileged and omnipotent race. [...] We are not in favour of the kind of catastrophic eugenics that pays homage to one or more of these preconceptions. We wish to use indirect means for the prophylaxis of neuropsychological illnesses, even if we lack faith in the automatic elimination of psycho-degenerated, of the unfit and the carriers of hereditary predispositions.[267]

As an alternative to the German "catastrophic eugenics," De Sanctis proposed the assistance activities of ONMI "with the aim of improving the stock." And his eugenic ambitions, more than Germany, looked to California, particularly to Lewis Terman's model of "supernormal classes": an effective means, according to the Italian psychologist, of "creating an elite of intellectually superior men, for the good of the community."[268]

In July 1936, in Paris, on the occasion of the Second International Congress of Mental Hygiene, Arturo Donaggio, director of the Neuropsychiatric clinics of Bologna and president of the Italian Society of Psychiatry, publicly contested Rüdin's theory in favor of eugenic sterilization:

265 "I Riunione Europea per l'Igiene Mentale (Parigi, 30–31 maggio 1932)," *L'igiene mentale* 12, no. 2 (1932): 20.

266 "I Riunione Europea per l'Igiene Mentale (Parigi, 30–31 maggio 1932)," 17.

267 "II Riunione Europea per l'Igiene Mentale (Roma, 27–28 settembre 1933)," *L'igiene mentale* 13, no. 3 (1933): 42–43.

268 "II Riunione Europea per l'Igiene Mentale (Roma, 27–28 settembre 1933)," 43.

> Prof. Rüdin, to justify compulsory sterilization, starts from premises based on a certainty. In fact, only absolute certainty can justify the intervention of authority, which decides the fates of a human personality, which imposes on his future in a decisive way. Now, does this indispensable condition of certainty in the premise of Prof. Rüdin really exist?[269]

According to Donaggio, Rüdin based compulsory sterilization on a "public health system inspired by eugenic principles," but actually eugenics was "not a real and proper science," but only a "complex of observations, which is trying to constitute a body of science." In second place, the "medical body excelling in the art of diagnosis" theorized by Rüdin, did not exist in reality, because "in fact, diagnoses are frequently disparate." Finally, "human hereditary biology" could not establish a sufficient basis for certainty, as it was also a "discipline still under formation." And all this was without even considering the possible "processes of regeneration," or the fact that "at times, beautiful minds and even geniuses—that is, propellant elements of human civilization, representative individuals or heroes in the sense of Emerson or Carlyle—had hereditary ancestors heavily burdened with mental illness." Donaggio's conclusive judgment admitted no doubts: "For the current state of our knowledge, the not so certain bases that we possess cannot allow a decision of authority regarding a disablement of human personality, that is, sterilization."[270]

A few years would pass before Italian psychiatry expressed what it was disposed to import from Nazi psychiatric eugenics: not the sterilization laws of 1933, but rather the model of a national center of genetic psychiatry, based on the example of Munich and Berlin.

---

269 "II Giornata genealogica," *Atti della Lega italiana di igiene e profilassi mentale* (1938): 106.

270 "II Giornata genealogica," 106.

CHAPTER IV

QUALITY THROUGH QUANTITY

EUGENICS IN FASCIST ITALY

The political rise of Benito Mussolini was followed with enthusiasm and trepidation by many mainline eugenicists. In December 1927, Raymond Pearl wrote to Corrado Gini: "I should like enormously to meet Mussolini. I have a great admiration for him. He seems to me to be the only really big figure of our times."[1] In 1928, thanks to Gini's intervention, the Norwegian Jon Alfred Mjøen, director of the Winderen Laboratorium in Oslo, obtained an interview with *il Duce*, during which he ardently admired his demographic policy.[2] In 1929 in Rome, during a meeting of IFEO, Eugen Fischer addressed a long memorandum to "the great statesman who, in the Eternal City, shows more than any other leader today, both in deed and word, how much he has the eugenic problems of his people at heart." Through Fischer, the IFEO appealed to Mussolini, asking *il Duce* to interest himself not just in the quantity of the population, but also in its quality:

> Here today, in the oldest capital of the world, we beg to express with the utmost solemnity our hope that those great men to whom the destinies of the highly gifted Italian nation are entrusted, will be first in setting a model to the world by showing that energetic administration can make good the damage which has already been done to our culture, by arresting the fall in population and

1 Raymond Pearl to Corrado Gini, 28 December 1927, Raymond Pearl Papers, American Philosophical Society (hereafter APS), Box 7.

2 ACS, SPD, CO, f. 210.802 "Mjøen, dott. Jon Alfred." Presidente del Comitato Norvegese per l'Eugenica.

> by preserving the best endowed. We pray that what was denied to earlier cultures may here be achieved in grasping fortune's wheel and controlling and turning it! Quality as well as quantity! The urgency brooks no delay; the danger is imminent.
> *Videat consul!* [3]

These hopes were soon to be deluded. Undoubtedly, many fundamental components of fascist ideology were able to justify the elective affinity with eugenics: the myth of the biological and spiritual regeneration of the nation; the technocratic and interclassist vision of social politics; a political language imbued with vitalism and social Darwinism.[4]

However, two important political and ideological factors prevented fascism adopting the "Nordic" example of a prevalently "qualitative" eugenics.

On 26 May 1927, with his famous Ascension Day speech, Mussolini introduced the fascist pronatalist population policy.[5] It is worth noting the fundamental role that Corrado Gini played in Mussolini's turnaround. Gini was president of the Central Institute of Statistics (*Istituto Centrale di Statistica,* known as ISTAT) from 1926, and of the Italian Society of Genetics and Eugenics (*Società Italiana di Genetica ed Eugenica,* known as SIGE) from 1924. Not only did the *Duce* repeatedly consult Gini regarding the technical details of the speech, but the relationship also influenced the timing of the natalist policy launch. From 1919, Mussolini had explicitly abandoned his youthful neo-Malthusian sympathies in favor of nationalist natalism. The turning point in 1927 was fuelled by his perception of change in the current Italian demography: Italy had also been hit with a slowdown in demographic growth. This different reading of the Italian demographic situation was probably influenced by Musso-

---

3 Draft with edits by Fischer in: MPG-Archives, Dept. I, Rep. 3, No. 23, pp. 262–65, cited in: Hans-Walter Schmuhl, *The Kaiser Wilhelm Institute for Anthropology, Human Heredity and Eugenics, 1927–1945. Crossing Boundaries* (Dordrecht: Springer, 2008), 116. See also Allan Chase, *The Legacy of Malthus. The Social Costs of the New Scientific Racism* (New York: Knopf, 1977), 345–46.

4 Mantovani, *Rigenerare la società,* 261–270; Aristotle A. Kallis, "Racial Politics and Biomedical Totalitarianism in Interwar Europe," in Turda and Weindling, eds., *Blood and Homeland,* 389–416; Roger Griffin, "Tunnel Visions and Mysterious Trees: Modernist Projects of National and Racial Regeneration, 1880–1939," in Turda and Weindling, eds., *Blood and Homeland,* 417–56.

5 On Mussolini's *Discorso dell'Ascensione,* see Treves, *Le nascite e la politica,* 126–39; Ipsen, *Dictating Demography,* 84–85.

lini's personal relationship with Corrado Gini. On the important public occasion of the inauguration of ISTAT, on 14 July 1926, Gini had in fact emphasized the danger of the "decadence of the white race" and also of the "Latin nations":

> Investigations on the population are always more convincing that the white race, or at least that part of the white race that gave rise to current occidental civilization, is at a decisive turning point in its history. After the marvelous development of the population seen in the previous century, we are now in a more or less stationary situation. [...]
> The other Latin nations, and perhaps also the Slavs, do not seem to be unconnected with this general movement, but follow it much more distantly; and naturally, it could be decisive, for the life of a Nation, proceeding through a turning point in history with a speed that is more or less slowed down, also because it is precisely in turning points that the more intelligent and most decisive runners better their position.[6]

In the Ascension Day speech, Mussolini indicated that the "discovery" of this new demographic situation implied the introduction of a pronatalist population policy in Italy: "For five years we have been saying that population is overflowing. It is not true! The river is no longer in flood, and is rapidly returning to its bed." Gini was obviously not the only source of Mussolini's populationism, but it is very probable that the *Duce* obtained his statistical evaluation of the growth of the population in Europe and in Italy from Gini, thus justifying the launch, exactly in 1927, of the fascist natalist population policy.[7]

In December 1930, Pius XI radically condemned birth control, premarital certificates, abortion and sterilization in his encyclical *Casti Connubii* [On Christian marriage]. In the following months, Agostino Gemelli, founder and dean of the Milan Catholic University of the Sacred Heart (*Università Cattolica del Sacro Cuore di Milano*) and vice-president of SIGE,

---

6 Corrado Gini, "Discorso di inaugurazione dell'Istituto Centrale di Statistica (14 luglio 1926)," *Annali di Statistica. Serie VI*, 2 (1929): 18–19.

7 Since 1928, Gini referred to Alfred Lotka, Louis Dublin and Robert R. Kuczynski researches in order to legitimize, on statistical grounds, the fascist pronatalist turning point marked by the Ascension Day speech: see Corrado Gini, "Il numero come forza," *Critica Fascista* 6, no. 19 (1928): 363; Corrado Gini, "The Italian Demographic Problem and the Fascist Policy on Population," *The Journal of Political Economy* 38, no. 6 (1930): 682–97.

actively defended and promoted the themes of the papal encyclical. Between March and October 1931, on the pages of the journal *Vita e Pensiero* [Life and thoughts], Gemelli responded to accusations of "medievalism" by the *Eugenics Review*, praising the "incalculable eugenic value" of Catholic sexual morals: chastity as a form of birth control; temperance as a safeguard against the damage of alcoholism; the sacrament of marriage as a remedy to the dysgenic risk of intermarriage and illegitimate births.[8] A year later, in October 1932, at the Florence Congress of Catholic Physicians (*Congresso dei Medici Cattolici di Firenze*), the principles of Catholic eugenics were again justified, not only by Gemelli, but also by Francesco Leoncini, professor of legal medicine at the University of Florence, and Giuseppina Pastori, professor of general biology at the Milan Catholic University.

Leoncini praised the new penal code, approved in July 1931, which, among the crimes against "the health and integrity of the stock," also considered the "procured impotence of procreation" through ionizing radiation: in this way, Leoncini believed, fascist juridical code demonstrated its full adherence with the "indefatigable principles of Catholic morals." Moreover, the penal code condemned any form of "eugenic sterilization," a measure—Leoncini commented—suggested by "a new civilization, which evidently is not our civilization, shaped by Latin genius and the spirit of Christianity."[9] For Giuseppina Pastori, the Church did not forbid "that we pursue eugenic aims." On the contrary, Catholic sexual morals were in themselves eugenic: "if one truly lives Christianly—Pastori confirmed—coercive legal dispositions with a eugenic aim would not be necessary." As for the rest, medicine itself seemed to confirm the eternal truth of Catholic sexual morals:

> Healing, today, is not amputating, but preserving: tomorrow, it will not be repressing, but preventing; therefore, even scientifically the physicians see in eugenics instructed by Catholic morals a great superiority in the face of immediate and violent means proposed by non-Catholic eugenics.[10]

8 Agostino Gemelli, "Le dottrine eugenetiche sul matrimonio e la morale cattolica," *Vita e Pensiero* 22 , no. 3–4 (March–April 1931): 195–99; Agostino Gemelli, "Ancora della condanna della eugenetica. Echi e critiche alla enciclica 'Casti Connubii' sul matrimonio cristiano," *Vita e Pensiero* 22, no. 10 (October 1931): 603–14.

9 Francesco Leoncini, "Relazione su la procurata sterilità di fronte alla morale e alla legge," *Studium-Quaderno dei Medici. Il II Convegno dei medici cattolici (Firenze, 16–18 ottobre 1932)*, suppl. no. 3 (March 1933): 38–64.

10 Giuseppina Pastori, "La relazione su l'eugenica e la morale cattolica," *Studium-Quaderno dei Medici*: 70.

At the end of the event, the Florence Congress of Catholic Physicians approved a resolution on eugenics, organized into three points:

> The physicians of Catholic Action (*Azione Cattolica*) [...]
> 1) invite Catholic physicians to keep abreast of scientific progress in genetics and invite Catholic scholars to cooperate with such studies and promote the good and healthy applications of this young and already greatly progressed science;
> 2) ask that the civil public authority prevent the diffusion in Italy of foreign propaganda of those eugenic methods that represent a violation of moral laws;
> 3) vote that Catholic physicians explain to the profane how the moral and physical improvement of humanity can not be obtained with the hurried and unjustified application of genetics to the human race, and neither with the propagation of those eugenic norms that contradict divine laws and are contrary to human dignity, but rather through the moral laws taught for centuries by the Catholic Church, norms that also govern the real progress of social hygiene and genetics.[11]

Undoubtedly, the institutional, ideological and political compromise between the fascist regime and the Catholic Church—sanctioned in 1929 by the signing of the Lateran Treaty —was decisive in the affirmation—in Italy as much as in the international context—of a natalist and populationist "Latin" eugenics.

The definitive adoption of "quantitative" eugenics was first announced by a rapid process of fascistization that in the second half of the 1920s overwhelmed the previous experiences of "qualitative" eugenics, and in particular, Aldo Mieli's SISQS and Ettore Levi's IPAS.[12]

In 1927, SISQS changed its name to Italian Society of Sexology, *Demography* and Eugenics (*Società Italiana di Sessuologia, Demografia e Eugenica*), and in the next year was incorporated into the Fascist Medical Union (*Sindacato Medico Fascista*).[13] In 1924, *Rassegna di studi sessuali* [Review of sexual studies] became *Rassegna di studi sessuali e di eugenica,* then *Rassegna di*

11 "Le deliberazioni del Convegno," *Studium-Quaderno dei Medici*: 100–101.

12 On the defeat of "qualitative" eugenics in fascist Italy, see also Mantovani, *Rigenerare la società*, 285–303.

13 See *Rassegna di studi sessuali, demografia ed eugenica* 8, no. 1 (January–March 1928): 25ff.

studi sessuali, demografia ed eugenica, and finally *Genesis,*[14] surviving a few years, until 1932.

In 1928 Aldo Mieli left his position as director to transfer to Paris as permanent secretary of the International Committee for the History of Science.[15] In 1930, a report of the Parisian division of the fascist Political Police named him an "adversary of the Regime and above all of demographic politics."[16]

In December 1928, IPAS was made dependant on the National Social Insurance Bank, (*Cassa Nazionale Assicurazioni Sociali,* or CNAS), later the National Fascist Institute of Social Security (*Istituto Nazionale Fascista per la Previdenza Sociale,* or INFPS). Struck by nervous exhaustion in 1926, Ettore Levi was replaced by Augusto Carelli, who was hostile to any form of "eugenic sterilization" and trusted in the effectiveness of the eternal mechanisms of nature.[17] In 1930, the direction of the review *Difesa sociale* passed into the hands of Cesare Giannini, who moved the interests of the journal to insurance medicine. In the meantime, guilty in the eyes of fascism of "carrying out propaganda for birth control,"[18] Levi was brutally expelled from every role within IPAS[19] and committed suicide in 1932.

The only person to commemorate him with sincere emotion[20] was the editor of *Pensiero sanitario,* Pietro Capasso. Even Capasso was under the watch of the fascist regime from 1925, when he had signed the Croce "Manifesto of Anti-fascist Intellectuals." Listed in the Register of political offend-

---

14 Starting from 1931, *Genesis* presented itself as an organ of an Italian Federation of Eugenics, which comprised SISQS, directed by Silvestro Baglioni, CISP and SIGE, both under the presidency of Gini. See *Genesis* 10, no. 1–2 (January–June 1931): 1.

15 See *Rassegna di studi sessuali, demografia ed eugenica* 8, no. 4 (December 1928): 240.

16 Report of the *Divisione Polizia Politica per la Divisione Affari Generali e Riservati,* 9 August 1930, in ACS, CPC, b. 24106, "Mieli Aldo."

17 See, in particular, Augusto Carelli, "Valore della sterilizzazione eugenica nel miglioramento della razza umana," *Difesa sociale* 7, no. 10 (October 1928): 341–45; Augusto Carelli, "A proposito di sterilizzazione eugenica," *Difesa sociale* 7, no. 11 (November 1928): 398; Augusto Carelli, "Quanti e quali individui dovrebbero essere sottoposti alla sterilizzazione eugenica ?," *Difesa sociale* 12 (1928): 436–40; Augusto Carelli, review of Charles Wicksteed Armstrong, *The Survival of the Unfittest* (1927), *Difesa sociale* 8, no. 3 (March 1929): 124–25.

18 See Ernesto Pestalozza to Levi, 10 January 1930, ACS, SPD, CO, b. 109005/2, "Levi Ettore." On the issue, see also Mantovani, *Rigenerare la società,* 300–303.

19 The numerous requests for the reintegration of Ettore Levi can be found in ACS, SPD, CO, b. 109005/2, "Levi Ettore."

20 See Pietro Capasso, "Ettore Levi," *Il pensiero sanitario* 14 (1932): 11.

ers (*Casellario Politico Centrale*) as "antifascist" and "opposer,"[21] Capasso was nonetheless able to keep direction of his review, restraining his veiled criticisms of the demographic campaign in the short space of the column *Spunti e punture* [Pricks and stings], which did not lack irony regarding the ridiculous excesses of fascist pronatalism.[22]

From the ashes of "qualitative" eugenics arose "quantitative" eugenics, linked more to the utopia of Tommaso Campanella and Leon Battista Alberti than to Galtonian gospels[23] and essentially founded on two scientific and ideological paradigms, influential on a national and international scale: the "integral" demography of Corrado Gini, on one side, and the medical constitutionalism of the endocrinologist Nicola Pende, on the other.

In the international context, Italian eugenicists expressed their unorthodox position by withdrawing the Italian Committee for Population Problem Studies (*Comitato Italiano per lo Studio dei Problemi della Popolazione,* known as CISP) from the International Union for the Scientific Investigation of Population Problems (IUSIPP), and SIGE from the International Federation of Eugenic Organizations (IFEO). At the World Population Conference of Geneva in 1927, the Italian delegation, led by Corrado Gini, clearly revealed its anti-Malthusian hostility.[24] Notwithstanding this, in 1928, at the creation of the IUSIPP, directed by Johns Hopkins biologist Raymond Pearl, Gini became vice-president, member of the executive committee and chairman of Commission III (Vital Statistics of Primitive Races).[25]

---

21 ACS, CPC, b. 19943, "Capasso Pietro." In 1941, the *Direzione Generale di Pubblica Sicurezza* suspended the surveillance given the subject's good conduct and the "sincere and effective contrition." In 1944, Capasso became undersecretary of State for Domestic Affairs during Badoglio government. See also Mantovani, *Rigenerare la società,* 296.

22 For Capasso's criticism of the demographic campaign, the regime and the encyclical *Casti Connubii,* see Mantovani, *Rigenerare la società,* 295–97.

23 In 1930, Michels praised, for example, Campanella's eugenic vision, in particular, as regarding the political order, with "the direction and the government of the State guaranteed of the high value of its principles, but also of the racial fusion that gives consistency and solidity to the population": see Roberto Michels, "Nei primordi della scienza eugenetica. Le utopie di Tommaso Campanella," *Rivista internazionale di filosofia del diritto* 10, no. 25 (1930) : 8–9, offprint. On the "myth" of Leon Battista Alberti, precursor of eugenics, see Mario Barbàra, "Leon Battista Alberti precursore di Galton," *Le Opere e i Giorni* 7, no. 11 (November 1928): 86–92.

24 Ipsen, *Dictating Demography,* 205.

25 Vice-president of Commission III was B. Malinowski. Other members were W. Schmidt, R. Pinto, G. Pitt-Rivers, O. Schlaginhaufen, R. Goldschmidt, E. Fischer, F. Boas, R. B. Dixon, H. B. Lundborg. The complete list is available in Raymond Pearl Papers, American Philosophical Society.

The Italian participation in the IUSIPP was mediated by the constitution of CISP, personally supported by Mussolini, who prepared, on instructions from Gini, a circular sent to the ministries and public entities, with the aim of constructing a broad network of financing for the Committee.[26] Gini did not hesitate to submit to Mussolini the drafts of the Union's statute, for the *Duce* to correct and elaborate.[27]

The biological determinism of Gini's demographic theory, and his direct relationship with the *Duce*, were quickly criticized by American social demographers, led by Harvard University mathematician Edwin B. Wilson. The hostility of the American social demographers was deeply connected with their commitment in New Deal reforming policies and social eugenics, based on family planning and birth control.[28] In 1930, Wilson's hostile activities towards the Union contributed to embittering the collaboration between Gini and Pearl, provoking a break.[29] In 1931, the International Population Conference was to be held in Rome, but Raymond Pearl established a separate conference in London to represent the Union. The Italian Committee did not recognize the legality of the decision of the Union and continued to organize the congress, which was held in Rome in September under the honorary presidency of Benito Mussolini. The IFEO and IUSIPP Committees of Argentina, Spain, France and Germany participated in the Rome congress. It was organized into eight sections, demonstrating Gini's "integral," multidisciplinary approach to the problem of population: biology and eugenics; anthropology and geography; medicine and hygiene; demography; sociology; economy; history; and methodology. The congress was not a meeting of pronatalists, but Wilson's concerns regarding Italy's political neutrality were not completely unfounded: Mussolini, for example, edited Gini's opening speech, instructing the latter to remove a passage praising Thomas Malthus[30] and opposed Gini's decision to invite Marie Stopes to Rome.[31]

---

26 ACS, PCM 1940–43, b. 2674, f. 1.1.16.3.5.27.000–7, sf. 2. For more details, see Cassata, *Il fascismo razionale*, 130.

27 Gini to Pearl, 11 February 1928, Pearl Papers, APS, Box 7.

28 On this topic, see in particular Edmund Ramsden, "Carving up Population Science: Eugenics, Demography and the Controversy over the 'Biological Law' of Population Growth," *Social Studies of Science* 32, no. 5–6 (October–December 2002): 857–99.

29 Gini to Wilson, August 14, 1930; Gini to Pearl, 20 August 1930; Gini to Pearl, 25 August 1930; Gini to C. E. McGuire, 16 January 1931; Gini to Pearl, n.d., but June 1931; Pearl to Gini, 13 June 1931. Pearl Papers, APS.

30 ACS, SPD, CO, b. 1172, f. 509560/III; see Ipsen, *Dictating Demography*, 205.

31 ACS, SPD, CO, b. 1172, f. 509560/III.

The rapport between SIGE and the IFEO, during the second half of the 1920s, was also beset with notable tensions. From 1926, the Italian eugenicists refused to pay the financing fee of the London Bureau of the organization, presided over by Cora B. S Hodson.[32] In 1928, at the IFEO meeting in Munich, Gini, member of a commission—also comprising Fischer and Mjøen—for the study of the internal organization of the Federation, proposed the elimination of the London secretary office, which the Italians claimed was only a "source of slowness, confusion and misunderstanding."[33] But SIGE's hostility to Ms. Hodson was not only a formal and bureaucratic question. The real problem resided instead in the "negative" eugenics proposals publicly supported by the IFEO secretary, above all regarding sterilization. These were clearly denounced in a 1931 letter from Gini to Charles Davenport:

> In the sitting of the Italian Congress of Genetis and Eugenics of 1929, in which M. Pestalozza presented his relation on sterilization, Ms. Hodson did not speak, although I, as chairman, had invited all the congress participants to join in the discussion three times. However, following, Ms. Hodson sent us in the minutes a long declaration that she never made. It was not completely regular. But more seriously was that in this declaration, she affirmed that the Federation, in the meeting in Rome, had voted in favor of sterilization, and it was in the name of the Federation that she made her declaration at the Congress![34]

On 20 August 1932, Gini communicated to Davenport SIGE's decision not to participate in the activities of IFEO, as the London secretary had not been abolished, as agreed upon at the Munich meeting in 1928.[35]

Therefore, when on 14 July 1933, national socialist Germany launched the most radical legislation on eugenic sterilization ever approved, fascist Italy had already assumed, in the eugenic field, a strongly critical position regarding both the IFEO and the IUSIPP.

In September 1933, Mussolini's newspaper *Popolo d'Italia* clearly indicated the directives against Nazi sterilization law:

---

32 Cora B. S. Hodson to Ernesto Pestalozza, 15 February 1932, Charles B. Davenport Papers, APS.

33 Gini to Davenport, 11 June 1931, Davenport Papers, APS.

34 Gini to Davenport, 11 June 1931, Davenport Papers, APS.

35 Gini to Davenport, 20 August 1932, Davenport Papers, APS.

> It could be that wanting to achieve qualitative perfection requires a series of successive sterilizations, which could lead to catastrophic consequences: that is, to the reduction of the race to a handful of men, too pure to remain men and make their living in this low world. Preserving the present and future health of the race is a duty, more, a fundamental duty of the State, this is the hinge of fascist doctrine; but our methods seem more suitable to the aim.[36]

The scientific community followed the indications of the regime, stigmatizing the Nazi eugenic extremism as "barbaric" and "anti-scientific" and countering German "Aryan" mysticism with "Mediterranean" and "Latin" equilibrium.[37] Sante de Sanctis, at the 2nd European Conference on Mental Hygiene in September 1933, defined coercive sterilization as "catastrophic" (see chapter III). The convention in Rome of the Society of Legal Medicine (*Società di Medicina Legale*) welcomed the conclusions of Salvatore Ottolenghi, in *Sterilizzazione del delinquente in rapporto alla medicina legale* [Sterilization of criminals in relation to legal medicine], which condemned sterilization as contrary to the spirit of the new fascist penal code.[38] As for demographers and statisticians, a crucial test came at the IUSIPP congress in Berlin in1935. Italy had not in fact definitively abandoned the Union in 1931. Notwithstanding Gini's opposition,[39] in 1935 the fascist government allowed Livio Livi, Gini's adversary and leader of the Italian social demographers, to become vice-president of the Union.[40] Nevertheless, on the eve of the IUSIPP Berlin congress in 1935, which,—Stefan Kühl has claimed—"marked the apex of international support of Nazi race policies and represented a great success for the Nazi propaganda machine,"[41] Mussolini ordered that Italy participate "with a delegation composed of few members, in the role of observers," since the congress would be concerned "also with the problems of the 'hygiene of the race' and of those inherent

36 "Popolo d'Italia," 14 September 1933.

37 Vincenzo Palmieri, *Denatalità. La grande insidia sociale vista da un medico* (Milan: Società Palermitana Editrice Medica, 1935; Lorenzo Ratto, "La sterilizzazone coattiva in Germania," *Avvenire Sanitario* 50 (1934): 1.

38 Salvatore Ottolenghi,"Sterilizzazione del delinquente in rapporto alla medicina legale," *Policlinico-Sezione Pratica* 43 (1933): 171.

39 Letter from the Ministry of National Education to ISTAT Presidency, 26 September 1935, ACS, PCM 1940–43, b. 2674, f. 1.1.16.3.5.27.000-7, sf. 3.

40 Franco Savorgnan to Mussolini, 26 September 1935, ACS, PCM 1940–43, b. 2674, f. 1.1.16.3.5.27.000-7, sf. 3.

41 Kühl, *The Nazi Connection*, 32.

'psychological' problems of the population, which can not be anything but the controversial questions of sterilization and the Aryan."[42]

An identical opposition came, in the same months, from the Catholic milieu. The news of approval of the Nazi "Law on the Prevention of Genetically Deficient Progeny" (14 July 1933) was reported by the *Osservatore Romano* [Roman observer] on 4 August 1933, with a brief note, which recalled the contents of the encyclical *Castii Connubii.* On 13 August, the daily paper of the Holy See summarized, in a lengthy article, the speech of Agostino Gemelli at the Florence Congress of Catholic Physicians in 1932: "Catholic morals—the article concluded—have greater eugenic value than all the rules of eugenicists."[43] In October 1933, Gemelli protested strongly against the instrumentalization of his thoughts by the Nazi propaganda, to make it seem as even the dean of the Milan Catholic University was a supporter of the July 1933 laws. Advised of this operation by Father Costantino Noppel, dean of the German College in Rome, Gemelli declared that he had never approved of the "infamous" German laws, and that he had always followed, in his role as "Catholic scientist" the directives of the Holy See in the field of eugenics.[44]

In October 1933, a letter of protest from Gemelli was published by the *Osservatore Romano*. Gemelli wrote: "The fact that many times, in my eugenic writing, I have demonstrated the gravity of moral error, not to mention biological, contained in the various sterilization proposals, should be enough to deny the affirmations [by the Germans]."[45] A copy of this denial was sent to the secretary of the Freiburg Caritas, in order to spread the news to German newspapers.[46]

Again in November 1933, the *Osservatore Romano* attacked the negative eugenics of the Nazi "advocates of death," reporting the detailed criticism by gynecologist Albert Niedermeyer (1886–1957) of the book *Von der Ver-*

42 Letter of the Cabinet of the Ministry of Foreign Affairs to ISTAT Presidency, 19 June 1934, ISTAT Archives, b. "Congressi internazionali. Partecipazione funzionari Istat."

43 "L'eugenica e la morale cattolica," *L'Osservatore Romano* (13 August 1933): 2.

44 Catholic University Archive (hereafter AUC), Agostino Gemelli Papers, Correspondence, b. 49, f. 70, August 28, 1933. On Gemelli's eugenics, see: Maiocchi, *Scienza italiana e razzismo fascista*, 75–76; Roberto Maiocchi, "Agostino Gemelli critico dell' 'eugenica' tedesca," *Vita e Pensiero* 83, no. 2 (2000): 150–69; Maria Bocci, *Agostino Gemelli rettore e francescano. Chiesa, regime, democrazia* (Brescia: Morcelliana, 2003): 421–24.

45 "Una smentita," *L'Osservatore Romano* (2–3 October 1933): 2. See also Gemelli's letter to Giuseppe Dalla Torre, editor of "L'Osservatore Romano," 29 September 1933, AUC, Gemelli Papers, Correspondence, b. 49, f. 70.

46 Gemelli to H. Höfler, 4 October 1933 AUC, Gemelli Papers, Correspondence, b. 49, f. 70.

hütung unwerten Lebens [The prevention of unworthy life].[47] In December 1933, in the monographic issue of the journal *L'Economia Italiana* [Italian economy] dedicated to the theme "Population and Fascism," Gemelli condemned yet again the Nazi legislation on sterilization, recalling the opposing conclusions reached by the Italian scientific community, at the two national eugenic congresses of 1924 and 1929, and, on a moral and religious side, by the Congress of Catholic Physicians in 1932.[48]

In 1935 and 1936, the International Congresses of Catholic Physicians were held in Brussels and Vienna, repeating the condemnation of negative eugenics. In Brussels, French physician Joseph Okinczyc attacked the materialistic logic which formed the basis of the negative eugenics of abortions and sterilizations, proposing instead a holistic medicine that cured the "person" rather than the "individual."[49] In Vienna, Gemelli underlined the importance of the Congress from the point of view of the Holy See: "The Pope expects us doctors to show him that the Catholic Church has not acted amiss when she condemned some eugenic trends. We shall propagate the doctrine among people as contained in the encyclical *Casti Connubii*." All participants agreed on the following: 1) The medical profession should reject sterilization as a method by which to eradicate the threat of hereditary disease; 2) Catholic physicians were warned of the "slippery slope" from eugenics to euthanasia; 3) Eugenic and penal castration were rejected outright, with the exception of castration in the cases of "psychopathic sex criminals"; 4) Positive eugenic methods should be reaffirmed, including the creation of Catholic counseling centers; 5) International cooperation by all Catholic medical associations should be favored in order to discuss the questions of eugenics and genetics.[50]

In the context of Italy's ideological, political and scientific opposition to negative and "nordic" eugenics, starting from the last half of the 1920s, it is

---

47 "Vita senza valore," *L'Osservatore Romano* (4 November 1933): 2. The review concerned Erwin Baur, W. E. Mühlmann, Friedrich Karl Walter, Paul Althaus, Ernst Heinrich Rosenfeld, Hans Meyer, Hans Duncker, *Von der Verhütung unwerten Lebens* (Brema: Halem, 1933). On Niedermeyer, see: Monika Löscher, "Eugenics and Catholicism in Interwar Austria," in Turda and Weindling, eds., *Blood and Homeland*, 310–12.

48 Agostino Gemelli, "La 'sterilizzazione coattiva e preventiva' nell'insegnamento degli studiosi italiani," *L'Economia Italiana* 11–12 (December 1933): 117–28.

49 Guido Lami, "Significati e moniti di un Congresso," *Studium* 31, no. 6 (June 1935): 362–65.

50 Guido Lami, "Il Congresso Internazionale dei Medici Cattolici a Vienna e il prossimo Congresso-Pellegrinaggio a Roma," *Studium* 32, no. 11 (November 1936): 628–631. See also: Löscher, "Eugenics and Catholicism in Interwar Austria," 311.

not surprising that, in 1935, it was SIGE, led by its president Corrado Gini and vice-president Agostino Gemelli, which promoted the constitution of a new organization in the international eugenics arena, an alternative to the IFEO: the Latin Federation of Eugenic Societies.

1. Corrado Gini's Hegemony: Demography and "Regenerative" Eugenics

Three factors essentially determined Corrado Gini's hegemonic role in fascist eugenics, at least starting from 1924. First, Gini was a relevant figure in the international scientific context, as statistician, demographer and sociologist. Secondly, he assumed, almost simultaneously, the presidency of the three most important Italian institutions in the field of population policy and eugenics: ISTAT (from 1926 to 1932), SIGE (from 1924) and CISP (from 1928). Finally, from a theoretical point of view, Gini's effective synthesis between populationist demography and biotypological constitutionalism provided a comprehensive framework for fascist "quantitative" eugenics, nationalism and pronatalism.

The First Congress of Social Eugenics, held in Milan from 20 to 23 September 1924 and promoted by SIGE and the Italian Royal Society of Hygiene (*Reale Società di Igiene*),[51] was already marked by the hegemonic presence of Corrado Gini, although the Italian eugenic debate was not at all monolithic.

Gini opened the first session of the congress, and his theory was presented as a profound critical analysis of the scientific legitimacy of the

51 *Atti del Primo Congresso italiano di Eugenetica sociale (Milan, 20–23 September 1924)* (Rome: Stabilimento Poligrafico dello Stato, 1927). In the executive committee, as well as Gini, were Luigi Mangiagalli (rector of the university and mayor of Milan); Icilio Boni, head physician at Milan's Ospedale Maggiore and president of the Royal Italian Society of Hygiene; Ernesto Pestalozza, senator and first president of SIGE; and Serafino Patellani. The promotional committee was made up prevalently of directors of university clinics of obstetrics-gynaecology, neuropsychiatry and dermo-syphilology and by directors of the institutes of zoology and comparative anatomy, and of hygiene. There were also economists, such as Attilio Cabiati, Luigi Einaudi, Maffeo Pantaleoni and Angelo Sraffa. Among the foreign guests participating in the conference, there was Leonard Darwin, president of the International Commission of Eugenics and the Eugenics Education Society; Lucien March, director of the *Statistique Générale de la France* (SGF) and representative of the *Société Française d'Eugénique*; Jon Alfred Mjøen, director of the Winderen Laboratorium (Oslo) and representative of the Consultive Eugenics Committee of Norway; Nikolai K. Koltsov, director of the Institute of Experimental Biology in Moscow and president of the Russian Eugenics Society.

biological "presuppositions" of "selective" eugenics: the heredity of some characteristics, the different modes of transmission of acquired and germinal characteristics, and the dominance of heredity over the environment in determining individual traits. Gini then remarked indifference of public opinion, in Italy and elsewhere, toward eugenic issues:

> Abroad, as here in Italy, while eugenics is alive and prospering as a discipline that interests the cultivators of biological and social disciplines, some political men, and several philanthropists, it is not however able—it would be in vain to deny it—to capture the conscience of the masses, who consider it with persistent skepticism, if not with evident mistrust.[52]

Faced with such a divergence from public opinion, the eugenicists were forced to "examine their conscience." They needed to clarify whether people's indifference was due to their lack of knowledge, or whether it was due to "an appreciation of the reality that many points remain to be clarified and many demands have to be contemplated before being in a position to move on, with a free conscience, to the application of a eugenics program."[53]

Gini did not hesitate to lead by example, listing the "significant doubts" that surrounded the theoretical assumptions of the "selective eugenics" movement.

First of all, the "resemblance coming from common descent" might not depend exclusively on heredity, but also "on the common environment during gestation or, earlier, during the development of the germ."[54] It was the so-called phenomenon of "induction": the germinal characteristics might not be permanent and hereditary across generations, but on the contrary, might be "induced in the germ" from the environmental influence, and be, as a result, temporary. If the germ's influential inductors of good or bad characteristics were (as with alcoholism or professional hazards) always recognizable, and if the effects of induction were permanent or irreparable, then "selective eugenics would always have a reason for existence and moreover it should complement a preventive eugenic intervention, aimed at impeding or favoring induction." When the influential

---

52 Corrado Gini, "Le relazioni dell'Eugenica con le altre scienze biologiche e sociali," in *Atti del Primo Congresso italiano di Eugenetica sociale*, 4.

53 Gini, "Le relazioni dell'Eugenica con le altre scienze biologiche e sociali," 4.

54 Gini, "Le relazioni dell'Eugenica con le altre scienze biologiche e sociali," 7.

inductors were recognizable, but the effects of the induction were short-lived, "selective eugenics faces its most difficult challenge, since the temporary effects of induction must be assessed in conjunction with the effects of heredity." When finally, as often happened, the influential inductors were not recognizable, selective eugenics was not effective, whilst preventive eugenics "maintained its justification for intervention, if this was facilitated by the understanding of environmental factors that could induce favorable or unfavorable characteristics."[55]

The existence of induction, in its diverse forms (parallel induction, mutual induction, continuation of induction) produced, therefore, a sort of "pseudo-heredity," which could mislead selective eugenics and favor the spread of hereditarily inferior individuals.[56]

As for the second assumption, that is, the diverse transmissibility of germinal and acquired characteristics, Gini posed a question mark, introducing the theme of "transmission of functional diathesis":

> The intense exercise of a function must have not only a mechanical effect on the development of the organ, but also a more subtle effect, probably biochemical, modifying the entire composition of the organism, even if barely perceptible or imperceptible to our means of observation [...]. The germs could in this way receive, due to the intense exercise of the functions, or of particular functions of the organism, biochemical modifications that render the products that derive from it predisposed to exercise the same functions.[57]

The transmission of functional diathesis could in this way modify the interpretation of the eugenic hierarchy of nations and social classes. If prolonged and intense exercise of the intellectual faculty on the part of the ascendants did in fact render the descendents more predisposed to exercise such faculties, then Italian, Russian or Greek emigrants could not be considered, as the American legislation would like, as "eugenically" inferior. Consequently:

> the eugenicist who, from the congenital value of the members of various classes and various nations, would like to judge [...] their eugenic value, with-

[55] Gini, "Le relazioni dell'Eugenica con le altre scienze biologiche e sociali," 8–9.
[56] Gini, "Le relazioni dell'Eugenica con le altre scienze biologiche e sociali," 10.
[57] Gini, "Le relazioni dell'Eugenica con le altre scienze biologiche e sociali," 10.
[58] Gini, "Le relazioni dell'Eugenica con le altre scienze biologiche e sociali," 11.

> out having first excluded that the congenital superiority of some could derive just from the major exercise of the faculty corresponding to his ascendants, would obtain, through selective action, radically erroneous and damaging consequences, rather than useful ones, for the progress of the race.[58]

As for the transmission of diathesis acquired through illness, if the illness produced immunization, then a function of illness could be identified in "evolution of the race," because the illness functioned in this case as an "immunizer of the germinal plasm":

> Abolishing illness in the present generations would mean exposing future generations, lacking immunization, to the risk of a serious crisis; eliminating sick people from reproduction would not have a vastly different effect, as reproduction would be left only to plasm that had not been recently immunized.[59]

Arriving finally at the problem of hereditariness of characteristics, Gini believed it did not explain, for example, how nations such as Australia and New Zealand sprung from colonies of deported criminals, or how "the distant descendants of great men vanish or degenerate."[60] It was therefore necessary to hypothesize—as maintained in the theories of Carl Nägeli, Theodor Eimer and Italian zoologist Daniele Rosa—an internal evolution of germinal characteristics:

> Germinal characteristics would evolve, at least for certain species, through internal forces, and the numbers of their population would evolve contemporarily, following a course which many compare to the course of individual development, with a period of gradual growth, a period of maximum development, and a period of decline which often finishes, sooner or later, in the extinction of the species.[61]

From the diverse varieties, races, populations, and families, "some would therefore become more advanced in the biological evolution of their germinal characteristics, but also consequently closer to decline and extinction; others more distant." The existence of a sort of parabolic evolution in

59 Gini, "Le relazioni dell'Eugenica con le altre scienze biologiche e sociali," 11.
60 Gini, "Le relazioni dell'Eugenica con le altre scienze biologiche e sociali," 13.
61 Gini, "Le relazioni dell'Eugenica con le altre scienze biologiche e sociali," 14.

germinal characteristics therefore reconciled the phenomenon of heredity with "the facts of great men coming from families of low origins and the successive decline of the descendants and similarly, those of normal development and excellent products coming from families of low extraction."[62]

On the whole, the three elements stressed by Gini—induction, transmission of functional and morbose diathesis and evolutionary tendencies of germinal characteristics—contributed to supporting the thesis of the mutability of the germinal plasm, from which he derived an inevitable condemnation of Anglo-Saxon eugenics. The nations, the classes or the families superior by wealth, culture or due to "individual congenital endowments," were not in fact "necessarily the nations, classes and families in which eugenics would favor proliferation, in view of the well-being of the race." On the contrary, claiming to improve the race through the selection of the elite would be like "improving a population through favoring the growth of adults, because they were stronger and trained, and opposing the growth of infants because they were weak and necessarily still lacking in all cultivation."[63] According to Gini, "selective" eugenics should not be completely excluded, but its field of action should be particularly reduced:

> Among families who in the past lived in the same environment, with an analogous amount of instruction, not yet elevated or elevated only recently from the lower classes, equally disposed to various diseases or immunized against these diseases, comparable from these various points of view, eugenics might yet exercise its selective action.[64]

Failing such conditions, "preventive" eugenic measures could be more effective: for example, "contracting [...] marriages at a young age, appropriately combining the characteristics of the spouses, avoiding crossings between unlike races, lengthening intervals between births, breastfeeding offspring, and reproducing preferably in determined seasons."[65] However, even in this sphere, there were many difficulties. What was, in fact, the best way to match marriages?

62 Gini, "Le relazioni dell'Eugenica con le altre scienze biologiche e sociali," 14.
63 Gini, "Le relazioni dell'Eugenica con le altre scienze biologiche e sociali," 15.
64 Gini, "Le relazioni dell'Eugenica con le altre scienze biologiche e sociali," 15–16.
65 Gini, "Le relazioni dell'Eugenica con le altre scienze biologiche e sociali," 17.

> Is it preferable to couple homozygotic individuals, and in this way separate the race into two categories, one of healthy homozygotes and the other of sick homozygotes, counting on the progressive decline and disappearance of the latter? Or is it preferable to let people cross, so that the crossings attenuate the damage of the sick forms, and perhaps favor the reproduction of heterozygotic individuals?[66]

According to Gini, the great part of the prescriptions of "preventive" eugenics had a consequence of demographic slowdown, and therefore, paradoxically, a reduction of the eugenic efficiency of the social organism. Even crossings between races that were very different should not be discounted, because the hybrids carried a major probability of individuals with "exceptionally favorable characteristics":

> We must ask ourselves if, and to what point, an on-average inferior population, but with a high frequency of people with exceptionally favorable characteristics might not be preferable, from a point of view of social efficiency, to an on-average superior population, but with more uniformly distributed characteristics.[67]

Additionally, "charitable," "egalitarian" measures, of prophylaxis, therapy, social medicine and labor medicine, inevitably contrasted with eugenics, because they impeded natural selection and the elimination of the weakest: between eugenics and euthenics it was therefore necessary "to find a compromise."[68]

Finally, from a long list of questions Gini arrived at the most serious. Eugenics implied a rationalization of births, which risked compromising the demographic power of the nation:

> It is here that we find perhaps the most serious doubt that perplexes eugenicists on the advantages of passing, in the current state of awareness and conditions, to practical action. It is the doubt whether the population or the classes, overcoming instinct and practicing eugenics, will rationalize the quantity of their offspring, not only from a point of view of quality, but even from a view of

66 Gini, "Le relazioni dell'Eugenica con le altre scienze biologiche e sociali," 17.
67 Gini, "Le relazioni dell'Eugenica con le altre scienze biologiche e sociali," 20.
68 Gini, "Le relazioni dell'Eugenica con le altre scienze biologiche e sociali," 22.

> advantage to the parents, and therefore reduce themselves to a number completely insufficient to maintain their place in the world.[69]

But the conclusion of the "long examination of conscience" did not suggest discouragement or surrender, but rather "prudence" and "persuasion." On one hand, according to Gini, eugenics had to recognize that it was still an "immature" science, not yet ready to go beyond the theoretical; on the other, it had to open up to the natural and social sciences, because from such synergy, eugenicists "could resolve the problems that constitute the basis of their science and the assumption of a future program of action."[70]

In effect, the First Congress of Social Eugenics seemed to faithfully follow Gini's call to scientific prudence. The final resolution, unanimously approved, was very moderate indeed:

> The First Italian Congress of Social Eugenics praises the scientific activities of the experts in genetics and eugenics and recognizes the importance of the resulting achievements. At the same time, we acknowledge that, in the face of the complex and delicate characteristics of the problems of applied eugenics, that which has been done is only little in the face of that which still needs to be done and, without excluding the possibility that from today we could draw useful results regarding the conduct of individuals and the action of public entities, we confirm that the greatest prudence will be imposed, and that in the meantime it is above all in the fields of research and observations that the eugenic specialists must focus their efforts.[71]

Not surprisingly, this resolution was signed, as well as by Gini and Patellani, by Agostino Gemelli. During the Congress, Gemelli synthesized the Catholic position toward eugenics, reprising the discussion contained in the document published in 1924 by the Secretariat for Morality of the Naples Diocese.[72] In this paper, Gemelli listed the reasons for the diffidence of the Catholic Church toward eugenics: particularly, the limited scientific

69 Gini, "Le relazioni dell'Eugenica con le altre scienze biologiche e sociali," 24.
70 Gini, "Le relazioni dell'Eugenica con le altre scienze biologiche e sociali," 25.
71 See "Nona seduta," in *Congresso Milano 1924*, in *Atti del Primo Congresso italiano di Eugenetica sociale*, LXIII.
72 Giuseppe De Giovanni and Mario Mazzeo, *L'eugenica* (Neaples: Pelosi, 1924).

grounds of eugenic precepts; the Catholic safeguard of human spirituality against the reductionist tendencies of science; and the defense of individual liberty against state intervention. This diffidence notwithstanding, Gemelli suggested the possibility of an "alliance" between eugenics and Catholicism, mediated by the assumption of the Catholic moral of chastity, defining this as a "subordination and rationalization of the sexual act,"[73] absolute before marriage, and relative after. Chastity would combat the possibility of illegitimate children, the transmission of venereal diseases and the conception of overly numerous or defective offspring. According to Gemelli, Catholic sexual ethics could lead to a progressive peaceful alliance between science and faith, in the name of eugenics: "We eugenicists must align ourselves to Catholicism in the battle against immorality and bad customs, and ask it to help us in our battle for the improvement of the race, availing ourselves of its weapons and making them our own."[74]

At the Congress, the Catholic rejection of negative eugenics was supported by a theoretical and scientific approach, which opposed rigid Mendelian–Weismannian hereditarianism with neo-Lamarckian faith in the heredity of acquired characteristics and in the modifiability of the "germinal plasm."[75] Only Gaetano Pieraccini's eugenics, with his deductions on the transmission of traits (in particular psychical ones) among the members of the Medici family,[76] could in some way be compared with the biological determinism of Jon Alfred Mjøen, who presented his pedigrees of families of criminals and geniuses. Mjøen wrote:

> Modern progress has [...] placed in doubt that the French revolution dogma of equality is based on incontrovertible circumstances as well as that men are born great or of no merit at random, independent of every law or organic relationship. We have been able to establish that there are families in reality formed by idiots, delinquents, perverts, idle people, and others instead with

73 Agostino Gemelli, "Religione ed eugenetica," in *Atti del Primo Congresso italiano di Eugenetica sociale*, 65. Gemelli's contribution was expressly requested by Gini in the organisational phase of the Congress, as demonstrated by Gemelli's reply of 25 April 1924: "At your insistence, I can do nothing but consent," in ACS, Gini Papers (hereafter AG), b. b4.

74 Gemelli, "Religione ed eugenetica," 66.

75 See Ugo Cerletti, "Necessità biologica delle malattie," in *Atti del Primo Congresso italiano di Eugenetica sociale*, 387–90.

76 See Pieraccini's monumental genealogical study, *La stirpe dei Medici di Cafaggiolo. Saggio di ricerche sulla trasmissione ereditariadei caratteri biologici* (Florence: Vallecchi, 1924).

> special attributes, composed of individuals who are eminent because of psychical, intellectual or artistic qualities, without being able to establish the diverse ways in which the conditions of the external world act on either of these.[77]

In fact, the four days of the Congress offered a composite picture of Italian eugenics, with several interconnections between social hygiene and social medicine. Eugenicists' contributions ranged from protection of maternity and infancy[78] to the fight against "social" illnesses;[79] from sexual education[80] to physical education;[81] from hydrotherapy[82] to the improvement of the work environment;[83] from the "prophylaxis of suicide"[84] to nutritional care.[85]

---

77 Jon Alfred Mjøen, "Delinquenza e genio alla luce della biologia," in *Atti del Primo Congresso italiano di Eugenetica sociale,* 170. See also N. Roll-Hansen, "Norwegian Eugenics: Sterilization as Social Reform," in Broberg and Roll-Hansen, eds., *Eugenics and the Welfare State,* 158–61.

78 See Camillo Pestalozza, "La natimortalità nei diversi periodi della vita italiana e milanese," in *Atti del Primo Congresso italiano di Eugenetica sociale,* 191–98 and 251–52; Emerico Biondi, "Il parto podalico e sua influenza sulla vita dei bambini," in *Atti del Primo Congresso italiano di Eugenetica sociale,* 202–10; Vittore Baldassari, "Alcuni dati statistici della Clinica ostetrica della R. Università di Genova," *Atti del Primo Congresso italiano di Eugenetica sociale,* 253–56; Giulio Calderini, "Sulla sorte dei feti nati da gravide albuminuriche," in *Atti del Primo Congresso italiano di Eugenetica sociale,* 273–80; Francesco Landucci, "Sul nuovo regolamento riguardante l'assistenza degli esposti," in *Atti del Primo Congresso italiano di Eugenetica sociale,* 415–18.

79 Giuseppe Antonini, "Alcoolismo ed Eugenetica," in *Atti del Primo Congresso italiano di Eugenetica sociale,* 117–20; Lanfranco Maroi, "Alcoolismo ed Eugenetica," in *Atti del Primo Congresso italiano di Eugenetica sociale,* 121–38; Eugenio Centanni, "La eredità dei tumori," in *Atti del Primo Congresso italiano di Eugenetica sociale,* 211–24; Andrea Pagani Cesa, "Dati statistici sull'influenza dell'ambiente famigliare come fattore di contagio tubercolare," in *Atti del Primo Congresso italiano di Eugenetica sociale,* 293–94; Giovanni Galli, "L'Eugenetica di fronte all'ereditarietà delle malattie cardio-vascolari," in *Atti del Primo Congresso italiano di Eugenetica sociale,* 307–10; Raffaele Jona, "Considerazioni cliniche e profilattiche sui rapporti fra tubercolosi ed Eugenetica," in *Atti del Primo Congresso italiano di Eugenetica sociale,* 311–18; Guido Rigobello, "L'ereditarietà nella tubercolosi," in *Atti del Primo Congresso italiano di Eugenetica sociale,* 319–24; Agostino Pasini, "La sifilide latente nei suoi rapporti con l'Eugenica," in *Atti del Primo Congresso italiano di Eugenetica sociale,* 325–32; Luigi De Berardinis, "La profilassi anticeltica nell'esercito," in *Atti del Primo Congresso italiano di Eugenetica sociale,* 333–40; Angelo Bellini, "Effetti vicini e lontani della blenorragia nell'uomo e nella donna," in *Atti del Primo Congresso italiano di Eugenetica sociale,* 345–54; Gaetano Dossena, "Il peso dei feti nati da madri tubercolose," in *Atti del Primo Congresso italiano di Eugenetica sociale,* 365–66; Giuseppina Pastori, "Sulla frequenza dell'eredolues nei fanciulli anormali," in *Atti del Primo Congresso italiano di Eugenetica sociale,* 425–30.

80 See Luigi Bellezza, "Educazione sessuale ed Eugenetica," in *Atti del Primo Congresso italiano di Eugenetica sociale,* 281–84; Emma Modena Camporini, "Eugenetica ed istruzione igienico-sessuale della donna," in *Atti del Primo Congresso italiano di Eugenetica sociale,* 363–64.

81 See Attilio Maffi, "L'educazione fisica delle masse altissimo fattore di Eugenetica sociale," in *Atti del Primo Congresso italiano di Eugenetica sociale,* 355–62.

82 Prassitele Piccinini, "Le fonti d'Italia," in *Atti del Primo Congresso italiano di Eugenetica sociale,* 419–22.

83 Luigi Devoto, "La famiglia del lavoratore del piombo," in *Atti del Primo Congresso italiano di Eugenetica sociale,* 409–10; Luciano Ermolli, "Un problema di Eugenetica operaia," in *Atti del Primo Congresso italiano di Eugenetica sociale,* 411–14; Giovanni Allevi, "Lavoro ed Eugenetica," in *Atti del Primo Congresso italiano di Eugenetica sociale,* 395–400.

84 See Vito Massarotti, "La profilassi del suicidio in rapporto all'Eugenica," in *Atti del Primo Congresso italiano di Eugenetica sociale,* 435–38.

As far as concrete eugenic proposals were concerned, Gini's moderatism encountered an almost unanimous chorus of confirmation from the other Italian eugenicists. Opposing Leonard Darwin, president of the International Commission of Eugenics and Britain's Eugenics Education Society, who called for segregation and sterilization of criminals,[86] was Leone Lattes, professor of legal medicine at the University of Modena, for whom "asocial tendencies" did not always derive from "heredity in its true sense" as much as from the consequences of some foetal illnesses. More than forbidding reproduction by criminals, Lattes believed that "eugenic practices" must turn their attention to the sanitary protection of pregnancy:

> Eugenic practices can, together with the remedy of impeding the reproduction of criminals and degenerates, be of valid assistance in curing germinal illnesses, in the period in which it is possible. Above all, it is necessary to turn the attention of physicians to the detection of hereditary syphilis in defective parents and the opportunity to cure it specifically during pregnancy, to prevent otherwise irreparable damage to the foetus.[87]

While Mjøen, defending society from immigrant "parasites," proposed the institution of an obligatory international identification card with all the relevant data of the subject,[88] Italian eugenicists, on the other hand, urged the eugenic value of national emigration. For Roberto Michels, the high qualifications of Italians workers emigrating to France, accompanied by an increasing birth control as a consequence of the improvement of their economic situation, would produce optimal results from a eugenic point of view.[89] For demographer Livio Livi, Italian repatriates represented both "rationally and morally a selected product." He declared: "I believe they and their offspring are more robust, healthier and more prolific examples compared to compatriots who don't emigrate."[90]

85 See Cesare Cattaneo, "Influenza della vitaminosi ed avitaminosi sul divenire della razza," in *Atti del Primo Congresso italiano di Eugenetica sociale*, 347–50.

86 See Leonard Darwin, "Eugenics and the Criminal," in *Atti del Primo Congresso italiano di Eugenetica sociale*, 151–58.

87 See "Quinta seduta," in *Atti del Primo Congresso italiano di Eugenetica sociale*, XXXVIII.

88 Jon Alfred Mjøen and Jon Bø, "The Norwegian System for Identification and Protection of the Individual," *Atti del Primo Congresso italiano di Eugenetica sociale*, 179–84.

89 Roberto Michels, "Taluni effetti dell'emigrazione nei suoi rapporti coll'Eugenica," in *Atti del Primo Congresso italiano di Eugenetica sociale*, 199–201.

90 Livio Livi, "Emigrazione ed Eugenetica," in *Atti del Primo Congresso italiano di Eugenetica sociale*, 50.

During the Congress, only Ettore Levi[91] and Felice Marta[92] declared themselves in favor of birth control. As for premarital examinations, despite the favorable position of several physicians,[93] the Congress voted for a rather moderate resolution, that mirrored the proposal of the Royal Society of Hygiene:

> The First Italian Congress of Social Eugenics approves the institution of a medical premarital certificate as simple eugenic information for the betrothed of the reciprocal conditions of health, and as a means of propaganda for an improvement of popular hygienic awareness. It is not a legal means upon which the permission to marry is granted by an authority, and we hope that, at least in the large urban centers, special public offices will be instituted to issue the certificate.[94]

The condemnation of sterilization as a eugenic practice was unanimous, although there were also veiled exceptions. The neurologist Eugenio Medea, professor at the Clinical Institutes of Improvement (*Istituti Clinici di Perfezionamento*) in Milan and leader of the Lombardy section of the League for Mental Hygiene (*Lega di Igiene Mentale*), was "waiting for our ability to realize the postulate that, as segregation should be imposed (and is already practiced) on those dangerous to society, so should sterilization be imposed on those dangerous to the species."[95] Meanwhile, he declared himself in favor of a "minimum program," that included the adoption of a premarital certificate and "health records."[96] Equally, law professor Domenico Medugno believed it was only a question of time and consensus:

---

91 Ettore Levi, "Le finalità eugeniche del controllo delle nascite," in *Atti del Primo Congresso italiano di Eugenetica sociale*, 257–72.

92 Felice Marta, "Eugenetica e neo-malthusianismo," in *Atti del Primo Congresso italiano di Eugenetica sociale*, 455.

93 See Carlo Francioni, "Le anomalie costituzionali e diatesiche dell'età infantile in rapporto coll'Eugenetica," in *Atti del Primo Congresso italiano di Eugenetica sociale*, 87–110; Romolo Costa, "Opportunità della reazione novocaino-formalinica prima del matrimonio," in *Atti del Primo Congresso italiano di Eugenetica sociale*, 295–96; Agostino Pasini, "La sifilide latente nei suoi rapporti con l'Eugenica," in *Atti del Primo Congresso italiano di Eugenetica sociale*, 325–32; Gian Angelo Ambrosoli, "Le malattie della pelle in rapporto all'Eugenetica," in *Atti del Primo Congresso italiano di Eugenetica sociale*, 341–44; Giuseppe Corberi, "L'ereditarietà nella epilessia," in *Atti del Primo Congresso italiano di Eugenetica sociale*, 431–34.

94 See "Nona seduta," in *Atti del Primo Congresso italiano di Eugenetica sociale*, LXIV.

95 Eugenio Medea, "Le malattie nervose e mentali in rapporto all'Eugenetica," in *Atti del Primo Congresso italiano di Eugenetica sociale*, 141.

96 Medea, "Le malattie nervose e mentali in rapporto all'Eugenetica," 143.

> Where education from experts of eugenics and related sciences is able to deeply permeate the social strata of the various classes, even the surgical instruments will be condoned, and sterilizing activities, carried out in accordance with the most recent scientific findings, will be proclaimed necessary and blessed. Currently, there are too many elements of a sentimental nature, too many customs that oppose, at least in Europe, any practice of the kind. This does not take away the fact that this is an aim that we must have, in order to set ourselves on the path to modern civilization.[97]

The condemnation of surgical operations was present also in the contribution of gynecologist and ex-president of SIGE, Ernesto Pestalozza:

> What I hope for eugenics is that, in the research of means to achieve its radiant ideals, it does not borrow from medicine any ancient, obsolete and repugnant operations. And, even if we do not believe in leaving the gradual elimination of appalling offspring to nature, the new science of eugenics can find in social hygiene promising rules to allow us to overcome single morbose conditions, focusing on every scientific research that extends the benefits of hygiene, that we are already able to offer to the individual and society, to the entire stock.[98]

Even Pestalozza, however, did not want eugenicists to be "driven by sentiment" and admitted "happily that if it was only through these operations that eugenics was able to cancel out, or al least limit, the hereditary transmission of illnesses that threaten the race, then the adoption would be justified without doubt, for the superior interests of humanity versus the individual."[99] In this way, voluntary abortion, although in general a "weapon both ineffective and dangerous,"[100] could be justified in the case of pregnancies in "subjects affected by hereditary nervous or mental degeneration," even if the justification would be limited to specific cases and carried out in public hospitals, after appropriate consultation.

97 Domenico Medugno, "L'azione dello Stato e l'Eugenetica," in *Atti del Primo Congresso italiano di Eugenetica sociale*, 147.

98 Ernesto Pestalozza, "Le indicazioni operatorie in rapporto all'Eugenica," in *Atti del Primo Congresso italiano di Eugenetica sociale*, 85.

99 Pestalozza, "Le indicazioni operatorie in rapporto all'Eugenica," 82.

100 Pestalozza, "Le indicazioni operatorie in rapporto all'Eugenica," 84.

In connection with the Milan Congress, between 20 and 22 September, the meeting of the International Commission of Eugenics was held, due to the initiative and contacts of Corrado Gini. On this occasion, the International Commission approved the Italian project of constituting an international library of eugenics, which had been proposed during the 1923 Lund meeting by Corrado Gini and Ettore Levi.[101] The first volume should have concerned Italian eugenics, but it was never published. The only book published in this series was, in 1930, *Le problème eugenique en Belgique,* edited by Albert Govaerts.

With the Milan Congress, Gini achieved complete hegemony over the Italian eugenic movement: starting from 1924 in fact, he was not only elected president of SIGE, but also undoubtedly became the Italian reference name in the international eugenics arena. In the second half of the 1920s, as well as intensifying the battle against birth control and eugenic selection of marriage, Gini specified, always in opposition to Anglo-American eugenics, his own interpretation of racial crossing. Significantly, Gini expounded his view on this topic during two international conferences: in 1927, at the Italian-Brazilian Institute of High Culture of Rio de Janeiro, and in 1929, at the Norman Wait Harris Foundation of Chicago.

Consistent with the positions expressed at the beginning of the century, Gini did not attribute a necessarily degenerative character to racial crossing.[102] In first place, according to Gini, the resurgence, in certain unions, of a pathologically latent character, did not imply in itself the negativity of crossings, but represented only "the necessary product of the gradual purification of the heterozygotes":

> When, in other words, an unfavorable trait appears in bastards, this does not actually signify degeneration, but is simply the effect of a scission typical of Mendelian laws, and is verifiable—given the presence of those unfavorable traits—also in the product of individual heterozygotes within the same race.[103]

---

101 See *Atti del Primo Congresso italiano di Eugenetica sociale,* LXVII.

102 On the centrality of the theme of racial crossing in 20th century eugenics, see Claudio Pogliano, *L'ossessione della razza. Antropologia e genetica nel xx secolo* (Pisa: Edizioni della Normale, 2005), 211–68.

103 Corrado Gini, *Le basi scientifiche della popolazione* (Catania: Studio editoriale moderno, 1931), 308.

Crossings, in Gini's opinion, could not be labeled as uniformly positive or negative, but instead produced—as demonstrated by Davenport and Steggerda, East and Jones, Hankins and others[104]—a major variability in the descendants, allowing for "products more favorable or more unfavorable, or intermediate as compared to the parent-races."[105] Racial crossing, Gini argued, produced frequent "physical, intellectual and moral disharmonies."[106] Above all in the case of "disharmonies in the moral sphere," Gini did not exclude the influence of social stigma:

> We must not forget that, especially in countries where the union between individuals of different races is the subject of general disapproval, if not legal penalties, the mulatto or hybrid derives generally from the illegitimate coupling of a white man and colored woman, both of low class and bad morals.[107]

But the reference to social contrasts did not change the priority which Gini gave in his explanation of the biological factor:

> It is also reasonable to admit that it [moral disharmony] may often be due to the even greater contrast between the psychology of the various races, as, for instance, between the ambition, the love of power, and the adventurous spirit of the whites and the idleness, the inconstancy, the lack of self-control and often adequate intelligence of many colored people.[108]

As for the low fertility of hybrids, according to Gini, the problem regarded only the "crosses of very different races, such as the white and the black, or the black and the yellow," but even in such cases "the results of observations are not in agreement."[109]

Gini constantly insisted on the need to consider crosses on a case by case basis. In fact:

---

104 C. Gini, "The Cyclical Rise and Fall of Population," in Corrado Gini, Shiroshi Nasu, Robert R. Kuczynski, and Oliver E. Baker, *Population* (Chicago: Harris Foundation Lectures, The University of Chicago Press, 1929), 116–17. The Italian version was Corrado Gini, *Nascita, evoluzione e morte delle nazioni* (Rome: Libreria del Littorio, 1930).

105 Gini, *Le basi scientifiche della popolazione*, 309.

106 Gini, *Le basi scientifiche della popolazione*, 311.

107 Gini, *Le basi scientifiche della popolazione*, 310. See also Gini, "The Cyclical Rise and Fall of Population," 122.

108 Gini, "The Cyclical Rise and Fall of Population," 123.

109 Gini, *Le basi scientifiche della popolazione*, 312.

> While some crosses, such as those between whites and blacks, have mostly damaged products, those between the colonial Dutch and Hottentot women in South Africa—studied with particular diligence by E. Fischer—resulted in several traits intermediate from the parent races and in others superior to both. Analogous results have been observed in the United States in crosses between whites and Red Indians, and in Oceania, between whites or Chinese with the Polynesians.[110]

Also in Brazil (and for the "Indian hybrids" in Canada) crosses did not, in Gini's opinion, present "very high quality," although in the Brazilian state of Cearà there was a population endowed with high fertility, "particular energy" and "physical characteristics of resistance," that justified the hypothesis that "a new ethnic type, destined to spread across the South-American continent,"[111] was developing. According to Gini, it was necessary to consider the multiplicity of factors that determined the eugenic quality of hybrids: the characteristics (physical, mental and moral) of racial crosses, the asymmetry of the relationship between the parent races, the surrounding social environment and the type of physical habitat.[112] In general however, with reference to the international literature on hybrids (Davenport–Steggerda, Fischer, Herskovits), Gini considered the "mixture of Whites and Negroes" particularly unfavorable. In Chicago, in 1929, he declared:

> It cannot be denied that mulattoes are generally intermediate between the Whites and Negroes, consequently superior on the whole to the latter and inferior as regards most of the traits in which the Whites are superior; superior to the former and inferior to the latter in those few traits in which Negroes excel.[113]

In spite of "isolated assertions due probably to unjustifiable generalizations," mulattos did not manifest any traces of heterosis, that is, "those manifestations of greater strength, precocity or vital resistance which char-

[110] Gini, *Le basi scientifiche della popolazione*, 312. See also Gini, "The Cyclical Rise and Fall of Population," 117.

[111] Gini, *Le basi scientifiche della popolazione*, 313. See also Gini, "The Cyclical Rise and Fall of Population," 102.

[112] Gini, "The Cyclical Rise and Fall of Population," 116–22.

[113] Gini, "The Cyclical Rise and Fall of Population," 125.

acterize many hybrids in the animal and vegetable kingdoms, and also [...] certain human hybrids."[114] On the contrary, while mulattos "present a higher percentage than Negroes of individuals who are unsuccessful at intelligence tests, they do not present an equal or higher frequency than do the Whites of particularly gifted individuals." Finally, Gini's conclusion was that "the crossbreeding of Whites and Negroes gives unfavorable results."[115]

But if crosses resulted generally in negative and disharmonious products, how could the fact that it was "historically and anthropologically ascertained that the great races and the great civilizations, just like the most progressive elements in a single nation, generally come from crossing" be explained?[116] In reality, the apparent contradiction could be justified with the selective mechanism represented by the struggle for life, sexual selection and emigration: these elements, according to Gini, "account for the fact that the most advanced nations, notwithstanding the fact that they owe their origins to the fusion of anthropologically heterogeneous elements and that they must probably in their beginnings have presented very considerable and marked diversity of forms, grow more and more homogeneous, until in time they present [...] uniformity of type."[117]

Therefore, in Gini's view, all the "great races" could be seen as anthropological "fusions." This was the case of the "European races, or those of European origins," that is, "the best that the human species has so far produced":

> Now among these races the pigmentation of eyes and hair, which display—albeit with varying frequency—all gradations from blue to brown and from fair to black, respectively, and even the form of the hair, which varies from absolute straightness to the thickly curled variety, are indisputable evidence of the fusion of diverse racial elements.[118]

Even the "most advanced of the yellow races," the Japanese, was probably a cross between the Chinese and the Malaysians or Polynesians. In the same way, among the Malaysian races, the Javanese dominated and were a combi-

[114] Gini, "The Cyclical Rise and Fall of Population," 126.
[115] Gini, "The Cyclical Rise and Fall of Population," 127.
[116] Gini, *Le basi scientifiche della popolazione*, 316.
[117] Gini, "The Cyclical Rise and Fall of Population," 97.
[118] Gini, "The Cyclical Rise and Fall of Population," 97–98

nation of diverse anthropological elements. Meanwhile, "the demographic decadence of many African populations" was contrasted with the expansion of the Bantu group in South Africa, a product of crossings between "Negroes and Hamites," which "causes anxiety to the white supremacy in South Africa."

As far as Italy was concerned, Gini's 1912 inquiries on the cephalic indices of Italian soldiers, the results of which had been confirmed by Franz Boas in 1913,[119] had revealed that "the greatest degree of variability is found in Central Italy, where the fusion between the Mediterranean dolichocephalic and the Alpine brachycephalic races has been very extensive."[120] It was no wonder that the Italian Renaissance had historically developed here. Therefore, as only those combinations that had been victorious in the struggle for existence were known, it was possible to hypothesize that racial crosses only "sometimes" gave rise to "populations endowed with superior characteristics to those so-called pure."[121]

The genetic dynamic of crossbreeding and successive "isolation" would reconcile, in Gini's view, the cyclical theory of nations with what happened in nature, in the domestication or rational breeding of plants and animals. Nature also gave rise to crossing and selective isolation:

> Apart from the appearance of mutations, not only the dominating races of mankind [...], but all races, derive their origin from crossbreeding. The group feeling determined by physical, or social, or cultural, or administrative factors (race, cast, city, state, etc.) and the hostility of neighboring groups, acts as an isolating factor, and in isolation the complete fusion of races which have been thus mingled gradually takes place. In this consists the biological function of the group feeling.[122]

In conclusion, for Gini, "pure" races did not exist, but were instead "purified" races, which however, could not survive indefinitely in their national-biological isolation, because, reaching a certain level of homogeneity, they

119 See Franz Boas, Helene M. Boas, "The Head Forms of the Italians as Influenced by Heredity and Environment," *American Anthropologist* 15, no. 2 (April–June 1913):163–88. See also Gini to Boas, 6 September 1913, APS, Franz Boas Papers.

120 Gini, "The Cyclical Rise and Fall of Population," 98–99.

121 Gini, *Le basi scientifiche della popolazione*, 317.

122 Gini, "The Cyclical Rise and Fall of Population," 136.

would decline if they were not reinvigorated with new crosses. Consequently,

> the cyclical process of evolution which occurs in the human races, if at first it may seem a wasteful system, inasmuch as it implies periodical recovery and dispersion of energy, really, under the biological laws governing organic life, corresponds to the ideal system suggested by the most modern results of genetics.[123]

In the cyclical theory of nations, the explanatory role of crossbreeding was fundamental in justifying both the birth and the "revival" of nations. In the first aspect, the concept of the germinal plasm was again central. If the inter-breeding involved individuals "in whom the germinal plasm has different variations, sometimes opposed, sometimes even complementary," the plasm of the hybrid could present a "plasticity that allows the start of a new vital cycle, which could lead to the formation of a new race."[124] This would also explain how "many times, new nations arise from the crossing of a superior, civilized and dominating race, with a race still primitive in its mode of life and its culture: that is, from one race specialized [...] in an intellectual sense, with one specialized in a physical sense, muscular."[125]

But aside from new races, born from crossings between native races with immigrants, or between immigrants of diverse origins, history, Gini underlined, offered many cases of "revival" of nations that had been stagnant for centuries, "without the change seeming to be provoked by an immediate external racial influence."[126] This was the case, for example of the Renaissance in Italy and France, or the transformation of Japan in the second half of the nineteenth century. Even these phenomena of revival could be explained, according to Gini, by crossbreeding, which occurred "not between a subject population and invaders, but between internal stocks that have previously remained more or less separate":[127]

> The populations in which these phenomena occur are, generally speaking, those in which different races have lived side by side, sometimes for long periods of time, whose amalgamation has hitherto been hindered by political barri-

123 Gini, "The Cyclical Rise and Fall of Population," 137.

124 Gini, *Le basi scientifiche della popolazione*, 318.

125 Gini, *Le basi scientifiche della popolazione*, 319. See also Gini, "The Cyclical Rise and Fall of Population," 106.

126 Gini, *Le basi scientifiche della popolazione*, 321. See also Gini, "The Cyclical Rise and Fall of Population," 110.

127 Gini, *Le basi scientifiche della popolazione*, 321–22.

> ers, or by psychological resistance, or by legal prohibitions, or by differences of culture or of language. The time comes when these obstacles which kept them apart are eliminated, when they assimilate their respective cultures, intermix [*sic*] on a large scale, and come to form indeed a single nation.[128]

The "fascist revolution" was, in Gini's view, the result of the "biological unification" of the Italian nation, which had its initial moment in the *Risorgimento*:

> Our Italy, from the start of the previous century, has finally started to show undoubted signs of revival, which have accompanied the *Risorgimento* and the reconquest of independence; in the current century, this phenomenon seems to have quickly undergone intensification, accentuated all the more by the last war, from which the fascist revolution is the recent fruit. [...] The *formation of the Italians* that D'Azeglio hoped for from a moral point of view has partly happened and is still progressing, even in the anthropological field; and we are starting to see the fruit.[129]

Gini's eugenic interpretation of crossbreeding and "revival" culminated, therefore, in a theory of fascism as the biological completion of the *Risorgimento*:

> Not only from the point of view of political psychology, but also from the racial standpoint, Italians had to be unified, and that unification, now hastened by the centralizing policy of the government, is beginning to bear its fruits. If this be the case, then the hope—and more than the hope, the intimate feeling which many have—that the Italian nation is now reviewing itself to write new and glorious pages in its history is not without biological foundations.[130]

The connection between eugenics and natalism theorized by Gini found its consecration between 1929 and 1931 with the organization of two important congresses: the Second Italian Congress of Genetics and Eugenics (1929) and the International Congress for Studies on Population (1931), respectively. In his inaugural discourse in 1929, Gini focused first of all on the new title of the congress, which presented the word "genetics":

128 Gini, "The Cyclical Rise and Fall of Population," 111.
129 Gini, *Le basi scientifiche della popolazione*, 322.
130 Gini, "The Cyclical Rise and Fall of Population," 114.

> This signifies that the study of factors, susceptible to social regulation, which might improve or worsen the physical and psychical characteristics of the human race—study that constitutes the object of eugenics—is indissolubly connected with the laws of heredity and the variability of all the animal and vegetal world, laws that form the contents of genetics.[131]

Five years after the First Eugenics Congress in Milan, Gini declared the low level of improvement in the general situation: in Italy, as in all the "Latin countries," the problems of eugenics interested only a "small group of scientists" and were not shared by a larger public audience. Certainly, the Italian spirits were not agitated by the "questions of race that worried every part of the Anglo-Saxon world,"[132] but eugenics was inevitably important for them, both due to the "contact with different races" in the lives of emigrants, and for the "effects of internal migration and crosses between like racial stocks" within the peninsula.

According to Gini, the skepticism of Italian eugenics toward theories that were "dear to the Anglo-Saxon and Nordic eugenicists"—"the theory of prevalence of heredity over the environment in the determination of human traits, the theory of the superiority of the Nordic race, the theory of the progressive degeneration of modern nations due to the increased reproductiveness of the lower classes" [133]—, was clearly "proof of the Latin balance" and was justified by numerous scientific doubts on the mechanisms of heredity.

In the face of the "complexity of the laws of heredity" and the difficulty of predicting the effects of crossbreeding, Gini repeated his conviction that time was not yet ripe for the practical application of eugenics.[134] Additionally, it was not necessarily true that the development of eugenics was indissolubly linked to a prevalence of heredity over the other factors:

> If eugenics concludes that the factors that, under social direction, can improve or impair the racial characteristics of future generations are a bit less heredi-

[131] Corrado Gini, "Discorso d'apertura," in *Atti del Secondo Congresso italiano di Genetica ed Eugenica (Roma, 30 settembre – 2 ottobre 1929)* (Rome: Failli,1932), 17–18.

[132] Gini, "Discorso d'apertura,"18.

[133] Gini, "Discorso d'apertura," 18–19.

[134] Gini, "Discorso d'apertura," 20.

> tary than was believed, and a bit more of a different nature, no one can say that eugenic science is any less than originally aimed for.[135]

According to Gini, an overly vast meaning had been contributed to the concept of heredity, "comprising, in this denomination, every similarity between parents and children that cannot be attributed to environmental conditions during individual development, or, in other words, every similarity between germinal characteristics of the successive generation."[136]

As in 1924, Gini once again underlined the importance of environmental influence on individual characteristics and stressed the role of induction and its consequences. Finally, in the last part of his inaugural discourse, he indicated "two directives" for the future development of Italian eugenics.[137] The first was "that it was not right to limit the study of similarity to immediate ascendants and descendants, but should be systematically extended to an examination of many successive generations," so as to distinguish with major precision the influence of heredity from that of induction or the evolution of the "family stock."[138] The second, on the other hand, was directed toward identifying factors that determined "the development and rise of new stocks."

In contrast to "conservative" eugenics, such as Anglo-Saxon or German, which focused on the defense of the biological elite and the elimination of defectives, Gini proposed a "regenerative" eugenics (*eugenica rinnovatrice*), prevalently interested in the study of biological factors of the birth, evolution and death of nations:

> How do new stocks grow? Admitting that they definitely come from the obscure mass of population, what are the circumstances that determine their rise? Evidently, this cannot come from the heredity of superior factors, which in the past did not exist. Could the origin be found in fortunate combinations; sorts of crosses between stocks not overly different and favored by natural selection? Could the change of environment caused by migration contribute? And what importance does the selection that operates within migration have?[139]

135 Gini, "Discorso d'apertura," 20.
136 Gini, "Discorso d'apertura," 21.
137 Gini, "Discorso d'apertura," 26.
138 Gini, "Discorso d'apertura," 26–27.
139 Gini, "Discorso d'apertura," 26–27.

In particular, Gini believed that eugenicists would find in migration and crossbreeding "the key to the generation or regeneration process that allows humanity to perennially renew its hereditary patrimony throughout the centuries." At the 1929 Italian Congress of Genetics and Eugenics and the 1931 International Congress for Studies on Population, this new paradigm of Italian eugenics assumed an undoubted hegemonic role. The first characteristic of "regenerative" eugenics was a very different approach to the classic issues of European and American eugenics such as racial crossing and sterilization.

Charles Davenport, director of the Eugenics Record Office,[140] claimed at the 1929 Congress that there was sufficient "proof of disharmony in human hybrids" and concluded that it was "bad for race crossing to happen on a large scale."[141] But Italian eugenicists had a different opinion. The biologist Cesare Artom saw in "hybridism" and genetic mutations two phenomena able to produce "new organisms with completely new biological and morphological properties,"[142] while the zoologist Alessando Ghigi highlighted the importance of "the devastating influence on the human species" of consanguinity. The problem of the "constitutional fertility of the mulatto" did not seem so obvious, but required deeper and more accurate statistical research, which could resolve "one of the most important problems of humanity, because it is linked to the possibility of a regression in the average intelligence of those populations that are being colonized by Africa, and those that have founded their very agricultural richness on the use of Negro workers."[143] Similarly, at the 1931 International Congress for Studies on Population, the positive value of some crosses was stressed by Luisa Gianferrari's and Giuseppe Cantoni's paper on the "demographic and genetic effects of inbreeding,"[144] while, in the section dedicated to racial crossing, the strongly hereditarian posi-

140 On Charles B. Davenport, see Kevles, *In The Name of Eugenics*, 41–56. See also Jan A. Witkowski and John R. Inglis, *Davenport's Dream. 21st Century Reflections on Heredity and Eugenics* (Cold Spring Harbor, NY: Cold Spring Harbor Laboratory Press, 2008).

141 Charles B. Davenport, "Sono utili gli incroci di razza?," in *Atti del Secondo Congresso italiano di Genetica ed Eugenica*, 60.

142 Cesare Artom, "Costituzioni genetiche nuove per mutazionismo e per incrocio," in *Atti del Secondo Congresso italiano di Genetica ed Eugenica*, 77.

143 Alessandro Ghigi, "Fecondità e sterilità nell'ibridismo e nella consanguineità," in *Atti del Secondo Congresso italiano di Genetica ed Eugenica*, 172.

144 Luisa Gianferrari, "Effetti demografici e genetici della consanguineità," in Corrado Gini, ed., *Atti del Congresso internazionale per gli Studi sulla Popolazione (Roma, 7–10 settembre 1931)* (Rome: Istituto Poligrafico dello Stato, 1934), vol. 2, 295–308; Giuseppe Cantoni, "Su la consanguineità nelle valli alpestri della Venezia Tridentina," in Gini, ed., *Atti del Congresso internazionale per gli Studi sulla Popolazione*, vol. 2, 309–14.

tions of Eugen Fischer—director of the *Kaiser Wilhelm Institut für Anthropologie* of Berlin—and Jon Alfred Mjøen, were contrasted by more circumspect contributions offered by Americans Stanley D. Porteus, based on the application of "Maze tests" on the population of Hawaii, and Harry L. Shapiro, coordinator of a study on Polynesian crossing financed by the Rockefeller Foundation.[145] As for the burning issue of sterilization, just as had happened in 1924 in Milan at the First Eugenics Congress, at the 1929 Second Congress, the gynecologist Pestalozza condemned it.

For "a similar violation of the liberty and physical integrity of the individual" to be justified—Pestalozza declared—"serious social damage" derived from the "free procreation of psychopaths and deficients" would have to be demonstrable, together with a list of the various forms of mental psychopathy that were hereditarily transmissible. But, regarding the latter point, it was "very probable that psychopaths throughout the generations would undergo auto-elimination, due both to their sterility or low fertility, and to the difficulties that would oppose their marriage." Regarding the former, he declared that "such a demonstration is far from being given, nor is it possible to give it with the current state of our eugenic knowledge."[146]

This was without taking into account the difficulty of correctly evaluating the consensus of the patient. Considering, rather, the central role of environmental conditions in the transmission of psychical characteristics, "the improvement of the race is to be looked for in the fields of prenatal care and education" rather than in sterilization.[147] In conclusion, according to Pestalozza:

> The improvement of the stock is not to be expected from neo-Malthusianism, nor from the limitation of offspring, nor from compulsory sterilization; in sum, not from limitation and prohibition, which represent only the negative part of the eugenic program. It is a program of positive eugenics that we should value, like the program implemented by our national Italian govern-

[145] See Eugen Fischer, "Die gegenseitige Stellung der Menschenrassen auf Grund der mendelden Merkmale," in Gini, ed., *Atti del Congresso internazionale per gli Studi sulla Popolazione*, vol. 3, 179–88; Jon Alfred Mjøen, "Biologische und biochemische Untersuchungen bei Rassenmischung," in Gini, ed., *Atti del Congresso internazionale per gli Studi sulla Popolazione*, vol. 3, 199–202; Stanley D. Porteus, "Race Crossing in Hawaii," in Gini, ed., *Atti del Congresso internazionale per gli Studi sulla Popolazione*, vol. 3, 203–12; Harry L. Shapiro, "Race Mixture Studies in Polynesia," in Gini, ed., *Atti del Congresso internazionale per gli Studi sulla Popolazione*, vol. 3, 213–20. On these contributions, see also Pogliano, *L'ossessione della razza*, 50–52.

[146] Ernesto Pestalozza, "Sterilizzazioni coattive," in *Atti del Secondo Congresso italiano di Genetica ed Eugenica*, 83.

> ment, with guaranteed assistance aimed at maternity and infancy, with prenatal care, social welfare, and the physical and moral education of the youth.[148]

The paper of neurologist and new president of IPAS, Augusto Carelli, was also in line with Pestalozza's position: as there was not "any serious proof to support the assumption of an increase in deficients and mentally defectives in current populations," the alarms regarding a presumed degeneration of the race were unjustified. Therefore, it was important to realize that "legal measures that are the direct consequence of such alarms [...] as well as being inhuman, do not have the least justification in the real facts."[149] Carelli proposed the constitution of a commission with the charge of studying the heredity of mental illnesses, their frequency and any eventual practical initiatives. At the 1929 Second Congress of Genetics and Eugenics, while Gaetano Pieraccini limited himself to supporting only the premarital certificate, as an "expedient of defense of the family and community,"[150] the physician Felix Tietze, president of the Austrian League for Regeneration and Heredity, and Mrs. Cora B. Hodson, secretary of the IFEO, declared themselves favorable to sterilization, the latter not hesitating to praise the humanitarian characteristics of Californian eugenic legislation.[151] But against these declarations, Pestalozza's reaction left no margins of debate:

> To Mrs. Hodson I would say that I reserve my enthusiasm for those surgical operations that tear the ill from their illness or from death, and not for mutilating surgical operations, that I as a surgeon would not deign to carry out, because there is no medical necessity, but only a social interest that has not been demonstrated.[152]

Next to the refusal of the "Anglo-Saxon" model of eugenics, a second characteristic of Italian "regenerative" eugenics was the importance attributed to the eugenic value of fertility and prolificacy. Several Italian contributions at the 1929 and 1931 Congresses focused on this problem: the physiolo-

---

147 Pestalozza, "Sterilizzazioni coattive," 85.

148 Pestalozza, "Sterilizzazioni coattive," 87.

149 Augusto Carelli, "Il presunto aumento dei deficienti e malati mentali fra le popolazioni," in *Atti del Secondo Congresso italiano di Genetica ed Eugenica*, 105.

150 See "Processi verbali," in *Atti del Secondo Congresso italiano di Genetica ed Eugenica*, 35.

151 "Processi verbali," 35–37.

152 "Processi verbali," 37.

gist Carlo Foà insisted, both in 1929[153] and in 1931,[154] on the priority of the economic and social causes rather than biological causes of the birth-rate decrease. In 1929, Silvestro Baglioni analyzed the parallels existing between the somatic and genetic functions,[155] while Agostino Gemelli, in 1931, proposed Catholic sexual ethics as a remedy for the psychological causes of sterility.[156]

The topic of the eugenic value of prolificacy was particularly based on the deep connection between natalist demography and medical constitutionalism. This is quite evident in the works of the statistician Marcello Boldrini, attempting to find a connection between the biology of social stratification and the demography of differential fertility.[157] Not surprisingly, it was Boldrini who, at the 1929 Second Congress of Genetics and Eugenics, advocated a synthesis between "quantity" and "quality" of population.[158] According to Boldrini, not only did the demographic power of a nation increase its "ethnic and somatic unity,"[159] facilitating "mixing and crossing of different groups,"[160] but it also contributed to attenuating the dysgenic consequences of the differential fertility of the social classes. Nothing could have been further from "Nordic" eugenics. Boldrini wrote:

> We are no longer looking at persuading the poorer classes to decrease their fertility, offering them a dream of greater well-being, but rather at resounding, as in the past, that internal voice in members of the higher classes, which encourages them to value paternity by the same standards of moral criteria.[161]

The links between "quantity" and "quality," or between populationism, on one hand, and biotypological constitutionalism, on the other, were clearly

---

153 Carlo Foà, "I fattori biologici della diminuzione delle nascite," in *Atti del Secondo Congresso italiano di Genetica ed Eugenica*, 173–94.

154 Carlo Foà, "I fattori biologici della diminuzione delle nascite," in Gini, ed., *Atti del Congresso internazionale per gli Studi sulla Popolazione*, vol. 2, 9–56.

155 Silvestro Baglioni, "Funzioni somatiche e genetiche," in *Atti del Secondo Congresso italiano di Genetica ed Eugenica*, 153–60.

156 Agostino Gemelli, "Le vedute della psicologia e della psichiatria nel problema della natalità," in Gini, ed., *Atti del Congresso internazionale per gli Studi sulla Popolazione*, vol. 2, 343–46.

157 See, in particular, Marcello Boldrini, "Biotipi e classi sociali," in Lucio Silla, ed., *Atti della SIPS. XX riunione (Milano, 12–18 Septtembre 1931)* (Rome: SIPS, 1932), vol. 1, 63–73.

158 Marcello Boldrini, "Qualità e quantità," in *Atti del Secondo Congresso italiano di Genetica ed Eugenica*, 379–404.

159 Marcello Boldrini, "Qualità e quantità," *Rassegna di studi sessuali, demografia ed eugenica* 10, no. 4 (October–December 1930): 262 (the article reproduces the text of the paper from the 1929 Congress).

160 Boldrini, "Qualità e quantità," 273.

161 Boldrini, "Qualità e quantità," 280.

expressed, at the Congresses of 1929 and 1931, by the results of the demographic and anthropological study of Italian large families, carried out by Corrado Gini.

From 1928 onward ISTAT had organized, on the initiative and under the direct responsibility of Gini, a scientific inquiry into Italian families with more than seven children. Based on the data of the registry office and declarations from the heads of families, priorly advised by the Mayor, the research was a census of more than a million and a half large families (exactly, 1,532,206). The analysis of the data was carried out in successive stages: the results of the first 11 provinces were presented in Gini's contribution to the Second Congress of Genetics and Eugenics in 1929;[162] a second analysis, regarding another 23 provinces, was presented at a conference held by the National Institute of Insurance in Rome (27 February 1931)[163] and at the University of Geneva (23 March 1931); finally, Gini made the other results public during the 1931 International Congress for Studies on Population.[164]

In January 1931, the demographic inquiry directed by ISTAT was completed with an anthropometric and constitutionalist investigation, coordinated by CISP and aimed at the biotypological study of parents of large families. The teams of physiologists, anthropologists and biologists who joined the initiative dealt with a series of municipalities, subdivided into homogenous regions "from an ethnic and geographical-climatological point of view."[165]

[162] Corrado Gini, "Prime indagini sulle famiglie numerose," in *Atti del Secondo Congresso italiano di Genetica ed Eugenica*, 289–338.

[163] Corrado Gini, "Nuovi risultati delle indagini sulle famiglie numerose," *Atti Istituto Nazionale Assicurazioni* 4 (1932): 7–46.

[164] Corrado Gini, Angelo Ferrarelli, "Altri risultati delle indagini sulle famiglie numerose," *Metron* 11, no. 1 (June 1933), then in Corrado Gini, ed., *Atti del Congresso internazionale per gli Studi sulla Popolazione*, vol. 8, 355–98. Two other papers from the conference linked to the inquiry on numerous families were: Corrado Gini, "Sulla nuzialità differenziale delle varie classi sociali," *Metron* 11, no. 1 (June 1933), then in Corrado Gini, ed., *Atti del Congresso internazionale per gli Studi sulla Popolazione*, vol. 7, 357–62; Corrado Gini, "Un nuovo fattore di selezione matrimoniale? L'ordine di generazione," *Metron* 11, no. 1 (June 1933), then in Corrado Gini, ed., *Atti del Congresso internazionale per gli Studi sulla Popolazione*, vol. 2, 245–60.

[165] The list was as follows: Alberto Aggazzotti (Modena, Formiggine, Concordia sulla Secchia), Mario Barbàra (Genoa), Carmelo Cafiero (Nola, Bacoli), Angelo Caroli (Bari, Monopoli, Mola, Polignano), Luigi Castaldi (Cagliari, Ales, Aritzo), Cristoforo Cuscunà (Nicolosi, Paternò),Umberto D'Ancona (Sienna, Grosseto, Monteroni d'Arbia, Abbadia San Salvatore), Filippo Dulzetto (Catania), Carlo Foà (Milan), Fabio Frassetto (Bologna, Imola, Riccione, Ferrara), Giuseppe Genna (Trapani), Carlo Jucci (Sassari, Tempio), Alberto Marassini (Parma), Aldobrandino Mochi (Florence), Osvaldo Polimanti (Perugia, Terni), Angelo Rabbeno (Camerino), Giuseppe Russo (Catania), Arturo Sabatini (Crotone, Catanzaro, Soverato, Chiaravalle, Cirò), Massimo Sella (Rovigno d'Istria, Pisino, Canfanaro, Dignano, Lussimpiccolo, Sanvincenti, Pirano, Gimino), Emilio Sereni (Naples, Vietri, Scafati), Sergio Sergi (Roma), Mario Tirelli (Olevano Romano, Bellegra), Gaetano Viale (Genoa, Imperia, Diano Marina), Velio Zanolli (Padua).

The inquiry included a "qualitative" analysis, based on a biotypological card, created by CISP,[166] a "quantitative" anthropometric analysis (stature, thoracic perimeter, length of the lower limbs, abdominal diameter, cephalic diameter, etc.) and, in some cases, the examination of blood groups. Every collaborator was required to analyze from 500 to 1000 families. Thanks to the mobilization of the municipalities, previously alerted by CISP, the examinations were carried out in municipal clinics or in specific locations prepared for the occasion, although home visits were also carried out, particularly in the big cities.

In June 1931, the first completed records arrived at CISP, while in the successive months most of the single collaborators' reports were consigned. In August, the inquiry could be said to be already concluded, and its results dominated the section of *Anthropology and Geography* at the 1931 International Congress.

A sort of bio-political recording of society, which involved public administrations, medical staff and the national academic system, the ISTAT-CISP inquiry had a double aim. In first place, large families had to become the fulcrum of the demographic and eugenic policies of the fascist regime, as Gini clearly confirmed at a conference held on 16 March 1928, at the Faculty of Law of the University of Bari:

> The most effective method to re-raise the birthrate, or to contain the decrease, is not to encourage the reproduction of small families and individuals that shun marriage, but rather that of those who have managed to remove every obstacle from their families that opposed their expansion and multiplication, who have preserved the generative power of earlier times intact.
>
> Keeping these families in the country by putting the brakes on emigration, facilitating their natural tendency to reproduce by appealing to the sentiments and considerations that could entice them, executing, wherever necessary, their transplantation to regions that have strong need of prolific elements, constitute the most effective measures.[167]

[166] The "qualitative" aspects included anamnestic traits (name, age, number of children, number of brothers and sisters, level of education, number of people in the residential complex, illness contracted, current state of health, menstruation, etc.) and natural descriptive traits (quantity and color of hair, form of the face and profile, dimensions of the head, aspect of the eyebrows, eyes and coloration of the iris, profile and dimensions of the nose, dimensions of the lips, development of the hair, state of the teeth, color of the face, etc.).

[167] Corrado Gini, "Problemi della popolazione," *Annali Istituto di Statistica dell'Università di Bari* (Bari: Tip. Cressati, 1928), 19–20.

Moreover the scientific observation of large families would help lead to an "anthropological type" of fertility, in order to verify the constitutional theories about the relationship between biotype and "genetic instinct." Gini once again confirmed in 1928:

> The inquiry could also be useful for science from another point of view, as it could ascertain the scientific grounds of the arguments of the constitutionalist school, or at least of several exponents of such a school, according to which the genetic instinct would be often particularly strong among macrosplanchnic or brevilinear individuals, whose biotype would therefore be favored by selective reproduction, while microsplanchnic or longilinear individuals would be particularly obstructed. If true, this theory would explain the persistence of the brevilinear type, so frequent in the population, compared to the longilinear type, that matrimonial selection should favor.[168]

Not surprisingly, the section of *Anthropology and Geography* at the 1931 International Congress reserved a significant space for the problem of the relationship between constitution and fertility and the identification of a "maternal biotype." While the zoologist Alessandro Ghigi insisted above all on the need to study the relationship between male heredity and fertility more deeply,[169] Nicola Pende summarized the results of an anthropometric investigation conducted on 250 women from the region of Liguria: "62% of hyper-fertile women belong to the brevilinear biotype, while 38% are the longilinear biotype: among brevilinear women 50.7% were hyper-fertile, while among the longilinear it is 23.5%."[170] The link between "brevilinear type" and fertility was reaffirmed shortly after by Piero Benedetti, of the medical clinic of the University of Bologna: "The brachytypic category possesses [...] in respect to the others, the greatest fertility; the longitypic the least."[171]

In the same section, a vast anthropometric and constitutional inquiry on an entire class of conscripts in the State Armed Forces was described.

---

168 Gini, "Problemi della popolazione," 23.

169 Alessandro Ghigi, "Costituzione e fertilità," in Gini, ed., *Atti del Congresso internazionale per gli Studi sulla Popolazione*, vol. 3, 75.

170 Nicola Pende, "Costituzione e fecondità," in Gini, ed., *Atti del Congresso internazionale per gli Studi sulla Popolazione*, vol. 3, 86.

171 Piero Benedetti, "Contributo alla ricerca dei rapporti tra fecondità e costituzione," in Gini, ed., *Atti del Congresso internazionale per gli Studi sulla Popolazione*, vol. 3, 116.

This research was organized by ISTAT, in collaboration with the Ministry of War,[172] and aimed at the identification of a hypothetical "Italian ethnic type," deriving from increased internal immigration.[173] The "anthropological" record adopted included the following titles: full name, criminality, vaccinations, infirmities, anthropometric data (weight, height, skin color, nasal, face and head shape, color and quantity of hair, eye color, profile of the face, nose, and chin, mouth shape, teeth, eyebrows, etc.) indices, blood group, and vocal range. Presenting the first results, relative to 1900 soldiers of the 1907–1909 classes, Gini came to a conclusion that only partly confirmed the biotypological theory, focusing on the centrality, in the relationship between fertility and constitution, of the "intermediate type" instead of the brevilinear.[174]

Again in 1933, outlining the comprehensive results of the inquiry on large families in Paris, at the 6th International Congress of the *Ligue Internationale pour la Vie et la Famille*, Gini went so far as to identify an anthropological and racial type of fertility, which was much closer to Quételet's "average man" than Nicola Pende's "brevilinear" type:

> The partial results obtained from different collaborators of the anthropometric inquiry carried out on the fathers and mothers of large families lead to the conclusion that the morphological type of the individuals examined, both men and women, for the major part of characters, oscillates around average values, as much for the fundamental measurements as for the index values.[175]

Following this, Gini summarized the bio-social characteristics of this "average" man:

172 The subcommittee of the study, nominated by Gini, was composed of: Livio Livi, president, member of the High Council of Statistics; Marcello Boldrini, Milan Catholic University; Pio Cartoni, the General Headquarters for drafted non-commissioned soldiers and troops, (*Direzione Generale Leva Sottufficiali e Truppa*) of the Ministry of War; Medical Captain Alfredo Corsi, the General Headquarters of Military Health (*Direzione Generale di Sanità Militare*) of the Ministry of War; Medical Colonel Giovanni Grixoni, director of the School of Military Health; Medical Lieutenant Colonel Gabriele La Porta, the General Headquarters of Maritime Military Health (*Direzione Generale di Sanità Militare Marittima*) of the Ministry of Marine; Aldobrandino Mochi, director of the Institute of anthropology and paleontology at the University of Florence; General Fulvio Zugaro, director general of logistical services at the Ministry of War; Medical Lieutenant Colonel Luigi De Berardinis, head of the ISTAT department of demography and vital statistics.

173 For a description of the initiative see Duilio Balestra, "La preparazione dell'indagine antropometrica sugli iscritti in una classe di leva in Italia," in Gini, ed., *Atti del Congresso internazionale per gli Studi sulla Popolazione*, vol. 3, 7–34.

174 Corrado Gini, "Alcuni risultati preliminari dell'indagine antropometrica sui soldati italiani," in Gini, ed., *Atti del Congresso internazionale per gli Studi sulla Popolazione*, vol. 3, 98.

> Environmental and economic conditions generally not very favorable; occupations for the father generally manual and tiring; for the mother, household duties; for the men, little adiposity, agile body shape, long limbs, large chest, normal abdomen, an on-average tall height; for the women, a more squat body shape, tendency to adiposity, narrow chest, average abdomen, short limbs, medium-short height, normal menstrual cycle.[176]

In 1932, as head of an Italian delegation to the 3rd International Eugenics Congress in New York, Gini once again presented the prospect of a eugenics based on a harmonic connection between the quantity and quality of the population:

> In the matter of population, as in other fields, the problems of quantity and quality are indissolubly connected. As I see it, they are indissolubly connected not only because in practice it is difficult to think of a measure affecting the number of inhabitants which does not also affect their qualitative distribution, or of a measure hindering or encouraging the reproduction of certain categories of people which does not also modify, directly or indirectly, the number of the population, but also and above all because population is a biological whole, subject, as such, to biological laws, which show us that mass, structure, metabolism, psychic phenomena, the reproduction of organic life are all indissolubly connected, both in their static condition and in their evolution, so that it would be vain to try to modify some of these characters without taking into account the stage of development attained by the other.[177]

Not surprisingly, at the 1932 International Congress, Gini's paper concentrated on one of the main themes of "regenerative" eugenics: the heterosis of hybrids.[178] According to Gini, empirical data did not seem to support the American geneticists East and Jones, who theorized a 50 percent diminu-

---

175 Corrado Gini, *Enquête démographique sur les familles nombreuses italiennes. Résultats des recherches* (Paris: Gembloux Imprimerie - J. Duculot Éditeur, 1933), 28.

176 Gini, *Enquête démographique sur les familles nombreuses italiennes. Résultats des recherches*, 28.

177 Corrado Gini, "Response to the Presidential Address," in *A decade of progress in eugenics: Scientific papers of the Third International Congress of Eugenics* (Baltimore: The Williams & Wilkins Company, 1934), 25–26. The Italian version of Gini's contribution was: Corrado Gini, "III Congresso internazionale di Eugenica (New York, 21–23 agosto 1932)," *La ricerca scientifica* 3 (1933).

178 Corrado Gini, "Remarks on the explanation of heterosis," in *A decade of progress in eugenics*, 421–24. (The Italian version was: Corrado Gini, "Osservazioni sulla spiegazione dell'eterosi," *Genesis* 1–2 (January–June 1932).

tion of heterosis between the first and second generation of hybrids and a progressive diminution over the successive generations:

> Then, granted that on the contrary this reduction does occur, we must conclude that the above explanation is insufficient and that it must either be completed by an additional explanation or replaced by another more in keeping with the facts.
> It would appear also, that a 50 per cent reduction of heterosis from the first to the second generation of hybrids is not always in keeping with experience, so that also from this side, the theory is not always confirmed by facts.[179]

In New York, the Italian delegation also participated in the exhibition, organized on the occasion of the Congress, providing exhibits which painted a picture of Gini's hegemony on fascist eugenics. In fact, the Italian contribution included three series of diagrams and cartograms organized by ISTAT; the proceedings of the 1924 and 1929 Eugenics Congresses published by SIGE; the issues of *Genesis* and *Metron*; the volumes published by CISP, and finally, Lidio Cipriani's African facial masks, exhibited at the American Museum of Natural History.[180]

Some years later, under the banner of "regenerative" eugenics and in opposition to the "Nordic" (Anglo-American and German-Scandinavian) component of the IFEO, Gini inaugurated the Latin Federation of Eugenic Organizations (*Federazione Latina delle Società di Eugenica*). The turning point came, not surprisingly, after the International Population Congress in Berlin, in the summer of 1935. On 26 September 1935, a letter sent by the Ministry of National Education to the Presidency of the Council of Ministers and the Presidency of ISTAT, based on a detailed report by Gini, explicitly stated the intention to draw back from the IFEO:

> The Italian scholars must abstain from collaborating with the International Federation of Eugenic Organizations, from which our representatives have distanced themselves in consideration of its program, which evidently contrasts with the Italian direction regarding the qualitative population policy.[181]

179 Gini, "Remarks on the explanation of heterosis," 423.
180 Gini, "III Congresso internazionale di Eugenica," 5.
181 The letter is conserved in ACS, PCM, 1940–43, b. 2674, f. 1.1.16.3.5.27.000-7, sf. 3.

In the speech Gini prepared for the first meeting organized by the Latin Federation, held in Mexico City, on 12 October 1935, the new *Latin* eugenics was characterized by three elements. First of all, the rejection of birth control and the search for a balance between the "quantity" and the "quality" of the population:

> The idea of a league of nations with low birth-rates could not originate among the Latins. Nor is it likely that Latins would ever grasp at the expedient of sending propagandists to countries with high birth-rates to spread the seeds of limitation of the birth-rate and mitigate their demographic pressure in this way. [...] This all shows that the fundamental eugenic problem of the relationship between quantity and quality of the birthrate can be objectively studied in the Latin Federation, in all its complexity, without postulating a contrast that needs to be demonstrated and without unilaterally taking into consideration only the facts that seem to bear witness in one sense.[182]

Similarly, in regard to migratory movements, the variety of situations within "Latin" countries and the absence of a policy that "defended the national market from the competition of foreign labor" favored "an impartial examination of the effects of immigration and emigration on the quantitative development of the population, such as the selective character of the emigrations and therefore their influence on the characteristics of the population of the country of origin and that of destination."[183] Finally, regarding the problem of race, and, in particular, the theme of crossbreeding, Latin eugenics could assume, according to Gini, a more balanced position, avoiding democratic egalitarianism, without however degenerating into national-socialist mixophobia:

> [Latin eugenicists] are not blinded with national sentiment to the point of believing, against history, that we can speak of a superiority of race for every time and place. It is, on the other hand, probable that, when crosses with another race appear inevitable, they can be kept from falling into the opposite extreme, judging all the races as absolutely equal from the point of view of their intellectual attitudes.[184]

182 Corrado Gini, "Parole inaugurali del Prof. C. Gini, lette alla riunione delle Società di Eugenica dell'America Latina tenutasi a Città del Messico il 12 ottobre 1935," *Genus* 2, no. 1–2 (June 1936): 78.

183 Gini, "Parole inaugurali del Prof. C. Gini, lette alla riunione delle Società di Eugenica dell'America Latina tenutasi a Città del Messico il 12 ottobre 1935," 78–79.

184 Gini, "Parole inaugurali del Prof. C. Gini, lette alla riunione delle Società di Eugenica dell'America Latina tenutasi a Città del Messico il 12 ottobre 1935," 79.

Several Latin nations found themselves at the peak of their economic and cultural power, others were rapidly developing, others, still "having a past superior to the present," were passing through a "phase of renewal with hopes for a grand arrival": Only Latin nations, therefore, could observe eugenics "without badly concealed concern," through the lens of Gini's cyclical theory of nations, "recognizing [...]—as in the evolution of other animal and vegetable species—the fundamental importance of the internal biological forces and mutations coming from variations of environment or crosses."[185]

The three cardinal points of "regenerative" eugenics, according to Gini, were very clear: the eugenic value of populationism, the renewing effect of migrations and the phenomenon of heterosis in crossbreeding. This was the Latin model. In Gini's view, all the eugenic measures, including "the most extreme and, for some of us, highly repugnant,"[186] had to be examined and discussed. And this neutral analysis could be provided only by Latin populations, who were in "favorable conditions" to address these problems "with scientific objectivity." In fact:

> As the Latin countries have never been used as colonies of deportation, they will not encounter those sources of degeneration that weigh on the economic and moral balance of other nations, nor do sexual perverts assume in their populations such importance as to suggest to scientists to constitute a third sexual category, or give rise to movements because this judgment is juridically recognized. These are circumstances that help to understand how suggestions of radical measures of elimination came to be listened to in other countries.[187]

In any case—Gini concluded—the "Latin" scientists would never forget the lessons of ancient Roman civilization and would never accept the practice of sterilization:

> It is very natural that the descendents of Rome, which [...] thousands of years ago imposed the abolition of human sacrifices, and then gradually achieved

185 Gini, "Parole inaugurali del Prof. C. Gini, lette alla riunione delle Società di Eugenica dell'America Latina tenutasi a Città del Messico il 12 ottobre 1935," 79.

186 Gini, "Parole inaugurali del Prof. C. Gini, lette alla riunione delle Società di Eugenica dell'America Latina tenutasi a Città del Messico il 12 ottobre 1935," 80.

187 Gini, "Parole inaugurali del Prof. C. Gini, lette alla riunione delle Società di Eugenica dell'America Latina tenutasi a Città del Messico il 12 ottobre 1935," 80.

> the abolition of slavery, feel complete reluctance in the face of a measure that deprives man of one of the most essential attributes of his personality and sacrifices one of the most salient manifestations of life.[188]

It was on this theoretical foundation that two years later, in August 1937, the First Latin Eugenics Congress was held in Paris, due to the strategic alliance between Gini's SIGE and the eugenic section of the French *Institut International d'Anthropologie*.[189]

French, Romanian and Italian physicians, hygienists and anthropologists participated at the Paris congress, emphasizing an ideological and scientific position markedly opposed to "Nordic" eugenics. The theme of birth control was almost nonexistent, replaced by Italian-French natalism, underlining the "eugenicity" of prolific families.[190] As for racial crossing, only René Martial, professor at the Institute of Hygiene of the Medical Faculty in Paris, celebrated the American eugenic fight against miscegenation, judging crossbreeding between the French and the "yellow" or "black" races negatively and calling for the introduction of a eugenic control of immigration.[191] Professor of veterinary medicine and agronomy, Étienne Letard, instead claimed that it was not possible to create a "hierarchy" of the biological

188 Gini, "Parole inaugurali del Prof. C. Gini, lette alla riunione delle Società di Eugenica dell'America Latina tenutasi a Città del Messico il 12 ottobre 1935," 80.

189 Members of the Latin Federation of Eugenic Organizations were, in 1937, in addition to Italy, Argentina, Belgium, Brazil, Spain, France, Mexico, Perù, Portugal, Romania, Switzerland. See *Bureaux des Sociétés Fédérées*, in Fédération Internationale Latine des Sociétés d'Eugénique, *Ier Congrès Latin d'Eugénique. Rapport* (Paris: Masson et C., 1937), 381–83.

190 See Raymond Turpin, Alexandre Caratzali and Gorny, "Contributions à l'étude de l'influence de l'âge et de l'état de santé des procréateurs, du rang et du nombre des naissances, sur les caractères de la progéniture," in Fédération Internationale Latine des Sociétés d'Eugénique, *Ier Congrès Latin d'Eugénique*, 240–61; Corrado Gini, "De quelques recherches sur les variations que présenteraient certains caractères suivant le nombre d'enfants de la famille," in Fédération Internationale Latine des Sociétés d'Eugénique, *Ier Congrès Latin d'Eugénique*, 262–69; Benjamin Weil-Hallé and M. Meyer, "La survie des enfants dans les familles nombreuses et restreintes," in Fédération Internationale Latine des Sociétés d'Eugénique, *Ier Congrès Latin d'Eugénique*, 270; Raymond Turpin, Alexandre Caratzali and Nicholas Georgescu-Roegen, "Influence de l'âge maternel, du rang de naissance et de l'ordre de naissance sur la mortinatalité," in Fédération Internationale Latine des Sociétés d'Eugénique, *Ier Congrès Latin d'Eugénique*, 271–77; Nora Federici, "Mortalité infantile et mortalité prénatale chez les familles nombreuses italiennes," in Fédération Internationale Latine des Sociétés d'Eugénique, *Ier Congrès Latin d'Eugénique*, 278–82; Raymond Turpin and Alexandre Caratzali, *Influence de l'âge maternel sur la mortinatalité des jumeaux*, in Fédération Internationale Latine des Sociétés d'Eugénique, *Ier Congrès Latin d'Eugénique*, 283–85.

191 See René Martial, "Métissage et immigration," in Fédération Internationale Latine des Sociétés d'Eugénique, *Ier Congrès Latin d'Eugénique*, 16–39.

validity of the human species,[192] while the physician Alfred Thooris, scientific consultant of the *Fédération Française d'Athletisme,* proclaimed the positivity of crossbreeding between the "Celtic race" and all the other stocks, with the exception of the Jews, whom he regarded as totally inassimilable.[193]

The attitudes toward the Nazi eugenic legislation differed: the law of 14 July 1933 was severely criticized, for example, by the French physician Franziska Minkowska,[194] but Georges Schreiber, vice-president of the *Société Française d'Eugénique,* highlighted the German example for the French, particularly with regard to the elements of the adoption of matrimonial loans to couples who had their eugenic efficacy certified.[195]

In general, "Latin" eugenicists at the Congress rejected rigid Weismannian hereditarianism and its socio-biological determinism. An entire section of the Congress was dedicated to the possible forms of healing the illnesses of the germ plasm[196] and several papers stressed the importance of environmental conditions, education and biotypological monitoring.[197] Called on

192 See Étienne Letard, "Les leçons de l'expérimentation animale dans le problème du métissage," in Fédération Internationale Latine des Sociétés d'Eugénique, *Ier Congrès Latin d'Eugénique,* 61–71.

193 See Alfred Thooris, "Considérations ethnologiques et démographiques sur la population française," in Fédération Internationale Latine des Sociétés d'Eugénique, *Ier Congrès Latin d'Eugénique,* 214–27.

194 Franziska Minkowska, "Eugénique et Généalogie," in Fédération Internationale Latine des Sociétés d'Eugénique, *Ier Congrès Latin d'Eugénique,* 341–50.

195 See Georges Schreiber, "Allocations familiales et Eugénique," in Fédération Internationale Latine des Sociétés d'Eugénique, *Ier Congrès Latin d'Eugénique,* 91–100.

196 See Edmond-Alexandre Lesné, "Influence des régimes carencés et déséquilibrés, suralimentation et sous-alimentation, sur la natalité et la mortalité des petits rats," in Fédération Internationale Latine des Sociétés d'Eugénique, *Ier Congrès Latin d'Eugénique,* 144–46; Oddo Casagrandi, "Tentatives microscopiques et biologiques en vue de l'identification de certaines tares organiques séminales, héréditaires et acquises," in Fédération Internationale Latine des Sociétés d'Eugénique, *Ier Congrès Latin d'Eugénique,* 147–49; Christian Champy, "L'importance des variations raciales de sensibilité aux hormones dans l'appréciation de la valeur sexuelle de l'individu," in Fédération Internationale Latine des Sociétés d'Eugénique, *Ier Congrès Latin d'Eugénique,* 150–53; Raymond Turpin, Alexandre Caratzali and H. Rogier, "Étude étiologique de 104 cas de mongolisme et considerations sur la pathogénie de cette maladie," in Fédération Internationale Latine des Sociétés d'Eugénique, *Ier Congrès Latin d'Eugénique,* 154–64; Henri Vignes, "De l'influence de l'intoxication alcoolique des procréateurs sur leur progéniture," in Fédération Internationale Latine des Sociétés d'Eugénique, *Ier Congrès Latin d'Eugénique,* 165–70; Gustave Roussy and René Huguenin, "Vues sur le rôle de l'hérédité dans le cancer humain," in Fédération Internationale Latine des Sociétés d'Eugénique, *Ier Congrès Latin d'Eugénique,* 171–86; Albert Brousseau, "De la viabilité et de la fécondité des insuffisants intellectuels," in Fédération Internationale Latine des Sociétés d'Eugénique, *Ier Congrès Latin d'Eugénique,* 187–97.

197 See Marcello Boldrini, "Constitution et Eugénique," in Fédération Internationale Latine des Sociétés d'Eugénique, *Ier Congrès Latin d'Eugénique,* 228–31; Georges Heuyer, "Constitution et Eugénique," in Fédération Internationale Latine des Sociétés d'Eugénique, *Ier Congrès Latin d'Eugénique,* 232–38; Giacomo Tauro, "La transmigration des classes sociales par l'éducation," in Fédération Internationale Latine des Sociétés d'Eugénique, *Ier Congrès Latin d'Eugénique,* 320–21; Giacomo Tauro, "Eugénique et pédagogie," in Fédération Internationale Latine des Sociétés d'Eugénique, *Ier Congrès Latin d'Eugénique,* 379–80.

to delineate a eugenic program for Romania, Gheorghe Banu, member of the Royal Romanian Society for Eugenics and Heredity, dedicated a large space to the questions of hygiene, the fight against social illnesses, and the protection of maternity and premarital certificates, leaving the proposal of limited sterilization of the chronically mentally ill, with consensus obtained from the families of the subject, to a brief concluding chapter.[198]

Following Gini's scientific paradigm, the Italian participants at the Congress focused their papers mainly on the problem of social metabolism produced by the cyclical evolution of nations, explicitly opposing the genetocratic social crystallization of Anglo-American eugenics.[199] An example was the relation of Giuseppina Levi della Vida, who criticized Karl Pearson's eugenic arguments, on the basis of Gini's theory. The biological decadence of the elite—maintained Levi della Vida—did not bring about the degeneration of civilization, as Pearson had claimed, but on the contrary, was absorbed by the parallel rise of the inferior classes:

> According to Gini's theories, social metabolism, far from representing a degenerative factor, constitutes a useful mechanism for society, in the sense that, continually renewing the ruling classes, for a certain period of time favors their development, and following this, prevents an overly rapid fall.[200]

Corrado Gini's contribution to the Paris Congress was centered on the problem of identifying a biological-statistical *medietas* as the fundamental criteria for racial biotypology. Entitled *Biotypologie et Eugénique*, Gini's paper claimed, first of all, the conceptual weakness of the "biotype" from a statistical point of view: since the frequency of the constitutional indices (thoracic index, ponderal index, etc.) were generally not in correspondence with the values that identified type, but rather had a relationship

[198] Gheorghe Banu, "Les facteurs dysgéniques en Roumanie: principes d'un programme pratique d'eugénique," in Fédération Internationale Latine des Sociétés d'Eugénique, *Ier Congrès Latin d'Eugénique*, 296–319.

[199] See Dino Camavitto, "Premiers résultats d'une recherche anthropologique sur les Zambos de la Costa Chica (Guerrero, Mexique)," in Fédération Internationale Latine des Sociétés d'Eugénique, *Ier Congrès Latin d'Eugénique*, 40–60; Paolo Fortunati, "Le métabolisme social d'après des recherches sur les étudiants de l'Université de Padoue," in Fédération Internationale Latine des Sociétés d'Eugénique, *Ier Congrès Latin d'Eugénique*, 79–90; Vincenzo Castrilli, "La nuptialité et la fécondité des diplômés de l'enseignement secondaire en Norvège," in Fédération Internationale Latine des Sociétés d'Eugénique, *Ier Congrès Latin d'Eugénique*, 110–19; Giuseppina Levi della Vida, "Le métabolisme social comme facteur de dégénération dans la société," in Fédération Internationale Latine des Sociétés d'Eugénique, *Ier Congrès Latin d'Eugénique*, 120–31.

[200] Levi della Vida, "Le métabolisme social comme facteur de dégénération dans la société," 129.

with the arithmetic mean of the same values, the "biotype" as defined by the constitutional school did not have a mathematical-statistical foundation, but represented only a sort of "mental category." It would therefore be better to define the "biotype" in terms of "constitutional form" or "constitutional morphology."[201]

Despite this criticism, Gini supported the need for a "statistical study of the constitutions" that would fit in the more general framework of the correlations between "the intensity of the same characteristic in two successive generations."[202] Nevertheless, according to Gini, a statistical-demographic approach to biotypology would bring two further problems with it: on one hand, the identification of a criteria of "normality," which Gini recognized in the geometric mean between linear or monotone relationships (for example, stature and thoracic perimeter);[203] on the other, deeper study into the problems of "heredity" of characteristics, aimed at defining the "inter-racial" or "intra-racial" origins of biotypes.

It was to inform this latter aspect that Gini reconsidered the data from the CISP-ISTAT inquiry on large families. This data showed that the brevilinear form was prevalent in the Po valley and that, on the other hand, the medium form was more frequent in Sardinia: couldn't the relationship between fertility and the brevilinear form—Gini asked—derive from a different reproductive capacity of the alpine and dinaric (brevilinear) races in comparison with the Mediterranean (longilinear)?

In conclusion, Gini repeated the necessity of reinforcing the scientific basis of biotypology:

> It is a delicate subject. We need to be clear about the terms, adopt the methods that are least susceptible to criticism and set up the research in a way that responds well to the questions to resolve. The difficulties regarding this last point are multiple, and the progress will consequently be slow.[204]

201 Corrado Gini, "Biotypologie et Eugénique," in Fédération Internationale Latine des Sociétés d'Eugénique, *Ier Congrès Latin d'Eugénique*, 200–04.

202 Gini, "Biotypologie et Eugénique," 204.

203 See Corrado Gini, "Une question importante pour la science des constitutions et pour la médecine militaire: comment juger si les proportions d'un individu sont normales?," *Revue de l'Institut International de Statistique* 5, no. 2 (July 1937): 107–14; no. 3 (October 1937): 203–11.

204 Gini, "Biotypologie et Eugénique," 211.

In Italy, "Latin" eugenics, largely shared by demographers and statisticians, nevertheless aroused the resistance of biological racists, who preferred to base fascist eugenics on the Nazi model.

In the field of colonial racism, for example, Gini's complex scientific evaluation of the problems of crossbreeding clashed, in 1937, with the introduction of the fascist laws against racial crossing. In 1932, presenting Lidio Cipriani's book, *Considerazioni sopra il passato e l'avvenire delle popolazioni africane* [Considerations on the past and future of African populations], published under the auspices of SIGE in the CISP series, Gini tried to reconcile the bio-demographic potential of racial crossings with the need to control them, above all the in Italian colonies:

> Recognizing the necessity of racial crossings for the conservation of the stock, and acknowledging that, according to the racial elements that are combined, the quality of the products will vary, there will be diversity, from a social point of view, in the value of these crossings in relation to the different environmental demands. However, this does not negate the importance of the eugenic problems of crossings. If anything, it accentuates it, insofar as, recognizing the inevitable characteristic of the phenomenon, the need to control it becomes more evident.[205]

Several years later in 1937, in an interview published in the journal *L'Azione coloniale* [Colonial action],[206] Gini explicitly approved the racist measures of the government, but repeated his arguments on the positive value of racial crossings as a factor in revitalizing the nation. The author

205 Corrado Gini, preface in L. Cipriani, *Considerazioni sopra il passato e l'avvenire delle popolazioni africane* (Florence: R. Bemporad & F., 1932). This text of Cipriani's summarised the much larger volume by the author, edited by the same publisher in 1932, with the title *In Africa dal Capo al Cairo*, published under the auspices of the Italian Geographical Society; Cipriani's racist ideas were expressed in chapter XI (*Alcune considerazioni generali sull'Africa e le sue popolazioni negre in rapporto al problema della colonizzazione*). Cipriani was one of the signatories of the Manifesto of Racial Scientists, in 1938. On Cipriani, see Paolo Chiozzi, "Autoritratto del razzismo: le fotografie antropologiche di Lidio Cipriani," in Centro Studi F. Jesi, ed., *La menzogna della razza. Documenti e immagini del razzismo e dell'antisemitismo fascista* (Bologna: Grafis, 1994): 91–95; Luigi Goglia, "Note sul razzismo coloniale fascista," *Storia contemporanea*, 19, no. 6 (December 1988): 1244; and Gianluca Gabrielli, "Prime ricognizioni sui fondamenti teorici della politica fascista contro i meticci," in Alberto Burgio and Luciano Casali, eds., *Studi sul razzismo italiano* (Bologna: Clueb, 1996): 80–82; on his activities as director of the Institute of Anthropology in Florence, see the interesting references in several essays contained in Enzo Collotti, ed., *Razza e fascismo. Le persecuzioni contro gli ebrei in Toscana (1938–1943)* (Rome: Carocci, 1999), in particular those by Camilla Bencini, Francesca Cavarocchi and Alessandra Minerbi.

206 The interview with Gini was published in two successive articles, signed by Genesio Eugenio Del Monte with the pseudonim "Eudemon": "Il fenomeno degli incroci nel pensiero di Corrado Gini" and "Il fenomeno degli incroci," in *L'Azione Coloniale* respectively on 25 February and 4 March 1937. *L'Azione Coloniale*, founded in 1931, was the official organ of the Fascist Colonial Institute, and was directed by Marco Pomilio.

of the interview, Genesio Eugenio Del Monte, during the purging trial against Gini in 1944–45, in an attempt to underline Gini's distance from official state racism, provided an interesting retrospective description of the whole affair:

> From 1928, I was introduced to studies on racial crossings by Father Mauro da Leonessa, capuchin missionary, currently in Rome at the Convent of the Capuchins of S. Lorenzo Fuori Le Mura.
> But only in January 1937 did the Italian press accept my article on the problem of racial crossing, since it was only then that the fascist government officially decided to follow the example that Great Britain, for some centuries, the United States of America since their constitution and the Colony of the Cape successively, had adopted, that is, racist policies, with results that even today are not denied.
> In Italy, Prof. Gini had for some years deeply studied these questions, and therefore I decided to ask him, in February 1937, for an interview for *L'Azione Coloniale*, which as has been noted, was the unofficial organ of the Ministry of Colonies. The interview, which was partly distant from my own ideas, was greeted with much enthusiasm by the Director of *L'Azione Coloniale*, Dr. Marco Pomilio, but although the first part came out, the following part was published after many difficulties due to an intervention from the government; and, unlike what had happened to similar articles, the Italian press completely *ignored* the highly important interview that was the synthesis of scientific research that since then has been acknowledged in that field.
> In the end, I myself was invited not to cite the studies of Prof. Gini in my writings on crosses, for reasons of appropriateness; I was made to understand that Prof. Gini was unpopular with some authorities, who, moreover, were irritated by his declarations in the interview, that in various points did not seem to be in accord with the racial policies of the fascist government.[207]

When, in 1939, Del Monte once again collaborated with *L'Azione Coloniale* for a series of articles on the bibliography of racial crossing, he was "categorically invited" to not cite Gini or the "Jewish" statistician Kuczynski.[208]

207 Declaration by Genesio Eugenio Del Monte, November 7, 1944, ACS, MPI, DGIS, Professori Universitari Epurati, 1944–1946, b. 16, f. "Gini".

208 Alberto Pollera, a colonial officer who served the colonial administration from his early twenties until his death in 1939, quoted the interview with Gini in an attempt to oppose, in his way, the introduction of the racial colonial legislation, to support the legitimacy and goodness of racial crosses: see Treves, *Le nascite e la politica*, 306–07. On Pollera, see Luigi Goglia, "Una diversa politica razziale coloniale in un documento inedito di Alberto Pollera del 1937," *Storia contemporanea* 16, no. 5–6 (December 1985): 1071–92; Barbara Sòrgoni, *Etnografia e colonialismo. L'Eritrea e l'Etiopia di Alberto Pollera 1873–1939* (Turin: Bollati Boringhieri, 2001).

In July 1938, the prominent journalist Telesio Interlandi, considered Mussolini's unofficial mouthpiece, attacked Gini's eugenics in the pages of the newspaper *Il Tevere*. After labeling Gini as a scholar "better known as a statistics expert than a pillar of eugenics,"[209] Interlandi interpreted the critical attitude toward national socialist racism, which several Italian scientists, symbolically represented by Gini, had adopted, as a "zone of dissidence" to be suffocated in order to obtain "greater political order":

> In this way science perpetuates a divorce that could be damaging to fascist society, denouncing in first place a deplorable political insensitivity. It is our work to signal the most scandalous manifestations of such insensitivity, because this way we can obtain the greatest political control in every zone of culture where dissidence flowers.[210]

Giovanni Preziosi, one of the most prominent Italian fascist anti-Semites, also heavily attacked Gini and his "infamous and antiracist eugenic Congress in Paris," describing the Latin Federation of Eugenics as an instrument "in the hands of Jews and Masons."[211]

Nevertheless, contrary to what Interlandi and Preziosi claimed, "quantity" once again prevailed over "quality" at the Third Congress of SIGE, held in Bologna in September 1938. In front of Luigi Cesari, delegate of the General Direction for Demography and Race, and Emil Witschi, professor at the State University of Iowa, Gini emphatically inaugurated the SIGE Congress, announcing the organization of a second International Congress of Latin Eugenics, scheduled for 1939 in Bucharest. Regarding communications and relations, the role of Italian genetics was on this occasion more important in comparison to the preceding congresses of 1924 and 1929. The Third Congress of SIGE was in fact characterized by two sections of genetics: general genetics, represented by Giuseppe Montalenti,[212] Claudio Barigozzi[213] and Adriano Buzzati-Traverso;[214] and animal and vegetal genetics,

209 Telesio Interlandi, "Cattolici sugli specchi," *Il Tevere* (23–24 July 1938).

210 Telesio Interlandi, "Zone di dissidentismo," *Il Tevere* (23–24 April 1938).

211 Giovanni Preziosi, "Per la serietà degli studi razziali in Italia (dedicato al camerata Giacomo Acerbo)," *La Vita Italiana* 28, 328, (July 1940): 74–75.

212 Giuseppe Montalenti, "I recenti studi sul problema della determinazione del sesso e dei caratteri sessuali secondari negli animali," *Genus* 3, no. 3–4 (June 1939): 193–214.

213 Claudio Barigozzi, "I nuovi orizzonti della citogenetica," *Genus* 3, no. 3–4 (June 1939): 35–72.

214 Adriano Buzzati-Traverso, "I nuovi orizzonti della radiogenetica," *Genus* 3, no. 3–4 (June 1939): 73–130.

represented by Alessandro Ghigi and the scholars of his Institute of Zoology in Bologna (where, not coincidently, the congress was held).[215]

The *human genetics* section was represented by Corrado Gini, Agostino Gemelli and Giuseppe Pintus, but it was above all the fourth session that was dominated by Gini's "regenerative" eugenics. While Gini's contribution was aimed at demonstrating that prolific women were no longer exposed to the danger of dysgenic twin births,[216] Marcello Boldrini, in opposition to the Anglo-Saxon position, emphasized the eugenic role of differential fertility. Boldrini referred in particular to the research of English neo-Malthusian eugenicist Raymond B. Cattell, according to whom the intelligence quotient was decreasing by one point every decade, due to differential fertility. On the contrary, according to the Italian statistician, the greater fertility of the lowest social classes did not necessarily have a dysgenic effect.

In first place, it was worth considering the low reproductiveness of "deficient and defective individuals."[217] Added to this was the fact that "the man immune from defects and the defective man, if not two abstractions, are at the least two relatively rare entities, while most people combine, coordinated in a system, both positive and negative qualities."[218] Human processes of adaptation determined, nevertheless, a "social neutralization of the defects and imperfections, which characterize every type and every non-anomalous combination of attributes."[219] Consequently, if it were true that the growing average number of children, from the top to the bottom of the social hierarchy, would favor, in future generations, both the positive and negative qualities of the inferior social classes, "the more advanced social neutralization of the most common psychical and physical imperfections in the higher and middle classes would cause—as regards negative traits—the opposite tendency."[220]

Finally—and it was Boldrini's last criticism of "Anglo-Saxon" eugenics—no one could know today the aesthetic ideal of the future: if the Spain of Philip IV had been preoccupied with eugenics, it would probably not

---

215 On SIGE third Congress, see also "Società italiana di genetica ed eugenica. Riunione di Bologna, 5–7 settembre 1938," *Genus* 3, no. 3–4 (June 1939): 369–70.
216 Corrado Gini, "Prolificità e frequenza dei parti plurimi," *Genus* 3, no. 3–4 (June 1939): 279–96.
217 Marcello Boldrini, "La fertilità degli individui deficienti e difettosi," *Genus* 3, no. 3–4 (June 1939): 301.
218 Boldrini, "La fertilità degli individui deficienti e difettosi," 304.
219 Boldrini, "La fertilità degli individui deficienti e difettosi," 303.
220 Boldrini, "La fertilità degli individui deficienti e difettosi," 305.

have fostered "a good number of those dwarves and buffoons" immortalized by Velázquez. In conclusion, therefore, "eugenic or dysgenic consequences could result from the differential fertility of the social classes; but not necessarily of the type that many eugenicists have foreseen."[221]

In addition to Boldrini's contribution, the demographer Nora Federici, a pupil of Gini's,[222] provided the first results from the ethnological research studies organized by CISP, of several "primitive" populations in a state of demographic isolation. These results were obviously in absolute agreement with Gini's "regenerative" eugenics. The data, regarding several anthropometric characteristics (stature, seated stature, weight and biacromial diameter) of populations studied by CISP—Karaites, Dauada and Berbers from Giado—confirmed Gini's theory on the negative effects of endogamy. Nora Federici wrote:

> All three populations examined behaved—notwithstanding the racial and environmental differences—in an analogous manner as regards development, demonstrating a visible slowing down in the development of all the considered characteristics compared to other populations that were not in a state of demographic isolation. These results therefore confirm the hypothesis that the regime of endogamy would have a detrimental influence on the corporeal development of the individual.[223]

Not surprisingly, the proceedings of the Third Congress of SIGE were published by *Genus*, the organ of CISP directed by Gini with the funds of the Italian National Research Council (CNR).

In the late 1930s, CISP's ethnological investigations represented the most relevant scientific contribution of Gini's "regenerative" eugenics. From 1928 to 1931, CISP had two principal initiatives: the demographic and anthropological inquiry on large families, and the collection of the archival sources of Italian demographic history, successively published in a monumental work of eleven volumes.[224]

---

[221] Boldrini, "La fertilità degli individui deficienti e difettosi," 307.

[222] On Nora Federici, see: Treves, *Le nascite e la politica*, 338–43; 459–65.

[223] Nora Federici, "La curva di sviluppo individuale presso alcune popolazioni isolate," *Genus* 3, no. 3–4 (June 1939): 343.

[224] CISP-Commissione di demografia storica, *Fonti archivistiche per lo studio dei problemi della popolazione fino al 1848* (11 vols. Rome: Tip. Luigi Proja, 1933–1941).

Anthropological and sociological research, financed and published by CISP, appeared massively influenced by Gini's cyclical theory of nations, focusing on particular issues dear to Gini, such as the mechanisms of social exchange or the different forces of expansion of various populations and social classes.[225]

Between 1933 and 1938, CISP organized ten expeditions, personally directed by Gini, which played a central role in Italian "Latin" eugenics: seven of these regarded populations considered "primitive" (the Dauada of Tripolitania, the Samaritans of Palestine, the Mexican populations, the Karaites of Poland and Lithuania, the Bantu of South Africa and the Berbers from Giado); the other three concentrated on the Italian "ethnic islands" (the Albanians in Calabria, the Ligurians in Carloforte and Calasetta in Sardinia).[226]

In every presentation of CISP's activities to the international scientific community, but above all in 1928 at the International Institute of Statistics in Brussels,[227] and in 1934 in Cleveland at the Hanna Lecture Foundation,[228] Gini explicitly linked the demographic and anthropological inquiries on primitive populations to the empirical testing of several aspects of the cyclical theory of nations, which were deeply connected with "regenerative" eugenics: in particular, the "revival" effect of crossbreeding and the dysgenic effect of demographic isolation.

"Primitive" populations represented, in Gini's view, the only anthropological source for a diachronic analysis of the different phases of the evolution of populations, almost a sort of snapshot that could restore the precise image of the mechanisms and causes of two demographic phases otherwise

225 See Carlo Valenziani, *Il problema demografico dell'Africa equatoriale* (Rome: Tip. C. Colombo, 1929); Paola Maria Arcari, *Le lingue nazionali della Confederazione Elvetica ed i loro spostamenti attraverso il tempo* (Rome: Tip. C. Colombo, 1930); Enrico Haskel Sonnabend, *L'espansione degli Slavi* (Rome: Failli, 1931); Reuben Kaznelson, *L'immigrazione degli Ebrei in Palestina nei tempi moderni* (Rome: Failli, 1931); Cipriani, *Considerazioni sopra il passato e l'avvenire delle popolazioni africane*; Dino Camavitto, *La decadenza delle popolazioni messicane al tempo della Conquista* (Rome: Failli, 1935); Enrico Haskel Sonnabend, *Il fattore demografico nell'organizzazione sociale dei Bantu* (Rome: Arti Grafiche Zamperini e Lorenzini, 1935); Radhakamal Mukerjee, *Le migrazioni asiatiche* (Rome: CISP, 1936); Wilton Marion Krogman, *L'antropologia fisica degli Indiani Seminole dell'Oklahoma* (Rome: Failli, 1936); Giuseppe Genna, *I Samaritani – 1. Antropologia* (Rome: CISP, 1938).

226 For a comprehensive synthesis, see Corrado Gini and Nora Federici, *Appunti sulle spedizioni scientifiche del Comitato Italiano per lo studio dei problemi della popolazione (febbraio 1933 – aprile 1940)* (Rome: Tip. Operaia Roma, 1943).

227 Corrado Gini, "Le Comité Italien pour l'étude des problèmes de la population," *Bulletin de l'Institut International de Statistique* 23, no. 1 (1928).

228 Corrado Gini, "Researches on Population," *Scientia* 55, no. 265 (May 1934): 357–73.

difficult to investigate, that is, the birth and death of the nation-organism. In 1928, Gini declared:

> One of the essential aims of the Committee is to gather the broadest data possible on these primitive and decadent populations, and to especially study the modality and, if possible, the cause of the decadence and gradual disappearance of certain races, and in the same way, the formation and blooming of new races, on which our ignorance is almost total.[229]

Regarding crossbreeding, CISP's scientific missions seemed to completely confirm Gini's theories: while demographic isolation and endogamy favored the senescence and decadence of a population, mixing produced a "revival" of nations.

During the 1940s, with reference to CISP's scientific missions, Gini developed a particular interpretation of "primitiveness" from the point of view of "regenerative" eugenics and the cyclical theory of nations.[230] For Gini, absence of culture, poverty, and "stationariness" were necessary, but not sufficient, characteristics for the definition of "primitive." A principle characteristic of "primitives" was technological backwardness, which in its turn impeded "those forms of culture and richness of a cumulative nature that are the essential causes of social progress."[231] But if, from a technological point of view, "primitives" were in an "infantile phase," from a biological and social point of view "primitiveness" was, for Gini, synonymous with "decadence" and "senescence":

> From a point of view of etiquette, of customs, social institutions, they are crystallized populations. Crystallized and often decadent. Lacking the capacity to progress, they are endowed with limited faculties of recovery: placed in difficult conditions, their social organization crumbles.[232]

Biologically, primitive populations were for the most part "worn, senescent, characterized by little variability, and therefore little adaptability, sometimes by degenerative characteristics, generally by limited and often insufficient repro-

---

229 Gini, "Le Comité Italien pour l'étude des problèmes de la population," 205.

230 Corrado Gini, "Le rilevazioni statistiche fra le popolazioni primitive," *Supplemento statistico ai Nuovi problemi di politica, storia ed economia* 3, no. 1–2 (1937); Corrado Gini, "I 'tradimenti' dei primitivi," *Genus* 5, no. 1–2 (1941); Corrado Gini, *Le rilevazioni statistiche fra le popolazioni primitive* (Rome: Manuali Universitari - Facoltà di Scienze statistiche, demografiche ed attuariali, 1940); Corrado Gini, "Caratteristiche e cause della primitività," *Genus* 5, no. 3–4 (1942).

231 Gini, *Le rilevazioni statistiche fra le popolazioni primitive*, 213.

232 Gini, *Le rilevazioni statistiche fra le popolazioni primitive*, 215.

ductive elements that make their demographic equilibrium unstable, or even determine their numerical decline."[233] In his analysis of the causes of primitiveness, Gini distinguished among "racial," "environmental" and "evolutionary" factors. As for the first, he did not deny the "low intellectual level," lack of inventiveness and some "physical deficiencies" in primitive populations, but was not disposed to generalize and claim that they were innate. Nevertheless, in a passage dedicated to "psychical deficiencies," Gini's discourse concluded with a justification of anti-Semitism as an "understandable reaction":

> There are some populations in which individuals spend the major part of their energies in emulative acts, which neutralize each other [...].
> If individuals of such populations were transplanted to other populations not habituated to emulative acts, they would make their fortune at the other's expense, *even if, in the long run, it causes understandable reactions.* This is the case for the Armenians and the Jews.[234]

If "racial qualities" did not appear to be necessary and sufficient conditions for "primitiveness," neither did environmental factors seem to exercise a determining influence. Even when transported "to the environment of civilized populations," the primitives did not lose their characteristics, and Gini cited, as a sort of apparent exception to the rules, the case of the "Negroes of America":

> There is—it is true—the example of the Negroes of America, who, introduced into Caucasian civilization several centuries ago, maintain evident characteristics that are inferior compared to the Whites.
> While that is undeniable, we must however recognize that the Negroes of America have made great strides on the path of civilization, so that it is difficult today to classify them as primitive.[235]

But the "civilization" of the African Americans was slow enough—Gini continued—to believe that the change was due not so much to environment as to the "progressive infusion of white blood and the progressive selection of individuals who had it."[236] Consequently, the "Negroes that emerge" were, in

233 Gini, *Le rilevazioni statistiche fra le popolazioni primitive*, 215–16.
234 Gini, *Le rilevazioni statistiche fra le popolazioni primitive*, 221; italics added.
235 Gini, *Le rilevazioni statistiche fra le popolazioni primitive*, 221.
236 Gini, *Le rilevazioni statistiche fra le popolazioni primitive*, 226.

reality "not true Negroes, but hybrids." Regarding individual environmental factors, not isolation, nor monetary exchange, not the scarcity of resources, nor even a temperate climate could help to clearly identify "primitiveness." Instead, it was the "evolutionary" factors, defined in the cyclical theory of nations, which furnished a "plausible explanation." As a result

> the more primitive populations are studied, the more we are persuaded that not only do they present an arrested development, but that very often they also present a qualitative and quantitative regression. [...] Primitive populations are, in the majority of cases, decadent populations, populations in the course of involution, senescent populations.[237]

The "primitives" were characterized by a substantial "physiological arrest": the "arrest of development that naturally waits for every living organism, individual or collective." Primitive populations were the forebears of "civilized" ones: they still survived and from them, thanks to the revitalizing power of crossings, some new, vigorous scion might arise.

So while the "primitives" therefore represented, in Gini's "regenerative" eugenics, the decadent and senile side of humanity, it was the hybrids—whether Bantu, inhabitants of Brazilian Cearà or the Black Americans—who would paradoxically announce, in the socio-biological transfusion between civilized and primitive, the rise of future populations.

2. Constitutionalism and "Latin" Eugenics: Nicola Pende's Biotypological Institute

The second pillar of Italian eugenics, between the 1920s and 1940s, was medical constitutionalism.[238] The Italian constitutional school had been founded, at the end of the nineteenth century, by Achille De Giovanni and Giacinto Viola. Italian constitutionalism was a neo-Hippocratic and holis-

[237] Gini, *Le rilevazioni statistiche fra le popolazioni primitive*, 240.

[238] On biotypology and constitutional medicine, see: Cristopher Lawrence and George Weisz, eds., *Greater than the Parts: Holism in Biomedicine, 1920–1950* (New York: Oxford University Press, 1998); see also J. Andrew Mendelsohn, "Medicine and the Making of Bodily Inequality in Twentieth-Century Europe," in Jean-Paul Gaudillière and Ilana Löwy, eds., *Heredity and Infection. The History of Disease Trasmission* (London and New York: Routledge, 2001), 21–80. On biotypology in United States, see: Sarah W. Tracy, "George Draper and American Constitutional Medicine, 1916–1946: Reinventing the Sick Man," *Bulletin of the History of Medicine* 66,

tic medical perspective, which stressed the relevance of "predisposition" in etiology and pathogenesis, shifting attention from causal agents of illness to the body's responses to such agents (the so-called "terrain"). It was based on three general principles: the primacy of the clinic; the individualized conception of illness; and natural treatment, aimed at aiding the body's own reaction to illness.[239]

Nicola Pende, a student of Giacinto Viola, can be considered as the principle exponent of Italian constitutionalism in the fascist period.[240] Born in Noicattaro, a small village near Bari, in 1880, Pende taught pathology and clinical medicine in Bologna, Messina and Cagliari, between 1907 and 1924. From October 1924 to 1925, he was the first chancellor at the Adriatic University of Bari. In 1925 he became the director of the Institute of Clinical Medicine at the University of Genoa. The year before, he had received *honoris causa* membership of the National Fascist Party. In 1933, he was appointed senator.

As regards De Giovanni's and Viola's constitutionalism, Nicola Pende introduced two important innovations. The first was the combination between medical constitutionalism and endocrinology. According to Pende, "constitutional hormonology" was based "on studies of the relationship between the endocrinal-vegetative system and biotypical aspects (morphological, humoral-functional, affective-volitive, intellectual)." In this framework, internal secretions became the "real fibers of the soul," that is, the fundamental connections between morphology and psychology. Pende's biotypological methods researched the "neuro-humoral" parameters (neuro-vegetative equilibrium, hormonal configuration) in order to define the relationships between corporeal and psychical nature,

no. 1 (1992): 53–89; and Heather Munro Prescott, "I was a Teenage Dwarf: The Social Construction of 'Normal' Adolescent Growth and Development in United States," in Alexandra Minna Stern and Howard Markel, eds., *Formative Years: Children's Health in the United States, 1880–2000* (Ann Arbor: University of Michigan Press, 2002), 153–82. On biotypology in Germany, see Michael Hau, *The Cult of Health and Beauty in Germany: A Social History, 1890–1930* (Chicago: University of Chicago Press, 2003). On biotypology in Latin America, see Yolanda Eraso, "Biotypology, endocrinology, and sterilization: the practice of eugenics in the treatment of Argentinian women during the 1930s," *Bulletin of the History of Medicine* 81, no. 4 (2007): 793–822.

239 On constitutional medicine in Italy, see: Giorgio Cosmacini, "Medicina, ideologie, filosofie nel pensiero dei clinici tra Ottocento e Novecento," in Corrado Vivanti, ed., *Storia d'Italia. Annali*, vol. 4, *Intellettuali e potere* (Turin: Einaudi, 1981), 1159–94; Giorgio Cosmacini, "Scienza e ideologia nella medicina del Novecento: dalla scienza egemone alla scienza ancillare," in Franco Della Peruta, ed., *Storia d'Italia. Annali*, vol. 7, *Malattia e medicina* (Turin: Einaudi, 1984), 1223–67.

240 See, in particular, Mantovani, *Rigenerare la società*, 225–33.

or, in other words, between, on one side, the anamnestic and biometric-descriptive level, and, on the other, the psycho-sociological and psychometric level.

Through endocrinology, Pende could provide an "integral biotypological profile" of the individual—the so-called *biotype*—geometrically defined as a quadrangular pyramid, the base of which represented individual, familial and racial inheritance, and the four sides of which indicated the different aspects of life: morphological individuality, physiological individuality, ethical and affective-volitive individuality, and intellectual individuality.

In Pende's theory, the individual was described as a "corporeal factory," whose structural-dynamic features were defined by four orders of factors: *hereditary* or *conceptional* factors, divided into racial factors and individual hereditary factors; *post-conception* conditional-environmental factors, which acted during the entire period of formation of the being and in the fulfillment of the hereditary program; *humoral* factors, both those that generated energy (nutritional material) and those that regulated the process of development of energy; and finally, the dominant *neuro-psychical* factors, that is, the nervous center of the life of relations and vegetative life, and psychical energy.

Pende's second twist to Italian medical constitutionalism—the interconnection between biotypology and politics—was based on his total scientific explanation of individual behavior. Since the hormones of the endocrine gland "influence the constitution and the harmonic form of the body" and were also "essential parts of the constitution and the form of the soul,"[241] it logically followed that the guiding principles of politics should be identified in biology. In 1921, Pende outlined an organicist theory of society, in which the "constitution of the State" was based on the collaboration between "the organs and the classes destined by nature to functions of vegetative life, that is, the production and distribution of common *pabulum* (nourishment) in all social activity," and "the classes destined by nature to functions of the life of relations, that is to coordinate the relationships between all the elements and the collective relationships with the external environment."[242] The "chain" that coordinated and unified the "nutritive cir-

[241] Nicola Pende, *Dalla medicina alla sociologia* (Palermo: Prometeo, 1921), 7.
[242] Pende, *Dalla medicina alla sociologia,* 72.

cle" and the "intellectual circle" of the social organism corresponded to that "neuro-hormonal chain that holds all the elements of the cellular state of the individual together." According to Pende, this chain came from the alliance between "intellectual aristocracy" and the "humble classes of manual workers." He continued:

> Such a chain [...] must be both double and single at the same time: on one hand, the influence of connection and control of individual activities, exercised by intelligence, that is, by an intellectual aristocracy; on the other hand, the influence of connection and control of the individualistic and egoistic tendencies [...] exercised by the real hormones of society, that is, by the social elements most evolved in the moral sense, more able to act as moral and altruistic restraints [...]. And since the great, inexhaustible mine of sentiment is the humble classes of manual workers—from whom the greatest moral genius, Christ, was born—the moral representatives, so to speak, of the government of the State, will rise, we hope, from this social class.[243]

In this 1921 essay, Pende's solution to the struggle of the classes lay in the alliance between the "aristocracy of the mind" and the "aristocracy of the heart," which had to prepare the way for the birth of "a future superior humanity."[244] With the advent of fascism, Pende's human biotypology soon took on the role of biological justification for totalitarian control of psychophysical individuality.

The Orthogenetic Biotypological Institute was inaugurated in Genoa in December 1926. In 1935, with the direct involvement of Mussolini, Pende was named director of the Institute of Medical Pathology and Clinical Methodology at the University of Rome, and in January 1936, the Biotypological Institute was also transferred to the capital. The Institute had organizational links with the Ministry for Public Instruction, the ONB (*Opera Nazionale Balilla*, the Fascist Party's youth group) and ONMI. In fact, the Institute carried out periodic examinations of the ONB members and students, acting as a diagnostic filter for youth destined to enter "differential classes," and was concerned with psychotechnics and professional orientation.[245]

243 Pende, *Dalla medicina alla sociologia*, 74.

244 Pende, *Dalla medicina alla sociologia*, 74–75.

245 Nicola Pende, *Anomalie della crescenza fisica e psichica* (Bologna: Cappelli, 1929), 2, 281–84.

Both in Genoa and Rome, the internal structure of the Institute was made up of different sections. The first room was dedicated to the anthropometric study of human morphology: here, the patients were photographed naked, with the photos placed in an archive, described as the richest in the world "as regarded anomalies of growth and constitution, endocrinopathic syndromes, etc."[246] Following this, the patients were weighed on precise scales, and measured using Viola's anthropometry, Pizzolni's craniometry, Thooris' body mass measurement, and Pende's "growth table." Finally, the morphological exam was completed with an evaluation of the level of development of the "five fundamental apparatuses": the muscular and ligamentary system, respiratory apparatus, hemopoietic apparatus and the sexual apparatus.[247] The second section was the "dynamic-humoral" section, which aimed at identifying the "individual somatic temperament." This section carried out the measurement of the basal metabolism, the "neuromuscular quality" (force, speed, resistance to fatigue, ability) and the "neuroendocrinic and electrolytic profile."[248] Psychology characterized the third section, where patients underwent a series of tests (Sante De Sanctis, Binet–Simon, Terman, Banissoni) to evaluate intelligence, memory, character and imagination.[249] The fourth section concerned psychotechnics and presented a series of analogical tests that reproduced work situations of different professional categories: drivers, construction workers, mechanics, mill workers. The psychotechnics section provided aptitude tests (proportional sense, combinatorial capacity, activity and motor force, motor skills) and examinations of the organs of sense and sensitivity (sight, hearing, touch, baric and muscular sense, sensitivity to heat and pain).[250]

All the information on heredity, morphology, psychology, and behavior of the subject was collected into a "biotypological card," a sort of "personality card," "the revelation [...] of the special type of human factory and special type of performance of the human psychical-physical motor,

246 Sellina Gualco and Antonio Nardi, *L'Istituto Biotipologico Ortogenetico di Roma* (Rome: Stab. Tip. Luigi Proja, 1941), 25. See also Pende, *Anomalie della crescenza fisica e psichica.*

247 Gualco and Nardi, *L'Istituto Biotipologico Ortogenetico di Roma*, 26–34.

248 Gualco and Nardi, *L'Istituto Biotipologico Ortogenetico di Roma*, 35–52.

249 Gualco and Nardi, *L'Istituto Biotipologico Ortogenetico di Roma*, 52–106.

250 Gualco and Nardi, *L'Istituto Biotipologico Ortogenetico di Roma*, 106–44.

which every individual represents."[251] It was an extremely complex classification that was difficult to apply on a large scale, but Pende recommended it to the fascist regime as a tool for the biological classification of the population. In 1934, a circular from the ONB instituted, for its millions of members, a simplified biological card with only four pages. But in Pende's hopes his biotypological card would substitute the citizens' and soldiers' "health passbook" (*libretto sanitario*), which was obligatorily introduced into schools in 1936.

Pende's biotypological card, moreover, was conceived to record and monitor the biopsychical state of the population, as well as to identify the symptoms of deviance within individuals, in order to correct them. This correction—the so-called *orthogenesis*—consisted of "opotherapy and organotherapy, stimulation and inhibition of internal secretion glands through the use of x-rays or phototherapy or special climates and nutrition; psychotherapy, special orthophrenic educational methods, and methods of correction of precocious amoral youth on biological bases, etc."[252]

The biotypological card, moreover, with its "complete diagnosis of the normal and post-illness or pre-illness psychophysical personality," was a true "individual document of identification, health and evaluation" of "citizens of the fascist regime," considered as "productive cells harmonically and consensually engaged in the complex cellular whole of Mussolini's State."[253]

In 1933, in the essay *Bonifica umana razionale e biologia politica* (Rational human reclamation and political biology) this organicistic analogy paved the way for a sort of biotypological totalitarianism. Not surprisingly, the essay was dedicated to Mussolini, the leader who "with the sound principles of a political biology weaves a new physical, moral and intellectual outlook, for a new, grand Nation." As single cells obeyed the fundamental laws of "cellular altruism," so—Pende argued—in the fascist state, individual liberty was "conditioned by collective liberty and interests." As in the

[251] Nicola Pende, "La scheda biotipologica individuale nella medicina preventiva e nella politica sociale," in Lucio Silla, ed., *Atti della SIPS. XXVI riunione (Venezia, 12–18 settembre 1937)* (Rome: SIPS 1938), vol. 5, 284–85.

[252] Nicola Pende, *L'indirizzo costituzionalistico nella medicina sociale e nella politica biologica* (Genova: Le Opere e i Giorni, 1926), 5.

[253] Pende, "La scheda biotipologica individuale nella medicina preventiva e nella politica sociale," 283.

human body, where "vital unity" derived from the "compenetration of the systems of organs of vegetative life and the systems of organs of the life of relations," so in the social organism "the two great classes can not evade the iron laws of fusion of the generating forces of prevalently muscular energy and the generating forces of prevalently creative and moral energy."[254] As "energetically differentiated cellular classes" could be distinguished in tissue, so in the national organism social classes coincided with biotypes and corresponded to the "biologically differentiated classes of workers and producers."[255] In this view, the biological system came to represent a sort of natural paradigm for fascist corporativism:

> [The fascist regime is] a truly biological political system, in which the central idea that individual liberty must be controlled, conditioned and limited by two immanent factors is implicit: that of the necessity and material interest and ideals from the corporative State to use the various forms of energetic value of individual citizens; and that no citizen must be able to cause damage, through his free will, to the collective life of the State.[256]

In Pende's biotypological totalitarianism, the deviant was comparable to the "malign cell of a tumor, which is removed for the good of the collective life of the human body, as it menaces its stability and validity."[257] On the contrary, the "biological and moral aristocracy of the nation" would originate from the "breeding ground" of fascist youth, called to carry out, in the social body, that work of "harmonization of the various productive categories of citizens," comparable to the "neurohormonal" mechanism of the "individual human organism."[258]

Starting from this organicistic analogy between the "vital unit" of the individual and that of the state, Pende went on to deepen the bio-political applications of the "science of orthogenesis," elaborating a sort of fascist biomedical architecture, structured on biotypological control.

According to Pende, orthogenetic and biotypological measures had to be systematically applied to the medical and sociological classification of

254 Nicola Pende, *Bonifica umana razionale e biologia politica* (Bologna: Cappelli, 1933), 38.
255 Pende, *Bonifica umana razionale e biologia politica*, 39.
256 Pende, *Bonifica umana razionale e biologia politica*, 40.
257 Pende, *Bonifica umana razionale e biologia politica*, 40.
258 Pende, *Bonifica umana razionale e biologia politica*, 40.

the four principal dimensions of the fascist state: children, women, workers and the race.

As for schools—"true workshops of the social personality of the individual"—"the study and the repeated testing of the individual biotype under formation" constituted, in Pende's view, the indispensable premise of an education that aspired to form "the total and harmonic man, that is, made of muscle, heart and brain, normally and harmonically developed, cultivated and oriented by the educator." Biotypology, above all, led to the "prior knowledge of what the scholar must cultivate," that is, to his "complete personality."[259] Biotypological anamnesis represented the scientific assumption of four biopolitical objectives connected to the scholastic sphere:

1) adapt "physical and moral education and instruction" to the different biopsychological phases of educational development: physical education, moral education, sexual "orthogenetic" education;

2) apply "differential" education to the subjects "who manifest retardation or precocity, defects or excesses, from the somatic and spiritual sides, in respect to the normal mass of companions of the same age";

3) correct and "normalise," with "modern physical, moral and intellectual orthogenetic means, the errors and deviations of normal physical and spiritual development, helping the disabled or mediocre in health, character or intelligence to achieve, as much as possible, the normal mean of the masses";

4) finally, select and orientate, that is, "reject those adolescents not suitable for certain scholastic careers capriciously, involuntarily or erroneously chosen, launching them in careers more suited to their capacities and attitudes, and orientating the normal adolescents, after having ascertained the special attitudes and inclinations and their pre-eminent psychophysical qualities, sending them to institutions adapted to introduction and learning of the type of school, trade or profession for which each appears to be best suited by his nature."[260]

In particular, regarding education, in the first phase of the "development of the body and the spirit," biotypology would evaluate the "instruments

[259] Nicola Pende, *Trattato di biotipologia umana individuale e sociale, con applicazioni alla medicina preventiva, alla clinica, alla politica biologica, alla sociologia* (Milan: Vallardi, 1939), 466–67.

[260] Pende, *Trattato di biotipologia umana individuale e sociale*, 466–67.

of intelligence" (capacity of attention, memory, mental stamina). It would also evaluate the "forms of thought," with a distinction between "tachypsychic" (speedy mentality) and "bradypsychic" (slow and analytical) individuals. In a second phase, that of puberty (from 15 to 18 years), two more were added to these two first "biotypes": the "empirical realists" and the "mixed."[261] Four mental types were therefore outlined, corresponding to as many professional orientations:

> From the first, the intuitive tachypsychics, intelligent artists and artisans and certain quick and able qualified workers, and the professionals of the natural, legal, or experimental sciences are more likely to develop. From the second, the analytical bradypsychics, it is more likely that technical professionals, engineers, constructors, mathematicians, philosophers, magistrates, academics, and certain workers of precision, patience and analysis, will develop. From the last, the empirical realist, business men, men of practical action, men of commerce, industrialists, bankers, agriculturalists, and sailors, will develop.[262]

In the "moral education" field, biotypology could identify the connection between deviant behavior and biological (endocrinal) or environmental causes, and prepare the appropriate therapy: adolescents with "hyperadrenal temperaments" could become aggressive, those with "hyperthyroid-hyperthymus" problems were prone to "lying and small thefts," and so on. Every anomaly had its own biotypological diagnosis and required a "differential" approach:

> We must be warned that the educators of the old mould are accustomed to treating undisciplined, rebellious students, or those of low morals, indistinctly, with the same primitive criteria with which, once upon a time, they beat and tortured the insane instead of curing their illnesses.[263]

The biotypological investigation of "individual moral dispositions" must therefore always be the premise of the "moral orthogenesis" of adoles-

261 Pende, *Trattato di biotipologia umana individuale e sociale,* 470.

262 Pende, *Trattato di biotipologia umana individuale e sociale,* 470. For a project of Pende on the reform of the scholastic system, see Nicola Pende, *La scuola fascista preparatrice dell'uomo totale ed orientatrice del cittadino produttivo (discourse of Senator Pende in the sitting of 25 March 1938)* (Rome: Tip. del Senato, 1938).

263 Pende, *Trattato di biotipologia umana individuale e sociale,* 472.

cents. As for sexual education, only biotypology could identify the endocrinal modality of sexual development and adequately advise the educators. Therefore, "sexual orthogenesis" had to substitute psychology:

> Sexual education must not still be based on moral pedagogy or purely psychological methods, which either achieve nothing or sometimes do ill to future parents: but we must pay heed to sexual orthogenesis, to the necessity that the psychophysical sexual development of adolescents happens normally and is not obstructed by educational inhibitions or moral and religious orders, which do not pay attention to the medical physiological control of the subject, his temperament; in sum, to his special sexual biotype.[264]

After childhood selection, the next focus of biotypological control was the hygienic and moral preparation of future mothers:

> [This is] carried out during their growth, correcting any possible anomalies of sexual development, and fortifying them according to the needs of the individual organisms, so that they later produce numerous and healthy children. The biotypological card will continue to follow married women and mothers to advise and cure them, preventing that infinite series of organic and psychical unbalances which are often linked to the various phases and activities of female sexual life and to the critical period of cessation of ovarian function.[265]

In *Bonifica umana razionale,* Pende developed a program of "education of females on bio-psychological bases,"[266] which was organized on three levels: the body, the character and the intellect. Regarding the first aspect, after having identified the aesthetic ideal for a woman as the "maternal type"—characterized by the development of the lower abdomen and pelvis—Pende theorized a physical education that harmoniously shaped the "lower half" of the body and favored the growth of "female fat":

> In adolescent women, not yet sexually mature, the real beauty of the body can be achieved only by favoring the development of normal sexual proportion

264 Pende, *Trattato di biotipologia umana individuale e sociale,* 473.
265 Pende, "La scheda biotipologica individuale nella medicina preventiva e nella politica sociale," 285.
266 Pende, *Bonifica umana razionale e biologia politica,* 115.

> and therefore promoting, through physical exercise, with suitable nutrition and hygienic practices [...] above all the regulated development of the lower half of the body, and preventing any irrational muscular exercise that arrests that development of the lower half or that exaggerates the largeness and thickness of the neck, thorax, arms and shoulders.[267]

While, regarding character, a woman should be constantly educated to have maternal sentiments toward men, the "intellectual pedagogy" of women had to necessarily promote "realistic and practical" thought more than "abstract." The true female working environment, according to biotypological rules, was nevertheless not represented by the factory or the office, but by primary school teaching, and in particular, by the manual and artistic activities together:

> And above all the so-called professions of the needle that include cutters, seamstresses, lace workers, milliners, doll dressers, and workers with artificial flowers and feathers. Here is the real and narrow field of female work, where women can reign sovereign and be truly in their right place.[268]

Indeed, work constituted the third field of application of Pende's bio-politics. Even in the choice of profession, liberty needed to be "severely controlled and regulated by the intervention of the State."[269] The biotypological approach, in this sense, aimed at a triple objective: understanding the psychophysical aptitudes or "individual productive capacities or deficiencies," in order to "guide every worker to his right place"; ascertaining the "predispositions to illness and constitutional weaknesses that cause accidents and workplace illnesses," in order to halt them through preventive therapy; and finally, to resolving "in the most fair and rational manner the medical-legal questions inherent in workplace illness and industrial accidents."[270] In the scientific organization of work, the constitutional physician therefore

---

267 Pende, *Bonifica umana razionale e biologia politica,* 115. On Pende's role in sports medicine, see Gigliola Gori, *Italian Fascism and the Female Body: Sport, Submissive Women and Strong Mothers* (New York: Routledge, 2004); Lucia Motti and Marilena Rossi Caponeri, eds., *Accademiste a Orvieto: donne ed educazione fisica nell'Italia fascista, 1932–1943* (Orvieto: Quattroemme, 1996).

268 Pende, *Bonifica umana razionale e biologia politica,* 133–34. On this topic, see in particular: Victoria De Grazia, *How Fascism Ruled Women: Italy, 1922–1945* (Berkeley: University of California Press, 1992), 48.

269 Pende, *Bonifica umana razionale e biologia politica,* 142.

270 Pende, *Bonifica umana razionale e biologia politica,* 162.

had to support the hygienist and the industrial engineer. Biotypology was, in fact, the "rational premise of every sane and fertile medical-social act of worker protection":

> Only men whose biotypological qualities are exactly known, and who are rationally oriented toward the office or the work most suitable to their biotype, can fertilize and maximize the productivity of the techniques of modern scientific organization of work. Only men aware of their organic weaknesses, and in time cured and corrected, can easily avoid the assault by infective, toxic and traumatic agents, or of meteorological morbose factors, to which their work exposes them, notwithstanding the efforts of modern hygiene.[271]

In a professional biotypological orientation, the evaluation of "varf" (velocity + ability + resistance + force) assumed a primary importance, because the four human biotypes were differentiated in their combination of the four respective qualities: muscular force, together with resistance to fatigue, was prevalent in the "brevilinear type, toned, muscular and sanguine,"[272] while the "toned longilinear type" had velocity, together with a sufficient level of muscular force. Finally, the "flaccid brevilinear type" and the "atonic and weak longilinear type," as they were not able to develop force or resistance, "could be perfect for work that required ability and ingenuity."[273] According to Pende, the evaluation of the biotype was useful not only in the field of "physiology of work," but also for the foreknowledge of certain predispositions to illnesses and accidents: for example, the "muscular and sanguine brevilinear type" would be exposed to cardiac illnesses, while the opposite longilinear "atonic and asthenic" type would more easily suffer from "tuberculosis of the lungs, pleura, peritoneum, and glands."[274] Consequently,

> we see that the knowledge of individual biotype of the worker permits us to carry out his hygienic protection, that is, for the rational utilization of his work according to the physical and mental qualities prevalent in him, and above all to fortify him, through the means of preventive medicine, in those organs in

271 Pende, *Bonifica umana razionale e biologia politica,* 163.
272 Pende, *Trattato di biotipologia umana individuale e sociale,* 518.
273 Pende, *Bonifica umana razionale e biologia politica,* 173–76.
274 Pende, *Trattato di biotipologia umana individuale e sociale,* 519.

> which he appears weakest and least endowed by nature, and therefore more likely to sicken in the work environment.[275]

In Pende's view, the National Fascist After-work League (*Opera Nazionale Dopolavoro*) had to be utilized for the "constitutional reclamation of workers, founded on biotypological principles": at the end of their working day, workers had not only to be reassured in the spirit, but also "overseen and helped in the fortification and restoration of their body from the latent alterations of organic functionality that fatigue and the work environment could cause."[276]

The final sphere of application of biotypology was represented by eugenics and racial policy. The first aspect of the "political-biological problem of the race" concerned pronatalism or, in Pende's words, the demographic "illness of low birthrate." Not in civilization in general, nor in urbanism, would the causes of the decline of the birthrate be found, but in "occidental Nordic industrialism":

> Industrial civilization has brought with it the elevation of the quality of life, but also a profound modification of customs, adoption of expensive habits, multiplication of costly needs, abuse of consumption and pleasures of every kind, a false comprehension of social wellbeing, an increase in selfishness, and above all the working of women and children and the decline of the concept of family.[277]

In particular, the working of women, "both manual and intellectual," had damaging consequences on the organisms of mother and children, creating "states of organic weakness or early stress of the maternal organism and disturbances of the development and constitution of the tender sprouts, suffering from malnutrition, both intrauterine and post-natal."[278] Additionally, certain professions—above all among city-dwelling female manual workers and office workers—directly exercised a "sterilizing influence." Next to work, the "second scourge" that induced women to limit their number of children and abandon the domestic hearth to satisfy "the craving for massages and sports," was constituted by the diffused conviction that maternity

---

[275] Pende, *Trattato di biotipologia umana individuale e sociale*, 519.
[276] Pende, *Trattato di biotipologia umana individuale e sociale*, 519.
[277] Pende, *Bonifica umana razionale e biologia politica*, 201.
[278] Pende, *Bonifica umana razionale e biologia politica*, 202.

compromised feminine beauty. On the contrary, according to Pende, a biological future of aesthetic deformation and psychical alteration awaited the childless woman:

> The persistent youthfulness of the body and spirit cannot be obtained through the unnatural limitation of fertility, as the poor woman deceives herself, but rather early senescence and flaccidity of the face and integuments, immediate expression of the ovarian insufficiency.[279]

Since, therefore, it was essentially the "modern woman" who had to "prevent the social illness of the declining birthrate that continues to worsen,"[280] the fascist state had to attain the bio-political objective of the preparation of future mothers, not so much through the campaign against urbanism as through an adequate and constant biotypological education:

> It is necessary to manage, with fascist wisdom, the forming of the Italian woman, starting from childhood, with a new educational direction, obligatory in primary and secondary schools. This education will aim to form the housewife and mother type, more than the science and sporting woman, and will give a new sexual education training, that will lastingly instill in the ingenuous and inexpert soul of the young girl the concept of the real meaning of the somatic and psychical attributes of her sex, destined on the whole by nature to the maternal function.[281]

Together with the decline of the birthrate, the second aspect of the bio-political problem of the race concerned, according to Pende, the preservation and the improvement of the "Italian stocks." In *Bonifica umana razionale,* Pende engaged with German biological racism, distinguishing between "physical somatology of the race" and the "psychology or dynamism of the race": "Within a race," he argued "physiologically and psychologically diverse stocks exist, and these are biological-social or historical-biological human groups, and not only ethnic or anthropological."[282] It was possible therefore to speak "realistically" only of the psychology of the stock, and not the psychology of the race.

279 Pende, *Bonifica umana razionale e biologia politica,* 207.
280 Pende, *Bonifica umana razionale e biologia politica,* 209.
281 Pende, *Bonifica umana razionale e biologia politica,* 210.
282 Pende, *Bonifica umana razionale e biologia politica,* 215–16.

To identify, in particular, the stock to which the "Romans owed their greatness," Pende presented the results of his "ethnic biotypological" survey, which he carried out himself in the Institute of Biotypology in Genoa, in collaboration with his assistants Vidoni, Gualco, Tamburri and Landogna-Cassone. In this investigation, in the cities of Sabina and Ciociaria, the population of ancient Rome appeared,

> hypervegetative and vigorous, with rounded and elliptical cranium, almost mesaticephalic, and a long, robust face, caustic, satirical spirit, cutting to the point of aggressiveness, sometimes bloody, with roughness and frankness of manner and language, impassive and unemotional toward events and phenomena of the ideal or abstract order.[283]

On the Tyrrhenian side of Lazio, in Abruzzo and Sannio, the stock of Campania Felix predominated, in which "a playful spirit, sentimentalism, aestheticism and idealism, serenity and religious mysticism perennially live."[284] Near the lower Adriatic, in Apulia, and partly in Lucania and Calabria, it was possible to trace the Iapygian-Messapian or Apulian stock, similar to the Calabrian-Sicilian. In Tuscany and Umbria it was still possible to find the "inexhaustible artistic-literary scientific sense" of the ancient Etruscans, while in Lunigiana, Garfagnana and Lucchesia, up until Liguria, there was evidence of the Atlantic-Mediterranean branch, visible in the Ligurian stock, with its "tall, dark and strong [men], with their mesaticephalic or subdolicephalic heads."[285] In Northern Italy, three "great psychological types of stock" could be identified, corresponding to the three great families of protohistoric populations that had invaded Italy: proto-Celts, proto-Umbrians and proto-Illyrians. The Piedmontese type stood out

> for his rather rough temperament, [...] for his attachment to his soil and his homeland, his tenacity and will, his military spirit, his disciplined respect to political and religious authority, the rather melancholic tone of the soul, not disconnected however, from simple serenity and festivity, [...] the type of realistic intelligence with little tendency to flights of fancy such as abstract thought.[286]

283 Pende, *Bonifica umana razionale e biologia politica*, 218.
284 Pende, *Bonifica umana razionale e biologia politica*, 218.
285 Pende, *Bonifica umana razionale e biologia politica*, 220.
286 Pende, *Bonifica umana razionale e biologia politica*, 222–23.

In the Lombard-Emilian type emerged

> gaiety and sociableness and innate joy of living, not disconnected from a certain unrest and mobility of the soul, a great industriousness, and above all a concrete mentality, at the same time associated with an exquisite aesthetic sensibility, artistic attitudes and an analytical type of intelligence.[287]

Finally, the Venetian type was characterized by "indomitable and bellicose [...] sentiment, exaggerated by their honor, value [...] and frankness."[288]

In Pende's view, race was the result of crossbreeding between different stocks. Therefore, the "Latin" race was not exclusively represented by the Romans, but by the "fusion of all the Italian stocks, and above all the stocks of the Mediterranean race, which Rome was able to harmonize and meld with its great realistic and political sense."[289] Following the example of ancient Rome, fascist racism had to pursue the objective of "juridical harmonization" of the Italian stocks.

On this basis, in 1933, Pende directly and explicitly criticized German *völkisch* and biological racism (in particular, Rosenberg's and Günther's theories):

> Once again we find men of high intelligence ignoring what our Chief does not ignore; and that is that a German race does not exist, and that the German population, like all the populations of the Earth organized into nations, are composed of many distinct biological races, who have lived side by side for millennia and collaborated for the economic and cultural progress of their State. Once again, we fascists, with our stance on political problems of race, demonstrate the realistic Mediterranean balance in the face of Nordic abstractness and mysticism.[290]

A racial policy such as the Nazi one, founded on "political prejudices, religious sentiment or a sectarian spirit" and not "on scientific, objective and realistic logic"—Pende argued— could only lead to "comic and illogical consequences":

287 Pende, *Bonifica umana razionale e biologia politica*, 222.
288 Pende, *Bonifica umana razionale e biologia politica*, 222.
289 Pende, *Bonifica umana razionale e biologia politica*, 225.
290 Pende, *Bonifica umana razionale e biologia politica*, 227.

> Must they be distanced from cohabitation and crossing with other non-Israelite dolichocephalic blonde Germans? And why must the dark, low brachycephalic Israelites, who are of the same blood as the German citizens of the alpine races, be excluded from crossings and political cohabitation with these other brothers of the race?[291]

Since, racial crossings notwithstanding, original stocks remained "always fixed,"[292] an effective racial policy had to value the "ethnic polyvalency of a single nation," starting from the perspective of biotypology and orthogenesis:

> Fascist Italy, instead of running behind the North-American, German and Scandinavian utopia of pure race, instead of aiming at homogenization and uniformity of the various stocks like the Soviet Republic, must jealously maintain intact this variety and ethnic polyvalency, which has been and will be the principal source of its renewed vitality and resurgent greatness.[293]

Concretely, according to Pende, the first step was to deepen the knowledge of the "ethnic balance of the Italian State,"[294] that is, of the "differential energetic values, in the somatic, moral, and intellectual fields, which most characterize the single ethnic stocks of the nation." Only starting from these premises would it be possible to develop a "State anthropotechnique,"[295] "differential for the various types of Italian people" and based on medical constitutionalism:

> Constitutionalism and hygiene, individual pedagogy, and bio-politics, strictly intertwined in this work of rational human breeding, will form the various selected types of the Italian of tomorrow. These new types will increasingly improve the mechanism of the corporative State, and they will move ever closer to that which we believe is the ideal of a perfectly organized human society [...], that is, one in which the unitary state results not from the social classes but from biologically selected classes of citizens.[296]

291 Pende, *Bonifica umana razionale e biologia politica*, 230.
292 Pende, *Bonifica umana razionale e biologia politica*, 231.
293 Pende, *Bonifica umana razionale e biologia politica*, 238.
294 Pende, *Bonifica umana razionale e biologia politica*, 232.
295 Pende, *Bonifica umana razionale e biologia politica*, 238.
296 Pende, *Bonifica umana razionale e biologia politica*, 239.

According to Pende, a first experiment in this direction could come from internal colonization and, above all, from the "reclamation of the stocks," which was being achieved in the swamps of the Agro Pontino:

> In this way, the internal colonies of Fascism will become, bit by bit, the true human breeding grounds of the nation, true centers of regeneration of the purest and most innate qualities of our ancient stocks [...]. And such breeding grounds, which today are humble, will perhaps create tomorrow the artistic, literary and political geniuses, and at any rate, truly aware citizens, because they are being raised in schools of work and sacrifice, to laboriously conquer, and not exploit, the earth that feeds them. And so from such breeding grounds, the nation will obtain new pure blood for its needs in peace and in war.[297]

It was on the opposition between "Latin" and "Nordic" that Pende based the cultural strategy with which, during the 1930s, he promoted the international diffusion of Italian biotypology. In particular, France and Argentina constituted the international network of Pende's "Latin science." In France, the Italian endocrinologist had contacts in the fields of Christian medical neo-humanism, homeopathy, neo-Hippocratism and cosmobiology: in particular, at the Paris Faculty of Medicine, Maurice Loeper, professor of therapeutics, and Maxime Laignel-Lavastine, psychiatrist and professor of history of medicine; Marcel Martiny, physician at the Leopold-Bellan Hospital in Paris; Georges Jeanneney, professor at Bordeaux Faculty of Medicine; and Maurice Faure, president of the Nice Society of Medicine and Climatology.[298] In 1934, in a conference at the Nice Mediterranean Academy, Pende praised the "Latin-Mediterranean spiritual unity," highlighting the physical robustness and fertility of the three "brunette" races (Mediterranean, Adriatic, Alpine) against the "civilization of machines and economic individualism" incarnated in the two "blond" races (Germanic and East Baltic).[299] Not surprisingly, Alexis Carrel, in *Man, the Unknown,*

297 Pende, *Bonifica umana razionale e biologia politica*, 241–42. On the projects of "State anthropology," linked to the zone of the Pontino swamps, and in particular the city of Littoria, see Sergio Sergi, "Antropologia di Stato. L'archivio comunale delle famiglie," *Razza e Civiltà* 1, no. 2 (April 1940): 183–89.

298 See the documentation in ACS, MPI, DGIS, Professori Universitari Epurati, b. 26, f. Pende.

299 Nicola Pende, *Biologia delle razze ed unità spirituale mediterranea*, in ACS, SPD, CO, b. 1005, f. 509057/509059.

cited the Biotypological Institute as a model,[300] and in July 1936, in a letter to Pende, underlined the importance of the defense of "Latin civilization":

> Nowadays the torch of Latin civilization has passed in Italy's hands. The Latins who live in the other nations of Europe and America put their trust in Italy. Thus, it is a happy circumstance that you in Rome will study one of the most important subjects for the future of mankind.[301]

As for Argentina on the other hand, in 1930, Pende held an important series of conferences, upon the invitation of Mariano Castex, professor of clinical medicine at the University of Buenos Aires. In the same year, the Argentinian President, General Uriburu officially sent the physicians Arturo Rossi and Octavio Lopez on an assignment to study Italian eugenic policies. Upon their return to Argentina in 1932, the *Asociación Argentina de Biotipología, Eugenesia y Medicina Social* was created, directed by Rossi. In 1933, the Council on Education and the Schools Department for the Province of Buenos Aires adopted, at Rossi's initiative, the school biotypological identity card.[302] On the basis of these international relations, Pende, in 1936, presented Mussolini with a project for a "Mussolinian University of High Latin and Mediterranean Culture in Rome," which would be a true "breeding ground for future creators of thoughts" for the Latin world.[303]

Pende's biotypology-based eugenics, like his racial theory, was very critical toward the "Nordic" model. Already in 1933, restating his doubts on compulsory premarital examinations, Pende proposed "orthogenesis" instead of negative eugenics:

> There is only the constant penetrating work of the physician, supported, as it is today in Italy thanks to the fascist State, by admirable laws of individual preventive hygiene, to create a somatic and psychical reclamation of the individual from infancy until the age of marriage; there is only the moral obligation

300 Alexis Carrel, *Man the Unknown* (New York: Harper & Brothers, 1935), 288. On Carrel's eugenics, see Andrés Horacio Reggiani, *God's Eugenicist. Alexis Carrel and the Sociobiology of Decline* (New York: Berghahn Books, 2007)

301 Carrel to Pende, July, 9, 1936, ACS, MPI, Professori Universitari Epurati, b. 26, f. Pende.

302 Nancy Leys Stepan, *"The Hour of Eugenics,"* 119. See also Andrés H. Reggiani, "La ecología institucional de la eugenesia: repensando las relaciones entre biomedicina y política en la Argentina de entreguerras," in Miranda and Vallejo, eds., *Darwinismo social y eugenesia en el mundo latino*, 273–309; Gustavo Vallejo, "Males y remedios de la ciudad moderna: perspectivas ambientales de la eugenesia argentina de entreguerras," *Asclepio* 59, no. 1 (January–June 2007): 203–38.

303 Pende to Osvaldo Sebastiani, July 14, 1936, ACS, SPD, CO 1922–43, b. 1005, f. 509057/509059

> on the part of parents to ascertain, commencing some time before marriage, the state of the future procreators. [...] Hygienic propaganda will be intensified by the registry office for the families that request a marriage [...]. Such propaganda, helped by the appropriate laws and State institutions of preventive medicine, such as the State biotypological-orthogenetic institutions, is the most rational and effective that the medical science and juridical conscience of a civil nation can provide.[304]

In 1938, at the annual reunion of SIPS, Pende criticized German negative eugenics, with its pretension to "liberate the race forever from those sorts of transmittable pests represented by hereditary illness."[305] Pende had two objections on this point: first, the major part of "subjects dangerous to the race are [...] carriers of latent defects that are apparently healthy and would therefore escape coercive anti-conceptional eugenics"; second, as German psychologist Walter Jaensch had also maintained, "the environment is more decisive than genetic factors, when we speak of superior strata of our psychical personalities, [that is] the most fleeting and the most recently acquired."[306]

In opposition to "Nordic, anti-conceptional selective eugenics," Pende proposed, in the first place, "familial or matrimonial eugenics," and in the second place, "post-conceptional orthogenesis" and the "constitutional reclamation of the individual." In respect to "matrimonial eugenics," Pende repeated (referring to the theories of Paolo Enriques) the positivity of crossbreeding between Italian ethnic stocks, but did not hesitate to base the racist and anti-Semitic fascist legislation on the principle of race-crossing among Italians (*Italiani con Italiani*):

> All this makes us believe that crossings between human races different not just in color, but also in level and type of mentality and different millennial environmental adaptations, even if they are both European populations, could instead produce degenerate descendents, or at least disharmonized ones, above-all mentally. And so, it seems to me possible to conclude that we Italians must value the principle *Italiani con Italiani*, in order to preserve and fur-

304 Pende, *Bonifica umana razionale e biologia politica*, 246–47.

305 Nicola Pende, "La profilassi delle malattie e anomalie ereditarie," in Lucio Silla, ed., *Atti della SIPS. XXVII riunione (Bologna, 4–11 September1938)* (Rome: SIPS, 1939), vol. 6, 70.

306 Pende, "La profilassi delle malattie e anomalie ereditarie," 70.

> ther improve the pure civilized characteristics of the progeny of Rome and the different ethnic components that in one sense or another have made a contribution of indisputable value to our supremacy.[307]

As for "post-conceptional orthogenesis" or "environmental eugenics," Pende stressed the relevance of the biotypological "natural" therapies, like sunshine, mountain air and mineral waters:

> Having refused, from both a practical and ethical point of view, prohibitive racist eugenics [...], we will give the preventive orthogenetic naturist and educative eugenic direction an increasingly greater value in achieving the glorification and continuity of the biological patrimony of the nation. We are aware that human reproduction can not be treated with the same means used in the selective breeding of beasts, and that the evolution of the body and above all the spirit of man obeys the physical chemistry of the genes only to a certain point, which is a part, but not all of the emerging evolutionary creator of man.[308]

Initially boycotted by Mussolini and the Ministry of Popular Culture, because of its distant position from the "Manifesto of the racial scientists" (July 1938), Pende's spiritualistic and biotypological interpretation of eugenics and racial policy emerged victoriously from the academic-scientific dispute for the management of fascist state racism, assuming an official character, above all in the period between 1939 and 1941.[309] In 1939, in the introduction to his essay *La scienza dell'ortogenesi* [The science of orthogenesis], Pende proposed a complete break between orthogenesis and "infamous," "Nordic" eugenics:

> Orthogenesis means regular, healthy and harmonious formation of men.
> What it must not be confused with is the infamous eugenics of certain eugenicists who believe that the race can be improved or purified by grafting the blood of individuals of distant or primitive races onto the trunk of decadent populations, or surgically sterilizing individuals of both sexes who have hereditarily transmittable illnesses.

307 Pende, "La profilassi delle malattie e anomalie ereditarie," 71. See also Nicola Pende, *Concetto e prassi della razza nella mentalità fascista* (discourse at the Cremona section of the Institute of Fascist Culture, 15 October 1938) (Cremona: Tip. Cremona Nuova, Cremona, n. d.)

308 Pende, "La profilassi delle malattie ed anomalie ereditarie," 11. See also Nicola Pende, "La scienza dell'ortogenesi. Principi e finalità," *La ricerca scientifica* 10, no. 4 (April 1939): 1–6, offprint.

309 See Maiocchi, *Scienza italiana e razzismo fascista*, 237–41.

> We propose—and here we find all the moral, scientific and social value of the Italian science of orthogenesis—instead of this utopia of creating better descendents through crossings with distant races or of selecting the fittest generators and excluding the unfit for the improvement of the race, the practice of putting the human being under scientific control from the moment of conception, from the beginnings of intrauterine life [...]; then, after this first post-conception and prenatal orthogenetic work, based on the hygiene of the gestating mother, we proceed with the protection and correction of development from the first days of birth, that is, the realization of post-natal orthogenesis.[310]

In 1940, Mussolini named Pende Chancellor of the Academy of Italian Youth of Littorio (GIL, *Gioventù Italiana del Littorio*). In 1938 the project for the Central Institute for Human Reclamation, Orthogenesis and Naturist Therapy (*Istituto Centrale di Bonifica Umana, di Ortogenesi e di Terapia Naturista*), desired by Pende from 1934 and financed by the *Pio Istituto di S. Spirito* and the *Ospedali Riuniti* of Rome, was approved by Mussolini, as part of the Universal Exposition E42. The architectonic profile of the model—a stronghold with four towers—symbolized the main pillars on which Pende's human reclamation was founded: children, women, workers, and race.[311] Within the stronghold, there was a green area for walks, and a naturalistic park of two to three hectares. In April 1939, Mussolini participated in the placing of the cornerstone. The works were carried out until 1943, and were then continued after the war.

Between December 1942 and May 1943 the Jesuits praised the "originality" and "ingeniousness" of Pende's biotypology, dedicating several articles in the review *Civiltà Cattolica* [Catholic civilization] to the exposition of the numerous affinities existing between orthogenesis and Catholic doctrine.[312] After the second world war, Pende reciprocated the attention, placing his biotypology, by now orphaned by fascism, at the service of Catholicism.[313]

---

310 Nicola Pende, *La scienza dell'ortogenesi* (Rome: CNR, 1939), 8. See also Nicola Pende, "Il principio biotipologico unitario," *Gerarchia* 11 (November 1940): 569–72.

311 See Maurizio Calvesi, Enrico Guidoni and Simonetta Lux, eds., *E42. Utopia e scenario del regime. 2: Urbanistica, architettura e decorazione* (Venice: Marsilio, 1987), 506ff..; Adolfo Mignemi, "Profilassi sanitaria e politiche sociali del regime per la 'tutela della stirpe'. La 'mise en scène' dell'orgoglio di razza," in Centro Studi F. Jesi, ed., *La menzogna della razza. Documenti e immagini del razzismo e dell'antisemitismo fascista* (Bologna: Grafis Edizioni, 1994), 65–72; Mantovani, *Rigenerare la società*, 330–31.

312 Mario Barbera, *Ortogenesi e Biotipologia* (Rome: La Civiltà Cattolica, 1943) (collected from articles published in *Civiltà Cattolica* from 19 December 1942 to 15 May 1943).

3. Demography and Biotypology: the Laboratory of Statistics at Milan Catholic University

A sort of synthesis between the two branches of fascist "Latin" eugenics—the demographic and the constitutional—seemed to come, in the last half of the 1920s, from the Laboratory of Statistics of the Milan Catholic University, and particularly from the contributions of the director of the Laboratory, the statistician and demographer Marcello Boldrini.[314]

After a first period still influenced by the typical sociobiological approach of the positivist anthropological tradition,[315] Boldrini progressively neared the constitutionalist school, attempting to verify, on a biometrical basis, the validity of the concept of "biotype" as a total explanation of the whole individual dimension: from the biological to the psychical aspects; from the demographic characteristics to the placement in the social stratification. In Boldrini's definition, the demonstration of the explanatory value of "constitutional type" came from the intercorrelation between different disciplinary approaches, summarizable in the following way:

1) Morphological-anthropometrical: this was the classical distinction between brevilinear type and longilinear type, based on the inverse correlation between dimensions in length of the human body and relative somatic mass.[316]

2) Biological-endocrinological: influenced by Pende's biotypology, Boldrini shared the idea of a connection between morphological structure and specific biological property. In particular, the brevilinear types presented, compared to the longilinear types, "a stronger biochemical activity, a higher blood

313 See, among others, Nicola Pende, *Corpo e anima* (Rome: SAET, 1947); Nicola Pende, *Il medico di fronte al Vangelo* (Milan: Il Giorno, 1948); Nicola Pende, *Medicina e sacerdozio alleati per la bonifica morale della società*, (Ancona: Tip. Flamini, n. d.)

314 For a brief profile of Marcello Boldrini and a bibliography, see in particular Giuseppe Locorotondo, "Boldrini, Marcello," in *Dizionario Biografico degli Italiani* (Rome: Istituto della Enciclopedia Italiana, 1988), vol. 34, 465–67. On Boldrini's eugenics, see also Maiocchi, *Scienza italiana e razzismo fascista*, 124–36.

315 See, for example, Marcello Boldrini, "I cadaveri degli sconosciuti. Ricerche demografiche e antropologiche sul materiale della Morgue di Roma," *La Scuola Positiva* 1, no. 7–8 (July–August 1920), 323–47; Marcello Boldrini, "Gli studi statistici sul sesso. Le traviate," *Rassegna di studi sessuali* 1, no. 2 (March–April 1921), 69–81.

316 Marcello Boldrini, "Tipi e attitudini costituzionali e sostituzione delle aristocrazie (XII Congresso dell'Istituto Internazionale di Sociologia, Bruxelles 25–29 August 1935)," in *Contributi del Laboratorio di Statistica. Serie IV* (Milan: Vita e Pensiero, 1936), 5, offprint; Marcello Boldrini, "Costituzione ed eugenica," in *Contributi del Laboratorio di Statistica. Serie V* (Milan: Vita e Pensiero, 1939), 185–89.

pressure, a greater physically active attitude, a prevalence of processes of accumulation over consumption and [...] a super-attitude to reproduction."[317]

3) Pathological: based on the intuitions of the ancient humoralists, Boldrini[318] and his students (Costanzo, Colloridi and Alberti)[319] proposed a necessary link between constitution and "morbose predisposition": the causes of illness had to be looked for not in external agents, but in the characteristics of biotypes.

4) Psychological: biotypes were distinguishable not only by their somatic aspects, but also by their psychical qualities, that is by "character," "temperament" and intelligence.[320] The Boldrini school (in particular, Mengarelli and Uggé) greatly developed these aspects, explicitly reconnecting them to the studies of the psychiatrist Kretschmer[321] and, in Italy, to Pende and Gemelli. In synthesis, the "brevilinear sthenic" variety presented "an open, frank, expansive, strong-willed, optimistic, malleable, achieving, euphoric character," while the longilinear type would be more "asthenic," that is "solitary, meditative, haughty of character," obstinate in temperament and with a "logical, hypercritical, profound, analytical" intelligence.[322]

Up to this point, Boldrini's analysis, although a systemization of the biotypological classification of the constitutional school, was not particularly original. The innovative contribution could be seen rather in the next conceptual step, that is, in the attempt to connect Pende's medical constitutionalism with Gini's biological demography, through the study of the relationship between constitutional structure and social class:

> Evidently, since they [the biotypes] differ in infinite points of view, from the pure form to the highest manifestations of personality, and as on such differ-

---

317 Boldrini, "Tipi e attitudini costituzionali," 5; Boldrini, "Costituzione ed eugenica," 189–90.

318 See Marcello Boldrini, *Sviluppo corporeo e predisposizioni morbose* (Milan: Vita e Pensiero, 1925); Boldrini, "Costituzione ed eugenica," 191–92.

319 Alessandro Costanzo, "Costituzione e mortalità," in *Contributi del Laboratorio di Statistica. Serie III* (Milan: Vita e Pensiero, 1934), 403–30; Alessandro Costanzo, *Costituzione e mortalità* (Milan: Vita e Pensiero, 1935); Franco Colloridi, "La donna media lombarda come campione antropometrico per le indagini ostetrico-ginecologiche in Lombardia," *Annali di Ostetricia e Ginecologia* (1934); Franco Colloridi, "Il tipo costituzionale nelle donne portatrici di fibromiomi uterini," *Annali di Ostetricia e Ginecologia* (1934); Salvatore Alberti, *La mortalità antenatale* (Milan: Vita e Pensiero, 1934).

320 For Boldrini's view on the measure of intelligence, with reference above all to the American psychological school (Sante Naccarati and H. E. Garrett), see, in particular, Marcello Boldrini, *La fertilità dei biotipi* (Milan: Vita e Pensiero, 1931), 167–70.

321 For a discussion of Kretschmer's theories, see Boldrini, *La fertilità dei biotipi*, 187–92.

322 Boldrini, "Tipi e attitudini costituzionali," 6.

> ences [...] natural, sexual and social selection is based, we understand that, due to the variety of external circumstances, from medical knowledge, tastes, social organization, certain types in different classes will be preserved over others, some constitutions and not others will be elected above all by the current that feeds the ruling classes.[323]

This "biotypological" theory of social mobility had its scientific consecration in 1935 in Brussels, in the Italian section of the 12th Congress of the International Institute of Sociology—led by no other than Corrado Gini.[324]

During the Congress, the conclusive results of a research program were presented, which had been undertaken by the Milan Laboratory of Statistics almost ten years earlier. Boldrini's investigation at the end of the 1920s focused on a group of 715 people from Padua, measured at twenty years of age, together with a respective "personal and family history." Regarding the relationship between constitution and social class, the results seemed to show a very strong connection. The longilinear types were found for the most part in the superior social classes, with the brevilinear types in the inferior ones:

> In 100 members of the superior category, there were 21.6 brevilinear types, 37.8 mesolinear types, 40.6 longilinear types. The percentages corresponding to the city's manual workers were different: 29.7%, 34.0%, 36.6%. The longilinear types were again in the majority, as in the superior class, but with a less intense occurrence. The situation was completely inverted for the following percentages relative to the countryside workers and farmers: 37.2% brevilinear types, 38.8% mesolinear types and just 28.5% longilinear types.
>
> There is no need for doubt, therefore, in considering this investigation as a confirmation of the high frequency of longilinear types in the superior classes, compared to the intermediate and inferior categories.[325]

The successive investigation on the "physical characteristics of the scientific personnel of Italian universities," conducted by Boldrini and Mengarelli, and presented at the 1931 International Congress for Studies on Population in Rome, constituted a confirmation: "the university body,

[323] Boldrini, "Biotipi e classi sociali," 71.

[324] For a collection of works, see *Contributi del Laboratorio di Statistica. Serie IV.*

[325] Boldrini, "Biotipi e classi sociali," 73.

taken as a whole, is tall and slim."[326] In the following years, the Laboratory of Statistics of the Catholic University in Milan continued to gather data and numbers to demonstrate the "biotypological" dimension of social stratification.

Mengarelli, for example, conducted broad research on the "physical characteristics of the Italians who have reached hegemonic positions in intellectual, artistic, political and economic-financial Italian life," confirming the biological difference between "active genius" and "contemplative genius":

> The most longilinear style of body are those who Mengarelli calls "men of theoretical life" and, in particular, the experts of the abstract disciplines. He considers these men, with their excess stature and low weight, as generally asthenic longilinear. Following them, with higher stature but also greater relative weight, and therefore a less outstanding longilinearity, are those who excelled in naturalist and technical research, and, at a notable distance, the "men of practical life" (political and economic-financial). These last [...] gravitate toward a brevilinear sthenic type.[327]

A second investigation of the "physical characteristics of nobility" demonstrated the "asthenic longilinear" constitution—a stature superior to the average, weight inferior, lighter pigmentation—of Italian aristocrats.[328] And while Mengarelli studied the "contemplative aristocrat," another student of Boldrini, Albino Uggé, was concerned instead with athletes, underlining their "brevilinear sthenic" constitution:

> In general, the sporting constitution is sthenic-brevilinear. It is comparable, therefore, with the physical form of men of practical life, but with a more accentuated body mass. The robust man, with a stout and brevilinear body, tends to emerge in sporting life, as well as in the political and business ones, according to whether he revolves his attitude of achievement toward purely physical activity, or intellectual.[329]

326 Boldrini, "Biotipi e classi sociali," 75.

327 Boldrini, "Tipi e attitudini costituzionali," 14. See also Carlo Mengarelli, "I caratteri costituzionali delle *aristocrazie* italiane," in *Contributi del Laboratorio di Statistica. Serie IV*, 157–82.

328 Carlo Mengarelli, "Su i caratteri fisici della nobiltà," in *Contributi del Laboratorio di Statistica. Serie IV*, 239–72.

329 Boldrini, "Tipi e attitudini costituzionali," 15–16; Albino Uggé, "Sul tipo morfologico degli atleti," in *Contributi del Laboratorio di Statistica. Serie IV*, 65–75.

Finally, Maggi's research presented data on the "new" aristocracy, such as the cinematographic artists: according to his survey, for example, the actors were part of the "sthenic longilinear" (stature and weight above average), while the actresses were of the "medium asthenic longilinear type."[330]

Closely connected to such analyses of the relationship between constitutional structure and social stratification, was the other problem dear to the eugenics of the Milan Laboratory of Statistics: that of the "differential fertility" of biotypes. For Boldrini, the difference in fertility between social classes did not depend in the first place on economic or social motivations (the *sociological* theory supported by the majority of Italian demographers), nor on the biological variations of reproductive capacity (Gini's *cyclical* theory), as much as on the biotypological composition of the social pyramid (the *constitutional* theory) with the less fertile longilinear types dominating the elite, and the highly sexually reactive brevilinear types crowding the lower classes:

> Since the longilinear type is proportionally frequent in the population, and more represented in the higher classes, and the natural, sexual and social selection in today's cycle favors it, it follows that the current members of the elite and those who would like to become members, elevated from the lower categories, are frequently hyper-evolved and, for this reason, possess an inferior fertility to the average population and above all less than the fertility of the middle and lower social classes.[331]

Compared to the sociological and biological theories of differential fertility, the constitutional theory—Boldrini stated—did not have an evolutionary dimension and could rather be conceived as a *static* image of a demographic "conjuncture":

> It [the constitutional theory] starts from the presupposition that the reproductiveness of the biotypes, linked to the morphological-functional structure, can be considered, over a brief period, as fixed, and that the differential fertility of the classes rises simply from the manner in which the biotypes, of which the population is formed, are divided among the classes. The phenomenon, in its

[330] Raffaello Maggi, "La costituzione degli attori dello schermo," in *Contributi del Laboratorio di Statistica. Serie IV*, 79–136.

[331] Boldrini, *La fertilità dei biotipi*, 203–04.

> intimate essence, has been caught in an "ontogenetic" moment, at a "conjuncture" that produces it, and does not admit, as a rule, an evolutionary process.[332]

Boldrini's constitutional theory did not completely negate the possibility of identifying an evolutionary tendency in the development of human society. It nevertheless placed the *primum movens* of such a process not in the biotypes themselves, but in the fluctuating selective mechanisms that derived, time after time, from the interaction between the external environment and the social system. In particular, Boldrini proposed a sort of philosophy of history, distinguished by a constant oscillation between two cycles: the phases of crises, change and revolution selected an elite of brevilinear types; the successive phase of stabilization and consolidation favored instead an elite of longilinear types. The "active genius," revolutionary and brevilinear, left his post to the "contemplative genius," intellectual and longilinear, and vice versa:

> If it is true that history assists in a rhythmic succession of phases of activity and of contemplation; that the craftsmen of one and the other are men of genius, who put, with their thoughts and their works, a personal seal on the political, social and religious life; finally, that the attitudes of creative activity are linked with the sthenic brevilinear structure and the theoretic predisposition with asthenic longilinearity, then we can conclude that the supreme power and superior direction of society are necessarily transmitted without pause by one type or the other.[333]

The "first contemplative period" of the modern era was the Renaissance, and in fact, Erasmus, "the most eminent and pure representative of humanism,"[334] was a "pure asthenic longilinear" type. A blow to Erasmus' theory however came from the "religious revolutionaries Luther, Zwingli, and Henry VIII, all well known as being of sthenic brevilinear structure."[335] Two centuries later, the Enlightenment saw the initial return of the "pure longilinear type" (Locke, Voltaire, Montesquieu, Diderot, d'Alembert, Rousseau, Wolff, Mendelssohn), soon dethroned "for a lack of practical capacity,"

332 Boldrini, *La fertilità dei biotipi*, 213.
333 Boldrini, "Tipi e attitudini costituzionali," 11.
334 Boldrini, "Tipi e attitudini costituzionali," 10.
335 Boldrini, "Tipi e attitudini costituzionali," 10.

by "true revolutionaries, such as Mirabeau, Danton, Robespierre, of the more or less sthenic brevilinear type." In the 1900s, the end of the First World War marked a new affirmation of activism: communism and fascism, although political adversaries, appeared to Boldrini to be united by the "brevilinear constitution of the leaders."[336]

Many research studies by the Laboratory of Statistics—such as the analysis of the relationship between biotype and social class—were produced to give statistical solidity to the hypothetical evolutionary tendency of the elite. Amintore Fanfani, for example, hypothesized a probable connection between the economic changes in Europe from the fifteenth century, and the formation of a new longilinear aristocracy.[337] Boldrini and Alberti's investigations into the transformation of the Italian elite in the last eighty years seemed to confirm the biotypological movement of the Italian ruling classes from a theoretical and longilinear type to a more active, practical brevilinear type.[338]

In Boldrini's view, the social stratification of biotypes and the constitutional theory of the elite represented the so-called "documentary or passive eugenics," which focused on the relationship between the constitutional characteristics, transmitted hereditarily, and the respective social and demographic consequences:

> Contrasting forces at the same time conserve and eliminate the types and the constitutional characters. As for the fundamental structure, the recessive is favored by homogamy. The longilinear type is additionally advantaged by the aesthetic evaluation, which facilitates marriage, but is impeded in diffusion by its lower natural fertility. Nor must we disregard, for the longilinear types, the disadvantage deriving from their frequent occurrence in the middle and higher classes, in which matrimonial rates are lower, the age of marriage higher, making the procreative will even weaker; and, by extension, offers the benefit of greater prosperity and a more comfortable and tranquil existence. [...] Nor must we disregard the negative sides of the constitution. The average duration of life, the mortality at various ages, the morbose propensities, are different for the two fundamental types, and as they are of a hereditary character, as is normal, their

336 Boldrini, "Tipi e attitudini costituzionali," 10.

337 Amintore Fanfani, "I mutamenti economici nell'Europa moderna e l'evoluzione costituzionalistica delle classi dirigenti," in *Contributi del Laboratorio di Statistica. Serie IV*, 137–56.

338 Marcello Boldrini and Aldo Alberti, "Il patriziato italiano nelle categorie dirigenti," in *Contributi del Laboratorio di Statistica. Serie IV*, 183–230.

> diffusion is positively or negatively influenced by the same factors that work to preserve or diffuse, or even eliminate the two typical opposing structures.[339]

Although he was a firm supporter of the hereditary nature of constitutional characteristics, Boldrini did not go so far as to accept the negative measures of what he called "active eugenics" (Anglo-American, German and Scandinavian eugenics). Not only for the reasons already listed—the natural harmony between the social system and human organism, the historical variability of eugenic ideals—but above all, for the recognition of the theoretical limits of a science that still had much to investigate and understand: "That the current world is the best, no one wishes to support; but no human mind is today capable of inventing another, at least not unless we talk of the mind of a novelist, such as Aldous Huxley, which we would not, however, aspire to realize."[340]

If therefore, in the future, the scientist had to content himself with continuing his studies, the politician could, in the meantime, "trust in the old instruments of hygiene, medicine, assistance, charity, and social legislation, with which defects, imperfections and illnesses are prevented, cured and rendered socially innocuous."[341] This cautious and moderate position, therefore, had not to induce pessimism about the eugenic hopes, but on the contrary, had to be interpreted, in Boldrini's words, as an honest scientific recognition of an immense field of work, which reserved "places and honors for all."

339 Boldrini, "Costituzione ed eugenica," 204.
340 Boldrini, "Costituzione ed eugenica," 208.
341 Boldrini, "Costituzione ed eugenica," 209.

CHAPTER V

EUGENICS AND RACISM (1938–1943)

Current historiography has completely dismantled the monolithic description of fascist racism in Italy. In fact, according to the most recent research, official racism developed in Italy, between 1938 and 1943, along three different lines, each distinct from an ideological, political and institutional point of view.[1]

Biological, or "Nordic" racism, characterized the publication of the most important scientific document, the so-called "Manifesto of the racial scientists."[2] The principle exponents of the biological current came from two different, although linked, groups: one, the journalistic lobby, headed by Telesio Interlandi, leading journalist of the regime and director of the daily newspaper *Il Tevere*, the weekly review *Quadrivio* and the twice-monthly journal *La Difesa della razza*;[3] the other, from the four main drivers of the "Manifesto" of July 1938. The latter group included the young anthropologist Guido Landra, main editor of the "Manifesto" and director from August 1938 of the Race Office of the Ministry of Popular Culture (*Ufficio Razza del Ministero della Cultura Popolare*); Lidio Cipriani, professor of anthropology at the University of Florence and director of the Florence national museum of anthropology and ethnology; Leone Franzì,

1 Mauro Raspanti, "I razzismi del fascismo," in Centro Studi F. Jesi, ed., *La menzogna della razza. Documenti e immagini del razzismo e dell'antisemitismo fascista* (Bologna: Grafis, 1994), 73–89.

2 See Aaron Gillette, "The Origins of the 'Manifesto of the Racial Scientists,'" *Journal of Modern Italian Studies* 6, no. 3 (2001): 305–23.

3 On Telesio Interlandi, see Francesco Cassata, *"La Difesa della razza." Politica, ideologia e immagine del razzismo fascista* (Turin: Einaudi, 2008); Meir Michaelis, "Mussolini's Unofficial Mouthpiece: Telesio Interlandi, Il Tevere and the Evolution of Mussolini's anti-Semitism," *Journal of Modern Italian Studies* 3, no. 3 (1998): 217–40.

assistant professor in the pediatric clinic of the University of Milan; Lino Businco, assistant professor of general pathology at the University of Rome and Marcello Ricci, assistant professor of anthropology also at the University of Rome.

Nationalist, or "Mediterranean," racism centered around historical and geographical considerations about race. It assumed an institutional relevance in February 1939, when Landra was replaced at the Race Office by Sabato Visco, director of the institute of general physiology at the University of Rome, and founder of the National Institute of Nutrition. The most famous exponent of this current was Giacomo Acerbo, president of the High Council of Demography and Race (*Consiglio Superiore della Demografia e Razza*), which was the principle institution of nationalist racism.[4] One of the most relevant initiatives of the General Council was the April 1942 document on the "Italian race," which among its stated intentions, aimed to become a new "Manifesto" on race.

The rise of Alberto Luchini at the head of the Race Office in May 1941 was an evident sign of the growing influence of the esoteric-traditionalist racist current, politically supported by Giovanni Preziosi and Roberto Farinacci,[5] and represented above all by the biological-metaphysical theories of Julius Evola.[6] Between 1941 and 1943, esoteric-traditionalist racism developed two particular projects: an investigation of the racial components (biological, psychological and spiritual) of the Italian population; and the constitution of a bilingual Italian–German review, entitled *Sangue e Spirito*

4 The High Council of Demography and Race included: Giacomo Acerbo; Filippo Bottazzi, professor of human physiology, University of Naples; Alessandro Ghigi, professor of zoology, University of Bologna; Raffaele Corso, professor of ethnology, University of Firenze; Vito De Blasi, lecturer of obstetrics and gynaecology, University of Genoa; Cornelio Di Marzio, journalist; Cesare Frugoni, professor of general clinical medicine, University of Rome; Livio Livi, professor of statistics, University of Florence; Biagio Pace, professor of topography of ancient Italy, University of Rome; Antonio Pagliaro, professor of glottology, University of Rome; Umberto Pieramonti, assistant professor of racial biology, University of Naples; Ugo Rellini, professor of palaeontology, University of Rome; Giunio Salvi, professor of human anatomy, University of Naples; Sergio Sergi, professor of anthropology, University of Rome; Francesco Valagussa, lecturer of Clinical Paediatrics, University of Rome.

5 On Giovanni Preziosi, see Renzo De Felice, *Storia degli ebrei italiani sotto il fascismo* (Turin: Einaudi, 1961); Michele Sarfatti, ed., *La Repubblica sociale italiana a Desenzano: Giovanni Preziosi e l'Ispettorato generale per la razza* (Milano: Giuntina, 2008). On Roberto Farinacci, see Matteo Di Figlia, *Farinacci: il radicalismo fascista al potere* (Donzelli: Roma, 2007).

6 For a complete biography of Julius Evola, see Francesco Cassata, *A destra del fascismo. Profilo politico di Julius Evola* (Turin: Bollati Boringhieri, 2003).

[Blood and spirit], in which the planned contents were supposed to constitute—at least in Luchini's intentions—the basis for a new and more radical racial policy.[7]

Eugenics represented a fundamental component of the opposition among the three directions of fascist racism. In particular, the nature/nurture debate acted as a boundary tool, mapping the different currents of fascist racism and defining their relationships with Nazi racial ideology and politics.

1. Biological Racism and Hereditarian Eugenics

Nazi negative eugenics was the point of reference for fascist Italian biological racism from 1934, as demonstrated between 1934 and 1938 by the journalistic campaigns carried out by Interlandi's journals, *Il Tevere* and *Quadrivio*.[8]

Starting from August 1938, the principle organ of diffusion of biological racism became *La Difesa della razza*. This journal was financed by the Ministry of Popular Culture and by a variety of banks, industrial concerns and insurance companies. Its foundation was closely linked with the publication of the "Manifesto of the racial scientists": the first issue of the magazine carried articles by the eight members of the Racial Manifesto committee, five of whom were on its editorial board (Landra, Cipriani, Franzì, Businco and Ricci).

In 1938, from their first issue, the hereditarian eugenicists of *La Difesa della razza* attempted to demonize and dismantle the neo-Lamarckian basis of Italian eugenics. Guido Landra was a driving force behind the anti-Lamarck campaign. Race—Landra declared in December 1938—essentially meant heredity. The environment did not exert any influence on ethnic types, described essentially as immutable and immortal:

> Commonly, we speak of the youth, maturity or age of a population. These terms, used mostly by historians, have value when they are used to refer to a

---

7 On these two projects, see Cassata, *"La Difesa della razza." Politica, ideologia e immagine del razzismo fascista*, 79–82.

8 On this topic, see Cassata, *"La Difesa della razza." Politica, ideologia e immagine del razzismo fascista*, 6–55.

> population, but not when they refer to a race. Racial qualities do not trace this fatal parabola: indeed, they always stay the same. And this holds true for physical qualities and, in a yet more outstanding manner, for psychical qualities. Racial qualities really have an immortal character, and are maintained as long as the men of a particular race are living.[9]

In May 1940, the journalist Willi Nix declared that heredity was "destiny." And further: "progenitors and descendants are inseparably linked to each other; the one is only a link in the chain, and completes the other like a new link."[10] At the beginning of 1941, in reference to the diverse vocational aptitudes of the races, the brother of Guido, Silvio Landra, insisted on the hereditarian paradigm:

> Man may change country, clothes, education, language, but from his deep interior there is always something connecting him to his racial origin, which, at any given moment and under determinate conditions, can bloom and manifest itself. [...] We can verify in men those things that we can verify in the entire animal kingdom. A hound is always a hound, a greyhound always a greyhound, a dachshund always a dachshund, not only in exterior form, but in its different ways of seizing and catching, which cannot be modified by his master.[11]

For the physician Giuseppe Lucidi, the biological purity of race lay in the "identity of blood" transmitted through the generations.[12] Blood groups therefore, had to be considered as "constitutional factors":

> From various research studies it appears evident that the blood groups must be considered as constitutional factors: recent research puts them in strict relation with the various anthropological characteristics, since the blood group of each individual is nothing less than the expression of the biological substrate of the individual. In fact, agglutinogens and agglutinins similar to those contained in blood are also contained in tissue. To be clearer, it is scientifically

9 Guido Landra, "L'ambiente non snatura la razza," *La Difesa della razza* 2, no. 3 (5 December 1938): 17.

10 Willi Nix, "Eredità e destino," *La Difesa della razza* 3, no. 13 (5 May 1940): 14.

11 L. S. [Silvio Landra], "Ambiente razza e attitudini professionali," *La Difesa della razza* 4, no. 5 (5 January 1941): 13–14.

12 Giuseppe Lucidi, "Il sangue. Individualità biologica di razza," *La Difesa della razza* 1, no. 5 (5 October 1938): 37–38.

> proven that if an individual has a blood group, their blood is different because the flesh is different, different from others.[13]

While Lino Businco underlined the need to deepen the hereditary etiology of diseases,[14] Luigi Castaldi, director of the Anatomic Institute of Cagliari, referred to "Galton's law" in November 1938 to demonstrate the heredity of the cephalic index:

> Through the germ plasm, something of our own substance passes to our descendents, reproducing in them our image, our attitude and abilities, our virtues and our weaknesses. And this sensation of living in them, and therefore of continuing in some way through them, is one of the principal causes of the affection and care that grandparents and parents have for us and that we have for our children, and in this long line, they will continue to be the basis of social life.[15]

In August 1938, the attack against neo-Lamarckism assumed the shape of a true "return to Galton," fuelled by concern over the progressive "decadence of the upper classes." The journalist Elio Gasteiner, for example, wrote:

> Leafing through magazines and journals we can observe the very large allocation of space dedicated to the various types of sport, and certainly almost all the readers have the firm conviction that it is in order to create a maximum of effort toward the future of the Nation.
>
> Racism—that is, the ensemble of sciences that deal with eugenics, human biology, and social anthropology—must, however, promptly disabuse them of this gratifying opinion. This immense work for the physical education of the youth has no effect on quality or on a desired hereditary racial improvement. For the individual there will certainly be constitutional advantages, but these improvements are paratypic; that is, they are not hereditary and therefore cannot change the race [...]. If it were truly possible to change man through exterior forces then the human race would not exist; its invariability through the millennia has been indisputably ascertained.[16]

---

13 Giuseppe Lucidi, "Rapporti fra gruppi sanguigni e caratteri antropologici," *La Difesa della razza* 2, no. 7 (5 February 1939): 8.

14 Lino Businco, "Individuazione e difesa dei caratteri razziali," *La Difesa della razza* 2, no. 10 (20 March 1939): 15–17.

15 Luigi Castaldi, "Nonni, figli e nipoti. Eredità dell'indice cefalico," *La Difesa della razza* 2, no. 2 (20 November 1938): 12.

16 Elio Gasteiner, "Un pericolo per la razza. La decadenza dei ceti superiori," *La Difesa della razza* 1, no. 2 (20 August 1938): 26.

The architect and art critic Giuseppe Pensabene did not hesitate to evoke the criticism aimed by the Jesuit periodical, *La Civiltà Cattolica*, against Lamarckian theories in the second half of nineteenth century.[17] In this ideological context, even Lino Businco, in other cases sympathetic to the idea of environmental influences, placed Francis Galton at the origins of fascist eugenics:

> Galton did not limit his works to scientific research. Convinced that his own ideas provided both model and warning for those scientists wrapped up in themselves, he enthusiastically acted. He wrote propaganda books and brought the people of London an institute of eugenics to which all who were preparing to celebrate marriage could turn for consultation. [...]
> In the climate created by fascism, with its renewed racial pride and the duties it brings, this fertile science can go among the people and make a great contribution to the increasingly strong new life born in imperial Italy.[18]

Marcello Ricci stressed the relevance of Mendelism as the theoretical lynchpin of hereditarian eugenics.[19] Under Ricci's interpretation, the laws referred not only to human anomalies and pathologies, but normal human characteristics, like eye and hair color:

> We can therefore conclude that all of human heredity explicates itself like that of plants and animals, dependent on Mendel's laws. The generalization appears justified by the fact that we cannot see why a diversity of transmission should exist among the various characteristics of an organism.[20]

Rather, it was above all the evident validity of Mendelian mechanisms in transmitting hereditary pathological and abnormal characteristics that gave rise to the hope, according to Ricci, of an "opportune application to the field of racial eugenics."[21] In October 1938, after focusing on the Mendelian transmission of hereditary diseases, the same author discussed the practical consequences more deeply:

17 Giuseppe Pensabene, "L'evoluzione e la razza. Cinquant'anni di polemiche ne 'La Civiltà Cattolica,'" *La Difesa della razza* 1, no. 2 (20 August 1938): 31–33.
18 Lino Businco, "Salute della famiglia, forza della razza," *La Difesa della razza* 2, no. 4 (20 December 1938) 37–39.
19 Marcello Ricci, "Le leggi di Mendel," *La Difesa della razza* 1, no. 2 (20 August 1938): 16–17.
20 Marcello Ricci, "Il mendelismo nell'uomo," *La Difesa della razza* 1, no. 3 (5 September, 1938): 19.
21 Ricci, "Il mendelismo nell'uomo," 19.

> It is therefore necessary to recognize that, ultimately, the single greatest benefit for racial improvement could come from the elimination of defectives. If rationally conducted over successive generations, the lack of continued inlet of new heterozygotes, hidden carriers of diseases, [...] would lead to an always greater rarification of abnormal and pathological cases. [...] A true racial improvement, based on effectively diminishing genetic defects through the application of means aimed at limiting active reproduction of the harmful individuals, appears to be the logical inference of a simple and serene reflection on what we have written.[22]

The article concluded, not by chance, with a reference to one of the most celebrated cases of international eugenic literature: Ada Juke and her "degenerate" descendants.

Galton and Mendel were not the only illustrious names invoked in this "invention of tradition" by *La Difesa della razza*. Guido Landra salvaged, for example, the hologenetic theory of Daniele Rosa, and its two functional applications to Italian racist ideology:

> 1) the common origin of racial elements that have contributed to the anthropological substrate of Italy with those of the other European populations, which today reveal physical and psychological affinities with our population, to higher or lower degrees;
> 2) the formation of a unique race on the soil of our homeland, a formation begun in a remote era, and accompanied by a continual evolution; as the centuries pass, the Italian race is increasingly differentiated from the other similar races, accentuating and developing determinate physical and psychological characteristics.[23]

Rosa's hologenesis, as elaborated by Georges Montandon in 1928,[24] was invoked to demonstrate the evolution of species "by internal forces," in opposition to Lamarckian environmentalism. Theoretical references included the utopian narrations of Tommaso Campanella[25] and Leon Bat-

22 Marcello Ricci, "Ereditarietà ed eugenica," *La Difesa della razza* 1, no. 5 (5 October 1938): 31.

23 Guido Landra, "La razza italiana nella teoria dell'ologenesi," *La Difesa della razza* 2, no. 11 (5 April 1939): 11. See also Guido Landra, "L'ologenesi del Rosa," *La Difesa della razza* 2, no. 10 (20 March 1939): 11–14.

24 On Montandon's hologenesis, see Georges Montandon, "La formazione delle razze umane," *La Difesa della razza* 4, no. 22 (20 September 1941): 9–12.

25 Fortunato Matarrese, "Demografia ed eugenica di Tommaso Campanella," *La Difesa della razza* 3, no. 15 (5 June 1940): 40–41; Paolo Nullo, "Il razzismo nella 'Città del Sole' di Tommaso Campanella," *La Difesa della razza* 4, no. 14 (20 May 1941): 13–15.

tista Alberti;[26] Vincenzo Giuffrida-Ruggeri's "monogenism;"[27] constitutionalism, with its contributions to the relationship between biotypes and fertility;[28] Georges Vacher de Lapouge's[29] *anthroposociologie* and genealogical researches, primarily that of the socialist physician Gaetano Pieraccini on the family pedigree of the Medici of Cafaggiolo.[30]

But it was above all the contributions of German and American eugenics that furnished the most solid scientific support for the biological racism of *La Difesa della razza*. This emerged most strongly in the writings of Guido Landra. The common thread running through Landra's contributions to the biweekly could be seen in his shift from nineteenth-century physical anthropology to twentieth-century "science of heredity" and racial genetics. He did not disdain traditional anthropometric methods[31] or the descriptions of the different taxonomies developed from international "racial studies,"[32] but favored the analysis of hereditary processes that characterized a single racial

26 Fortunato Matarrese, "Leon Battista Alberti, studioso di problemi razziali," *La Difesa della razza* 3, no. 9 (5 March 1940): 37–41.

27 Guido Landra, "Poligenismo e monogenismo," *La Difesa della razza* 3, no. 21 (5 September 1941): 27–29.

28 Giuseppe Lucidi, "Costituzione e natalità," *La Difesa della razza* 2, no. 9 (5 March 1939); Guido Landra, "Le razze europee e il problema delle aristocrazie," *La Difesa della razza*, 4, no. 13 (5 May 1941): 12–15.

29 Georges Montandon, "Vita e opere di Vacher de Lapouge," *La Difesa della razza* 4, no. 12 (20 April 1941): 24–26.

30 Luigi Castaldi, "Eredità delle attitudini psichiche," *La Difesa della razza* 3, no. 3 (5 December 1939): 26–31.

31 See Guido. Landra, "I metodi per lo studio delle razze umane," *La Difesa della razza* 3, no. 20 (20 August 1940): 29–35; Guido Landra, "Antropologia – Forme esterne del corpo umano, variazioni nel sesso e nell'età," *La Difesa della razza* 4, no. 12 (20 April 1941): 18–20; Guido Landra, "Antropologia – Ricerche e dottrine craniologiche," *La Difesa della razza* 4, no. 14 (20 May 1941): 26–29; Guido Landra, "Lo scheletro facciale nelle razze umane," *La Difesa della razza* 4, no. 15 (5 June 1941): 24–26; Guido Landra, "Antropologia – Ricerche craniologiche," *La Difesa della razza* 5, no. 3 (5 December 1941): 24–26; Guido Landra, "Antropologia – Studi razziali sulle differenze razziali della faccia," *La Difesa della razza* 5, no. 5 (5 January 1942): 22–23; Guido Landra, "Antropologia – Morfologia facciale," *La Difesa della razza* 5, no. 6 (20 January 1942): 28–29.

32 A review of "racial studies" was curated by Landra between June 1939 and June 1940. See Guido Landra, "Gli studi razziali nell'Europa balcanica," *La Difesa della razza* 2, no. 16 (20 June 1939): 32–34; Guido Landra, "Gli studi razziali in Polonia e in Russia," *La Difesa della razza* 2, no. 18 (20 July 1939): 14–17; Guido Landra, "Studiosi americani di problemi razziali," *La Difesa della razza* 2, no. 20 (20 August 1939): 13–16; Guido Landra, "Razza e nazionalità in Romania," *La Difesa della razza* 2, no. 21 (5 September 1939): 10–13; Guido Landra, "Studi razziali in continenti extraeuropei," *La Difesa della razza* 2, no. 23 (5 October 1939): 34–37; Guido Landra, "Gli studi razziali in Ungheria e in Bulgaria," *La Difesa della razza* 3, no. 3 (5 December 1939): 32–33; Guido Landra, "Studi razziali in Transilvania," *La Difesa della razza* 3, no. 14 (20 May 1940): 16–19; Guido Landra, "Studi sulle mescolanze etniche della popolazione," *La Difesa della razza* 3, no. 15 (5 June 1940): 12–13. On anthropological taxonomy, see Guido Landra, "Sistematica antica e moderna delle razze umane," *La Difesa della razza* 3, no. 19 (5 August 1940): 23–28; Guido Landra, "La classificazione delle razze umane secondo von Eickstedt," *La Difesa della razza* 4, no. 6 (20 January 1941): 12–15; Guido Landra, "Antropologia – Problemi di metodo per la definizione dei tipi razziali," *La Difesa della razza* 4, no. 10 (20 March 1941): 22–25; Guido Landra, "Le razze dell'Asia meridionale e orientale," *La Difesa della razza* 4, no. 11 (5 April 1941): 18–20.

factor: from finger and palm prints[33] to facial shape;[34] from the "integumentary system"[35] to blood groups;[36] from the heredity of illnesses[37] to the factors of corporeal optimization.[38] Mapping the quotations in Landra's articles it is possible to explicitly demonstrate the influence exercised on the Italian anthropologist by National Socialist eugenics. Furthermore, as head of the Race Office and then as a journalist, Landra visited some of the more active and relevant institutions of German *Rassenhygiene.* In particular, he made contacts within the laboratories of Eugen Fischer and of Othmar von Verschuer, respectively first and second director of the Kaiser Wilhelm Institute for Anthropology, Human Heredity and Eugenics in Berlin; the Institute of Anthropology and Ethnology of Breslau, headed by Egon von Eickstedt; and the Institute for Race and Heredity, directed by Heinrich Wilhelm Kranz at the Faculty of Medicine of the University of Giessen.

La Difesa della razza frequently ran articles—especially those signed by Guido Landra—dedicated to studies on hereditary diseases (Lange, Lenz, von Verschuer, Weitz),[39] twins,[40] and growth factors (Boas, Davenport, Dunn, Rodenwaldt, Fischer).[41] The 5 November 1939 issue contained, in prime position, a long essay by Eugen Fischer on the concept of race.[42]

References to the Nordic eugenic paradigm were accompanied by a proposal of eugenic practical measures, which seemed notably close to the Nazi legislation. There were certainly many discordant voices. The physiologist Silvestro Baglioni believed that eugenics must "elevate and cultivate

33 Guido Landra, "Studio razziale delle impronte digitali," *La Difesa della razza* 3, no. 12 (20 April 1940): 40–41; Landra, "Studio razziale delle impronte palmari," 36–37.

34 Guido Landra, "La forma del viso nelle razze umane," *La Difesa della razza* 3, no. 21–22 (5–20 September 1940): 51–54; Guido Landra, "Caratteri fisionomici identità razziale," *La Difesa della razza* 4, no. 9 (5 March 1941): 18–20.

35 Guido Landra, "Le variazioni del sistema tegumentario nelle razze umane," *La Difesa della razza* 3, no. 23 (5 October 1940): 11–16.

36 Guido Landra, "Ricerche moderne sui gruppi sanguigni," *La Difesa della razza* 4, no. 2 (20 November 1940): 34–37.

37 Guido Landra, "Gli studi di patologia ereditaria in Germania," *La Difesa della razza* 3, no. 17 (5 July 1940): 18–22.

38 Guido Landra, "I fattori ereditari dell'accrescimento," *La Difesa della razza* 3, no. 19 (5 August 1940): 36–39; Guido Landra, "Studi sull'aumento della statura in Scandinavia," *La Difesa della razza* 4, no. 5 (5 January 1941): 10–12.

39 Guido Landra, "Gli studi di patologia ereditaria in Germania," 18–22.

40 Guido Landra, "Il metodo dei gemelli," *La Difesa della razza* 3, no. 18 (20 July 1940): 28–31.

41 Guido Landra, "I fattori ereditari dell'accrescimento," 36–39.

42 Eugen Fischer, "La realtà della razza," *La Difesa della razza* 3, no. 1 (5 November 1939): 11–17.

the love of children," using the tools of preventive medicine and hygiene, rather than repressive methods.[43] In an article from July 1940, Giovanni Marchiori somewhat paradoxically interpreted the Nazi policy of sterilization as "the greatest racial experiment since Lycurgus,"[44] while radically rejecting negative eugenics: "segregation, like sterilization, is a coactive measure that clashes against our moral and juridical norms, that prohibit among other things the use of sanctions against those who have not committed any crime and are *compos mentis*."[45] These same doubts also clouded the potential efficacy of premarital certificates:

> In a conscious population, premarital examinations, eugenic certificates, and the prohibition of marriage in the case of serious situations should have their value. Today their effectiveness is dubious and such prohibitions could result in worse evils and open unions. Besides, marriage sometimes has only emotional aims, or offers mutual assistance, or regulates an earlier union: particularly for the elderly.[46]

In the pages of *La Difesa della razza*, Renato Semizzi, professor of social medicine at the Universities of Padua and Trieste, wrote a column on social medicine entitled *Salute della razza*, in which he constantly repeated the "euthenic" refrain:

> The State must impose a rigid euthenics. They must provide for the improvement of the environment in which all national activities are developed, [...] encourage prolific marriages in every way, to obtain a considerable number of hereditary combinations until the most favorable emerge [...], and above all combat the decreasing birth rate.
> The State must provide for the prophylaxis and the correction of the disabled in the fight against all social illnesses; as it must prevent, and possibly eliminate, all the causes suspected of facilitating the rise of determinate regressive mutations.[47]

43 Silvestro Baglioni, "Continuità della razza," *La Difesa della razza* 3, no. 4 (20 December 1939): 6–12.

44 Giovanni Marchiori, "Propaganda eugenica o misure coercitive?, " *La Difesa della razza* 3, no. 18 (20 July 1940): 21.

45 Marchiori, "Propaganda eugenica o misure coercitive?," 21.

46 Marchiori, "Propaganda eugenica o misure coercitive?," 21.

47 Renato Semizzi, "La medicina della masse," *La Difesa della razza* 4, no. 9 (5 March 1941): 13–15; see also Renato Semizzi, "La medicina sociale attraverso i tempi e le idee," *La Difesa della razza* 4, no. 11 (5 April 1941): 21–26; Renato Semizzi, "Eugenica e terapia razziale," *Critica medico-sociale* no. 7–8–9 (July–September 1940): 34–39.

The same author went on to hypothesize a genetic influence of "civilization" (that is, of a "perfected social organization") on the laws of heredity:

> Civilization also has a great importance in the field of heredity because its influence—whether or not it can be evaluated—together with euthenics and the evolution of human thought, modifies the complicated structure of genes, by altering, interweaving or dividing them, accelerating or slowing the reciprocal influence. In this way, new psychological constructions and somatic adjustments are created through temporal hereditary modifications [...] which, attentively observed, controlled, compared and followed, indicate the unmistakable and prolonged influence of the civilized environment.[48]

Opposing these limited voices critical of negative eugenics, most contributors invoked a more radical eugenic interventionism. In particular, the creation of a national index of Italian biological characteristics was frequently suggested by many collaborators of *La Difesa della razza*. From the first issue, the president of ISTAT, Franco Savorgnan, hoped for an update of the anthropometric inquiry into the Italian population, first realized by Ridolfo Livi in 1896:

> An anthropometric inquiry conducted on a vast scale seems, today, more than pertinent. It could demonstrate which physical characteristics present with major frequency in the Italian race, measuring the deviation from the average and normal type, the variability and the extremes (field of variation) and determine the differential characteristics of the small Italian race compared to others that make up the great Indo-European family.[49]

A few months later, Giuseppe Lucidi proposed a "census of blood" through blood typing, with the aim of achieving two objectives, the first "scientific-racial," and the second "practical":

> 1) *For a scientific-racial end,* an exact study of blood groups, beyond giving documental substance to our racism, would determine the biological characteristics of our race, placing our science at the *avant-garde* of all relevant research, considering that abroad they are actively working while here almost nothing

48 Renato Semizzi, "L'influenza della civiltà sui popoli," *La Difesa della razza* 4, no. 14 (20 May 1941): 10–12.

49 Franco Savorgnan, "I problemi della razza e l'opportunità di un'inchiesta antropometrica sulla popolazione italiana," *La Difesa della razza*, 1, no. 1 (5 August 1938): 18.

> is being done, nor is likely to be done, to develop a solid basis for a racial science. [...]
> 2) *For a practical end,* for an optimization beyond defense of our race, such research would permit us to know the precise blood group of any individual, which in times of war, more so than peace, could save thousands and thousands of lives, or simply make the practice of blood transfusions more practical.[50]

In March 1941, Giulio Silvestri favored the creation of "national race archives, rich in all the genealogical branches":

> We could reconstruct a genealogical tree of single families, or better, single individuals. It would be utterly interesting for observing various characteristics in the light of statistics. It would deal with [...] a work of extreme size that only the State could undertake, but which would give the exact measurements of the racial composition of the nation and, for every individual, the key to many apparently inexplicable inclinations that can be conveniently observed in daily life. It would also make the concept of race clear to the public, since everyone would clearly discover their own position within the thick network of relations and blood relations, which form the nation and the homeland.[51]

Besides national anthropological mapping, *La Difesa della razza* eugenicists proposed two radical practical measures against racial degeneration: the prohibition of race-crossing and the elimination of defectives.

As regards the first aspect, undoubtedly mixophobia and the denunciation of hybrids was a recurring theme in the pages of the biweekly. From the first issue, Guido Landra drew on Eugen Fischer's data regarding the so-called "bastards of Rehoboth" and the "Rhineland bastards" to demonstrate the degenerative effects of race-crossing[52]. Leone Franzì believed that the "lack of constitutional affinity between the maternal and paternal germ plasm," at the base of hybridism, produced racial damage that were both quantitative (increases in miscarriages and sterility) and qualitative (the "biological disharmony" that determined a greater frequency of patholo-

50 Giuseppe Lucidi, "Gruppi sanguigni e nuclei razziali. Necessità di un censimento del sangue," *La Difesa della razza* 2, no. 5 (5 January 1939): 15.

51 Giulio Silvestri, "Per un archivio genealogico nazionale," *La Difesa della razza* 4, no. 9 (5 March 1941): 24–27.

52 Guido Landra, "I bastardi," *La Difesa della razza* 1, no. 1 (5 August 1938): 16–17.

gies and mental diseases).[53] As most eugenic literature—and Franzì cited Davenport, Lundborg and Mjøen—demonstrated the negatives of hybridism, political intervention was desired in order to "avoid any type of racial crossing that might provoke the very efficaciously named 'racial chaos,' laying a dangerous trap for both the moral and physical hygiene of the population."[54] Franzì continued:

> This becomes much more evident and legitimate because there are already existing norms, not only secular but also religious, which prevent interbred unions and have purely eugenic aims. Racial crossings are dangerous and damage is certainly not a minor risk, but rather a major risk of interbred unions.[55]

There was therefore a need to impede not only marriages between "blood-relatives," but also "bastardisation," extending the ban to unions between "disparate racial elements, especially if inferior."[56]

In the article *Il meticciato, morte degli imperi* [Race-crossing: the death of empires], signed by the physician Giuseppe Lucidi, hybridism was synonymous with sterility, on one hand, and "physical and spiritual disharmony," on the other:

> Almost all of the major scholars: Davenport, Lundborg, Myoln [sic], Tillighart, agree that bastards have impaired physical qualities, due to alteration of the endocrinal equilibrium. Arassaz has particularly studied Brazilian hybrids, finding people without physical or moral energy.
> The physical disharmony in fact strikes again in the spiritual camp, as science continues to reveal how intimate and profound the relationships between the material and the spiritual are in the human body. This is even more valid for the Italo-Abyssinian crosses, who seem in a particular way inferior to the two progenitor races both spiritually and biologically.[57]

The physician Raffaele D'Anna Botta was of the same view, believing "racial crossings to be disastrous, especially for the superior races that

53 Leone Franzì, "Il meticciato. Insidia contro la salute morale e fisica dei popoli," *La Difesa della razza* 1, no. 4 (20 September 1938): 29–30. See also Ada De Blasio, "Frenastenie e meticciato," *La Difesa della razza* 6, no. 1 (5 November 1942): 17.

54 Franzì, "Il meticciato. Insidia contro la salute morale e fisica dei popoli," 33.

55 Franzì, "Il meticciato. Insidia contro la salute morale e fisica dei popoli," 33.

56 Franzì, "Il meticciato. Insidia contro la salute morale e fisica dei popoli," 33.

57 Giuseppe Lucidi, "Il meticciato, morte degli imperi," *La Difesa della razza* 2, no. 13 (5 May 1939): 18.

immediately lose their exceptional psycho-physical qualities and intellectual dowry."[58]

An entire special issue of *La Difesa della razza*—that of 20 March 1940—was dedicated to the problem of mixed blood, "with the aim—asserted Landra—of increasingly disseminating such studies among the Italians, contributing to the formation of racial pride."[59] The articles, signed by "specialists on matters of unquestioned authority and scientific seriousness,"[60] included a significant repertoire of "classics" from the international eugenic movement: Eugen Fischer on the "bastards of Rehoboth,"[61] Charles Davenport and Morris Steggerda on "mulattoes of Jamaica,"[62] Wolfgang Abel on "hybrids of Renania,"[63] Yun Kuei Tao on European–Chinese crossings,[64] Johann Schaeuble on hybrids in South America,[65] and Rita Hauschild on "Negro–Chinese crosses."[66] Moreover, a number of passages drew attention to American eugenics and their fear of "racial suicide" caused by the diffusion of miscegenation. The theories of Madison Grant and Lothrop Stoddard were cited as confirmation of a biological danger so real it could not be negated, even by the "tolerant and liberal Americans."[67] Also for Giuseppe Pensabene, responsible from 20 February 1941 for a special column on race-crossing, "mixing" was "a crime against God" and those who did not possess this "natural religious sentiment" should be judged and condemned as "abnormal."[68] Against the "upsetting" number of hybrids in the world, calculated by Pensabene to be in the order of 67 million—"an eighth of the Aryan population"—a rigid control of migratory flows was

58 Raffaele D'Anna Botta, "Meticciato," *La Difesa della razza* 5, no. 12 (20 April 1942): 22.

59 Guido Landra, "Studi italiani sul meticciato," *La Difesa della razza* 3, no. 10 (20 March 1940): 8.

60 Landra, "Studi italiani sul meticciato," 8.

61 Eugen Fischer, "I bastardi di Reoboth," *La Difesa della razza* 3, no. 10 (20 March 1940): 12–17.

62 Charles Davenport and Morris Steggerda, "Mulatti di Giamaica," *La Difesa della razza* 3, no. 10 (20 March 1940): 18–24.

63 Wolfgang Abel, "Meticci di Renania," *La Difesa della razza* 3, no. 10 (20 March 1940): 26–30.

64 Yun Kuei Tao, "Incroci fra cinesi ed europee," *La Difesa della razza* 3, no. 10 (20 March 1940): 33–38.

65 Johann Schaeuble, "Il meticciato nell'America del Sud," *La Difesa della razza* 3, no. 10 (20 March 1940): 46–49.

66 Rita Hauschild, "Gli incroci negro-cinesi," *La Difesa della razza* 3, no. 10 (20 March 1940): 52–53. On Hauschild's researches, see also Guido Landra, "Il problema degli incroci a Trinidad e nel Venezuela," *La Difesa della razza* 6, no. 2 (20 November 1942): 14–16.

67 Lorenzo Rocchi, "Razzismo nel Nord-America," *La Difesa della razza* 3, no. 8 (20 February 1940): 30. See also A. L, "Il razzismo nord-americano," *La Difesa della razza* 2, no. 1 (5 November 1938): 22–23; Giuseppe Ficai, "S.O.S. degli antirazzisti," *La Difesa della razza* 2, no. 11 (5 April 1939): 38–39.

68 Giuseppe Pensabene, "Il meticciato delitto contro Dio," *La Difesa della razza* 4, no. 8 (20 February 1941): 26–27.

needed, together with a high level of guardedness, to combat those "moral" hybrids making up the antechamber of biological hybrids.[69] According to Guido Landra, aggravating this problem during the war was the presence of "troops of colored people that democracy has unwittingly brought to fight on our continent."[70] From here the necessity—particularly supported by Landra—was to "in time identify hybrids and take the necessary preventive measures so that the blood of old Europe is not irremediably poisoned."[71] Embedded in the eugenic problem of hybrids, therefore, were the conflicting views of democratic egalitarianism and fascist racism: toward the "humanity of bastards" favored by the former, the latter advocated a "program of defense and optimization."[72] If liberalism and democracy opened the door to hybridism and biological confusion, creating—as the cases of France and Latin America demonstrated—this racial chaos which had its origins in social and political crises,[73] then Fascism was held up as a return to the natural (and divine) order of racial separation, source of biological health, not to mention social and political stability and security.

Regarding the problem of elimination of defectives through compulsory premarital certification and sterilization, the model was clearly German and American negative eugenics. In August 1938, a proposal for collaboration with *La Difesa della razza* was put forward by Germana Maulini—director, before the Spanish civil war, of the Meomenista Institute in Barcelona and then head of the Borgomanero Physiotherapy and Physical Re-education Studio—and her secretary, the physician Carlo Cosimo Borromeo, cultivator of "anthropo-biological aesthetics." Landra rejected the proposal with these words:

> The aim of racism is not to bring abnormal individuals to a normal level or to correct physical imperfections, but to increasingly defend and optimise the best elements of race. It seems to me therefore that the operations of

69 Giuseppe Pensabene, "Le due cause maggiori del meticciato nel mondo," *La Difesa della razza* 4, no. 10 (20 March 1941):10–12.

70 Guido Landra, "Il problema dei meticci in Europa," *La Difesa della razza* 4, no. 1 (5 November 1940): 15.

71 Landra, "Il problema dei meticci in Europa," 15.

72 Felice Graziani, "I meticci nella storia," *La Difesa della razza* 5, no. 15 (5 June 1942): 16–17.

73 On France, see: Nicola Marchitto, "Il meticciato e la Francia," *La Difesa della razza* 2, no. 10 (20 March 1939): 38–40. See also Giorgio Almirante, "Una razza alla conquista di un continente," *La Difesa della razza* 2, no. 1 (5 November 1938): 20–21; Roberto Raineri, "Il problema razziale brasiliano," *La Difesa della razza* 3, no. 7 (5 February 1940): 39–42; Ettore De Zuani, "Problemi razziali nell'America Latina," *La Difesa della razza* 2, no. 18 (20 July 1939): 11–13.

> Professor Maulini—though worthy from a humanitarian point of view—cannot find a place in the pages of "Difesa della razza," which has other precise objectives.[74]

Among the collaborators of *La Difesa della razza*, Lidio Cipriani was one of the most active advocates of the Nazi eugenic model. Although he underlined that the "Italian spiritual unity" was grounded on a specific "biological base,"[75] nevertheless Cipriani claimed, in any case, that the race-nation originated from a "melting-pot" of different human types and that, at the center, the elite represented the expression of "the most well-endowed ethnic element."[76] An article dedicated to this theme, titled *Miscugli di razza* [Mixture of race], was unsurprisingly censored by the Ministry of Popular Culture, because—as Landra wrote to Cipriani in August 1938—"for our politicians there exists in Italy only *one race*."[77] The theoretical basis of differential fertility was that "fecundity is in inverse ratio to the eminence of the physical and mental endowments, not to mention economic conditions."[78] Therefore, according to Cipriani, it was the fundamental work of fascist racism to "stimulate [...] the reproduction of the best through the publication of eugenic principles, with economic provisions and an appropriate exaltation of patriotic sentiments."[79] At last, to protect and favor "the diffusion of the best endowed," in the context of the national body, the Florentine anthropologist sent a memorandum to the Ministry of Popular Culture dated 15 July 1938, in which he proposed a proper project of "surveillance" of migration within the peninsula, with the aim of impeding to the utmost the "darkening" of Italian "racial groups."[80] Specifically, Cipriani advised "caution as regards secret policies of movement of ethnic groups onto Italian soil," hypothesizing an optimization of the most bio-

74 Guido Landra to Carlo Cosimo Borromeo, n.d. (August 1938), in ACS, Ministero della Cultura Popolare, Gabinetto, b. 151, f. "Collaboratori Ufficio Razza," sf. "Borromeo Carlo Cosimo."

75 Lidio Cipriani, "Unità spirituale degli italiani," *Corriere della Sera* (5 August 1938).

76 Cipriani, "Unità spirituale degli italiani."

77 Guido Landra to Lidio Cipriani, 24 August 1938, in ACS, Ministero della Cultura Popolare, Gabinetto, b. 151. The text of the censored article is conserved in ACS.

78 Lidio Cipriani, "La razza e la vita delle Nazioni," *Corriere della Sera* (3 December 1938).

79 Cipriani, "La razza e la vita delle Nazioni."

80 Lidio Cipriani to the Ministry of Popular Culture, 15 July 1938, in ACS, Ministero della Cultura Popolare, Gabinetto, b. 151.

logically favorable, accompanied by the "elimination of those ethnic groups judged undesirable." The reference to Germany and to the sterilization of the "Rhineland bastards" was implicit:

> The solution to an analogous problem, directly regarding the intention to eliminate certain ethnic groups judged undesirable, is being sought in secret in Germany, where it could be useful to inform ourselves about the exact methods adopted. From the other point of view, however, favoring the mixing of certain of our ethnic groups could signify the creation of new energies in the development of the country; but there exists a need to see clearly, in order to control the phenomenon in the most efficient way.[81]

With this point of view, it is not surprising that Cipriani considered Nazi eugenic legislation as a model to imitate. In an article published in *Gerarchia* in December 1939, and again in *La Difesa della razza* in January 1942, was an exaltation of the German *Rassenhygiene,* as much on a theoretical and scientific level as on a practical and political one.[82]

Cipriani's admiration for these negative eugenic methods was shared within the editorial board of *La Difesa della razza* by Marcello Ricci and Guido Landra. The former, in January 1939, dedicated a long review to the problem of premarital certificates, in which he analyzed the legislative situations in the United States, Germany, France, Sweden, Norway, Holland, Russia, Mexico, Argentina and Turkey.[83] The latter, in June 1941, after having also detailed the eugenic legislations of Europe and America, hoped that Fascism would also confront the "problem of race" in its "eminently qualitative aspect."[84] At the end of 1941, the Sicilian doctor Aldo Modica repeated that eugenic premarital controls were the foundation of the "proven transmittability" of "hereditary illnesses" and "degenerative psychical characteristics," of "damages ascertained as present in parents of those with serious constitutional illnesses," of "lesions that the sickness of one spouse could carry to the other, or to their generative capacity," of "dominance that has the characteristics of racial deterioration or inferior-

81 Cipriani to the Ministry of Popular Culture, 15 July 1938, in ACS, Ministero della Cultura Popolare, Gabinetto, b. 151.

82 Lidio Cipriani, "Le scienze antropologiche nella Germania hitleriana," *Gerarchia* (December 1939): 787–791.

83 Marcello Ricci, "Eugenica e razzismo," *La Difesa della razza* 2, no. 6 (20 January 1939): 22–23.

84 Guido Landra, "Il certificato prematrimoniale," *La Difesa della razza* 4, no. 16 (20 June 1941): 24–25.

ity as respects the characteristics of the race in which psycho-physical evolution has reached a superior refinement."[85]

In March 1942 it was again Landra who attacked the "purely quantitative population policy" of the fascist regime, calling in a loud voice for a political intervention aiming at "genetically improving the Italian population, impeding the increase of the worst, and instead favoring that of the best."[86]

As in other situations, however, it was the readers' column *Questionario*, the home of the campaign stirred up by *La Difesa della razza*, which supported a "negative" eugenics based on sterilization and obligatory premarital certificates. It was inaugurated by the opposing views of two "*camerati*," readers not well identified, named Vassetti and Falanga. For Vassetti, sterilization, in order to be effective, had to be obligatory; while for Falanga, the combination of voluntary sterilization and obligatory premarital certificates could be seen as a point of compromise between the fascist state and the Catholic Church:

> On the problem of instituting an obligatory premarital certificate, which involves the socially vast problem of love, the intervention of the State would certainly displease the individual; and the Church could not allow illegitimate unions to take place under its eyes, evading the civil laws and keeping individuals in a state of sin. From its side, the State, and here I refer to the Italian one, can not indefinitely remain a spectator in things of such vital importance for the health of the race.[87]

The invitation from the editorial board for contributions from readers discussing sterilization and premarital certificates, without "ignoring" or "undervaluing" the question,[88] was immediately taken up. While Aldo Modica and Lidio Cipriani dedicated long articles to premarital controls and sterilization respectively, citing Nazi legislation as a model to imitate, the *Questionario* was flooded with the views of its readers. Eleonora Villani underlined the "human side of sterilization," citing the pitiable case of two parents and their son affected by a "terrible defect"; Giambattista Volta

85 Aldo Modica, "Il certificato prematrimoniale," *La Difesa della razza* 5, no. 4 (20 December 1941): 30.
86 Guido Landra, "Fondamenti biologici del razzismo," *La Difesa della razza* 5, no. 10 (20 March 1942): 7.
87 "Questionario – Pro e contro la sterilizzazione," *La Difesa della razza* 5, no. 1 (5 November 1941): 31.
88 "Questionario – Pro e contro la sterilizzazione," 31.

proposed the "coupling of individuals with defects of antithetical character, in which case we might hope that in the product the defect of one might compensate for the opposite defect of the other"; Aurelio Migotto maintained that man "must be made innocuous" through sterilization, but repeated the eugenic importance of social policies to promote the birth rate and combat the development of large cities, which were "foci of infections that corrode the moral endowments of the race." Finally, Lorenzo Falanga called for a major spirit of collaboration on the part of the Catholic Church.[89]

The first critical note came in December 1941 from Claudio Del Bo: the weak debate in Italy on sterilization was not due to the lack of preparation of Italian scientists, but "to 'Italian' aspects that the problem assumes in our homeland, arising from the Mediterranean-Latin character of our people, who have always been well-balanced and rich in religious sentiments that directly or indirectly represent the Roman church." Pathological heredity remained a problem defined and clear from a scientific point of view. The Catholic Church had already expressed its total refusal in the encyclical *Casti Connubii*: the introduction of premarital certificates would favor illegitimate unions, threatening the "moral custom of the family [...] at the base of the social organism." Moreover, it was important not to forget the influence of the "environmental factor" and the effectiveness of a "quantitative" policy, more than a "qualitative":

> Therefore we must persevere in the means already taken by Fascism, intensifying the fight against social illnesses; encouraging the procreation of healthy and strong people; promoting, especially among the young, a healthy and sporting life; creating, in sum, a eugenic environment that is such as to eliminate, or at least limit, defects believed to be hereditary.[90]

This was an "environmentalist" position, provoking an immediate admonishment from the hereditarian Telesio Interlandi, editor of *La Difesa della razza*:

> We wish to advise *camerata* Del Bo to study the problems of heredity with greater attention. Since he wishes to give these factors a relative importance,

89 "Questionario – Pro e contro la sterilizzazione," 30–31.

90 "Questionario," *La Difesa della razza* 5, no. 4 (20 December 1941): 30.

> and in every case not a categorical one, he must realize that—should heredity fall—one of the pillars of racism falls with it: of true scientific racism, and not that which feeds on spiritualistic aims.[91]

In January 1942, Vassetti intervened again to dismantle Del Bo's argument: sterilization could not have "Italian" aspects, because biological defects did not change according to different races. Pathological heredity followed Mendelian mechanisms of transmission of characteristics, as the "Nazi laws on hereditary and degenerative defects" demonstrated, fruit not of "political or racial fanaticism," but of scientifically "controlled elements." The environment was a "concomitant and not determinate factor" in the development of the "hereditary germ."[92] Del Bo responded that if the transmission of characteristics was proven it was not however demonstrated that it respected fixed laws: "heredity exists but it is not provable case by case," in which case, what sense would it have to sterilize the carrier of an illness such as syphilis, transmittable not only through the sexual act? Would it not be better to isolate him? "Belief in a scientific program," Del Bo concluded, "is not a pure act of faith, but a duty to the evolution of the individual and civilization." With the practice of sterilization we would come to lack "that tenacity in the fight against illness that transforms a doctor into an apostle."[93]

Beyond the usual nature/nurture diatribe, the debate finished locked around a critical argument advanced by Raffaele D'Anna Botta in February 1942. The rejection of sterilization was justified on the grounds of anti-Semitic ideology:

> "Sterilization"—insidious weapon of scientific decadence—is nothing less than a practice that... every day practices, in the guise of science, pedantic and professional Judaism. The proof is the daily damage of Jewish and "Jewified" gynecologists, who, with the excuse of intervening to eradicate an ill... hypothetically serious, give hysterectomies to all the women who fall into their hands, without distinction, making them more ill than before, as well as rendering them sterile and infertile for all their lives.

91 "Questionario," *La Difesa della razza* 5, no. 4 (20 December 1941): 30.
92 "Questionario," *La Difesa della razza* 5, no. 6 (20 January 1942): 30.
93 "Questionario," *La Difesa della razza* 5, no. 9 (5 March 1942): 23.

> They could not be serving Israel better! In other terms, applying "sterilization" will promote the great destructive plan of Jewish Messianism which consists of the extinction of all the Nazarenes.[94]

Instead of sterilization, seen as a homicidal instrument in the hands of the international Jewish conspiracy, D'Anna Botta offered "desaprophytisation" (*desaprofitizzazione*), that is, the elimination of hereditary pathological pollution, invented by the "very Italian" doctor Pier Nicola Gregoraci."[95] According to D'Anna Botta, this practice had been introduced at the end of the first decade of the 1900s, when it had been unsurprisingly "attacked" by the "cabal of silence of the Jewish-Masonic sector."[96]

In the editorial board of *La Difesa della razza,* the campaign advocating the constitutional "Gregoraci doctrine" against "Jewish" sterilization did not find, however, a favorable hold. What was this "desaprophytisation," exclaimed Carlo Vassetti, in March 1942, refuting the theory of the "Jewishness" of sterilization:

> Certainly the accusation of Jewishness directed against the predominant medical science is one that will be particularly seductive to the observers of this debate [...]. Stating first that luminaries and scientists of undoubted Italianness as well as adamantine scientific faith belong and have always belonged to this predominant medical science, against which our contradictor strongly fights. I offer my opinion that a better polemic garb should at least be adopted in launching such fiery, intransigent censure, which is as generic as it is categorical. I object, if for no other reason, because the poisonous, symptomatological, localist, hyperscientific denunciations made by Professor D'Anna cannot touch the clear fame of scholars and scientists, who have been asked for

94 Questionario, *La Difesa della razza* 5, no. 8 (20 February 1942): 31.

95 In 1935, Giovanni Preziosi and Roberto Farinacci conducted a campaign of support for the neo-Hippocratic methods of Pier Nicola Gregoraci, against the growing political and scientific influence of Nicola Pende and against the "massonic conspiracy" of bacteriological medicine. See Giovanni Preziosi, "Parlo di pier Nicola Gregoraci," *La Vita Italiana* 23, no. 263 (February 1935):147–48; Pier Nicola Gregoraci, "La mia nuova Dottrina," *La Vita Italiana* 23, no. 263 (February 1935): 149–55; Giovanni Preziosi, "Fatti e commenti. Il caso Gregoraci," *La Vita Italiana* 23, no. 265 (April 1935): 515–18; Giovanni Preziosi, "Il caso Gregoraci," (with letters from Davide Giordano, Giacinto Viola and Benedetto Schiassi), *La Vita Italiana* 23, no. 267 (June 1935):788–92; Giovanni Preziosi, "Fatti e commenti. Scienza nuova?!...," *La Vita Italiana* 23, no. 268 (July 1935):99–100; Giovanni Preziosi, "Fatti e commenti. Probità scientifica," *La Vita Italiana* 23, no. 268 (July 1935): 100–01. On D'Anna Botta's positions, see also: Raffaelle D'Anna Botta, "La pseudo-scienza ebraico-massonica contro il genio italiano," *Il Tevere* (30–31 March 1939): 3.

96 Questionario, *La Difesa della razza*, 5, no. 8 (20 February 1942): 31.

> many decades of fatigue and faith, and were anything but obstinate, amphibious, or Jewish men.[97]

And for once Falanga was in agreement, in this case, with Vassetti. Declaring his sincere ignorance of the nature and etymology of "desaprophytisation," he explicitly maintained the neutrality of medical methodology in respect to the racial membership of the person applying it:

> The fact that Jewish doctors claim to exterminate the race of the goyim, or that a Jew invites the destruction of the German people, need not induce the gentiles to repudiate those methods to reach higher aims.[98]

With the exception of writings by Giuseppe Chiesa,[99] Gino Valisfanio[100] and "*camerata*" Giviani,[101] who all intended to demonstrate the denigrating campaign conducted by "Jewified" medicine against Gregoraci, the absence of a reply on the part of D'Anna Botta about the effective "worth of desaprophytising methods"[102] led the debate in *La Difesa della razza* to a dead end. An attempt to re-launch the argument was made, in August 1942, by the *camerata* Falanga. After a brief synthesis of the phases of the discussion and conclusions reached, Falanga repeated the importance of sterilization and premarital certificates in the political, moral, religious and scientific arenas:

> For politics, we are dealing with the necessity of defending the health of the race [...].
> For morality, it is a question of not offending that sense of human dignity that is in everyone: the drive to know they will continue through their children, beyond fleeting parentheses of individual lives.
> For religion it is a question of safeguarding the right to have offspring, avoiding a conflict between the conscience as citizen and the conscience as believer.
> For science, in the end, it is necessary to cure them through the most effective means, rather than prevent the propagation of hereditary ills, in order to guarantee healthy offspring.[103]

97 Questionario, in *La Difesa della razza* 5, no. 10 (20 March 1942): 23
98 Questionario, in *La Difesa della razza*, 5, no. 11 (5 April 1942): 22.
99 Questionario, in *La Difesa della razza*, 5, no. 14 (20 May 1942): 22.
100 Questionario, in *La Difesa della razza*, 5, no. 16 (20 June 1942): 22.
101 Questionario, in *La Difesa della razza*, 6, no. 6 (20 January 1943): 22.
102 "Questionario – Sterilizzazione," *La Difesa della razza* 5, no. 20 (20 August 1942): 22.
103 "Questionario – Sterilizzazione," *La Difesa della razza* 5, no. 20 (20 August 1942): 22.

Unexpectedly in April 1943, a letter appeared from Giovanni De Santis, a country medical practitioner from Rapagnano (Ascoli). Notwithstanding his reservations regarding the editorial staff of *La Difesa della razza*, De Santis supported the "Italianness" of "desaprophytisation" and the "Jewishness" of compulsory sterilization.[104]

The debate surrounding transmission of undesirable hereditary characteristics seemed, however, to find a new home in *Genetica* [Genetics], a column of *La Difesa della razza*, headed by Aldo Modica (under the pseudonym *Mod*) between August 1942 and June 1943. The tone of these "lessons" was maintained, for the most part, on a pseudoscientific theoretical level, but the rigid Mendelian-Weissmanian formulation, adopted to describe the "immortality" of the "germline,"[105] implicitly justified the foundation of negative eugenic methods. Neither prayers, nor the power of future post-natal prophylactic techniques—stated Modica in February 1943—could counteract these immovable, deterministic genetic theories:

> Neither the case under discussion, nor destiny and the imponderable divine are useful in this hard-and-fast genetic determinism. It is not possible, with a candle, no matter how big, or with a cycle of prayers to the sainted protector, to impede the genes of an illness or a specific abnormality, which might affect an entire line of descendants, when it does not emerge immediately in the direct descendent in a fatal way. It is not possible—not with medical solutions, post-natal prophylactics, or immense spending—to neutralize that gene or group of genes transmitted in associated or isolated forms, which "must" determine those pathological factors or malformations in the son, the daughter, or the near and removed descendents.[106]

If biology was destiny, then the only solution possible was a eugenic system that eliminated defective hereditary traits.

---

104 Questionario, *La Difesa della razza* 6, no. 11 (5 April 1943): 22.

105 Genetica, *La Difesa della razza* 5, no. 21 (5 September 1942): 21.

106 Genetica, *La Difesa della razza* 6, no. 8 (20 February 1943): 21.

2. Environmentalist Eugenics: Psychological and Anthropo-geographical Racism

La Difesa della razza interpreted racial eugenics in terms of discontinuity regarding previous social, health and demographic fascist policies. In contrast, *Razza e civiltà* [Race and civilization], a monthly journal of the General Council and General Directorship of Demography and Race (*Consiglio Superiore e Direzione Generale per la Demografia e la Razza),* provided a continuist interpretation.

This was evident from the first issue of the journal, in Carlo Bergamaschi's celebration of the virtues of the National Organization for the protection of Motherhood and Infancy, known as ONMI (*Opera Nazionale Maternità e Infanzia*): "The first time that the *Duce* decided on the terms of a concrete defense of the race was when, in his Ascension Day speech [26 May 1927], he mentioned the creation and the demise of ONMI."[107] In the "sector of racial defense"[108] represented by ONMI, it was not a matter of innovation or transformation, but simply of "intensifying work that has already been well begun, with new energy, and with criteria better adopted to the scope of the project," and to "accentuate the action in the pre-natal and post-natal sectors" in such a way as to obtain "an active reclamation of the race." The ethnologist Alfredo Sacchetti underlined the connection between sport and race. He identified a scientific relationship between the development of competitive activity and the "ascendant and culminating" phases of the evolutionary cycle of the history of nations as theorized by Gini. Sacchetti endorsed, on this basis, "a new program that should above all interest the new positive eugenicists and with advantages not only for current society [...], but also directly for the species, and therefore future society."[109] Praise for the demographic campaign as an instrument for quantitative and qualitative racial betterment was repeated and constant. In par-

107 Carlo Bergamaschi, "L'Opera Nazionale Maternità e Infanzia per la difesa sociale," *Razza e civiltà* 1, no. 1 (23 March 1940): 91; Bergamaschi was a Commissioner of OMNI. On the racist ideology expressed by *Razza e civiltà,* see M. Masutti, "La rivista 'Razza e civiltà': un aspetto del razzismo fascista," *Sociologia* I (2002): 83–100.

108 Bergamaschi, "L'Opera Nazionale Maternità e Infanzia per la difesa sociale," 97.

109 Alfredo Sacchetti, "Sport ed evoluzione dei popoli," *Razza e civiltà* 1, no. 2 (April 1940): 238.

ticular, the physician Giuseppe Tallarico reiterated, in all of his works, a precise link of continuity between fascist pronatalism and racism. In May–June 1940, he listed, for example, a long series of proofs of "experimental demography" in the agricultural and husbandry sectors, that demonstrated how in every natural context, whether vegetal or animal, an enhancement in "quantity" always led to an improvement of "quality." From here, he immediately criticized Anglo-Saxon eugenics as unacceptable from a fascist pronatalist point of view:

> And yet, does eugenics suggest that the firstborn or the first births will be the best and the most successful? Nothing confirms this implicit presumption... indeed, the biographies of grand men demonstrate the opposite, and Kattel [sic] has highlighted that often there is the need for a long line of children before the possible combinations of hereditary factors of genius and mental superiority are achieved and brought together in such a way, in a happy combination and in perfect harmony, to lead to men of genius and superior person. Only large families can improve the race, because only in the golden secret of their "number" can we find the most effective means of improvement, discovering in the hand of chance or destiny the greater number of hereditary combinations, the meeting and melding of chromosomes, the various affirmative and selective possibilities.[110]

In the following issue, Tallarico again wrote at length about the merits of fertility, referring both to the biological and psychological capabilities of the mother, and to the "birth-rate of a nation," in which "hypergenesis" and "eugenesis" coincided: "from numbers and mass come individuals of quality, while from lower numbers and limited fertility come modest physiologies and low constructive yields."[111] In March 1941, he investigated the relationship between human prolificacy and nutrition:

> The factors that influence human prolificacy and the birth-rate of the population are multiple, have higher or lesser importance, and are racial, environmental, social, psychical, moral and religious and also nutritional, because nutrition and man's birth-rate are linked more intimately than is commonly known.[112]

110 Giuseppe Tallarico, "Il numero è anche la qualità," *Razza e civiltà* 1, no. 3–4 (May–June 1940): 288–89.

111 Giuseppe Tallarico, "I pregi biologici della fecondità," *Razza e civiltà* 1, no. 5–6–7 (July–September 1940): 81.

112 Giuseppe Tallarico, "L'alimento e la prolificità umana," *Razza e civiltà* 2, no. 1 (23 March 1941): 81.

Together with this continuist interpretation of racism, a second element that distinguished the eugenic ideology of *Razza e civiltà* was undoubtedly the adoption of the environmentalist paradigm as a cornerstone of the mechanism of transmission of biological characteristics of the human species.

In October 1940, for example, the influence of hereditary predisposition as regards to alcoholism was particularly emphasized by Arnaldo Fioretti, head of the Doctors' and Nurses' Union. Alcoholism was essentially a "family and social" illness, and, as such, did not necessarily produce genetic degeneration and addiction:

> Until today, heredity, intuited by Bianchi, has been nothing but pure hypothesis. If this were the truth, we would despair of finding a cure for this illness, since heredity is something fatal, difficult to modify: chromosomes, and genes within chromosome's, resist strong force to alter their structure or modify their orientation, and we could hardly hope to effect a cure, when alcoholism could, in this way, be inherited by the entire population.[113]

This did not signify, however, that an anti-alcoholic policy should not be reasonably included in "a serious, prolonged, and wise attempt at racial reclamation":

> This means: curing the white race of those exogenic and endogenic factors that conspire actively against it. It is useless to defend it from contact or contagion from the inferior races if we must abandon it without defense to the vices that threaten its resistance and that compromise descendants.[114]

The Physician Luigi Cesari, in an article entitled *A question of racial reclamation: for the children of neuropsychotic defectives,* expressed many reservations on the possibility of a rigid application of Mendel's laws to the problem of hereditary transmission of human characteristics:

> With his laws, Mendel opened a horizon of some interest, but his concepts cannot find in man the scientific control for an exact evaluation of the transmissibility of hereditary characteristics. This is fundamental for genetics, which, with an increase in abundant scientific observations in recent years, has seen the addition of new laws to Mendel's classic ones.[115]

---

[113] Arnaldo Fioretti, "Lieo bifronte," *Razza e civiltà* 1, no. 8 (October 1940): 587.

[114] Fioretti, "Lieo bifronte," 591.

[115] Luigi Cesari, "Una questione di bonifica della razza: per i figli dei tarati neuropsichici," *Razza e civiltà* 1, no. 1 (23 March 1940): 75.

Cesari kept a clear distance from the practical applications of the eugenic hereditarianism, represented by the Nazi laws of sterilization:

> Gutt, Rüdin, and Ruttke have largely studied heredity in a series of illnesses and their deductions have resulted in Germany's laws of sterilization—the most complete, certainly, but also debatable because the hereditary prognosis, which should be absolutely certain, and not leave room for scientific scruples, or those of conscience [...], is based on the simple determination of facts derived from experience, on calculations of percentages of sickness in the descendants of the infirm, etc.; in other words, on empiricism.[116]

Cesari proposed an alternative to Nazi sterilization to resolve the problem of the inheritance of nervous and mental diseases, based on a "sure diagnosis" and "reconstruction of the genealogical tree." Even though he recognized the importance of the statistical frequency of mental disease and its degenerative impact on the "white race," Cesari believed the question could not simply be resolved through surgery, but must be confronted through the enhancement of "preventive and curative assistance for sick children."[117]

> Let us remember that progressive nations have, by now, adopted precautions to ensure childhood mental health services, both preventive and corrective. Let us remember that the mass of abnormal, different people has no precise limits, but is made up of diverse types not well classified. Let us remember that the problem is arduous and complex.[118]

Again in *Razza e civiltà*, Giuseppe Tallarico offered an analogous condemnation for the practice of sterilization in the name of "environmentalism":

> In expectation of the ideal remedy of pure eugenics committed to eradicating the bad from the germ plasm, the environmental remedy has again gained power and consistency: that is, the utilization of external factors; above all nutritional, functional, hygienic, economic, social, and educational, which have much value in the making of a man [...]. External factors with not only an indisputable individual action, but also a racial one, able, it seems, to change, through intense and constant action, the constitution of the genus.[119]

[116] Cesari, "Una questione di bonifica della razza," 76.
[117] Cesari, "Una questione di bonifica della razza," 80.
[118] Cesari, "Una questione di bonifica della razza," 81.
[119] Giuseppe Tallarico, "Il problema degli incroci," *Razza e civiltà* 3, no. 1 (January–March 1943): 479.

It was undoubtedly important to Tallarico to impede "crossings between defectives so that they would not harm the race," but it was above all necessary to "favor marriages between families who should be members of the race's 'golden book' by means of prizes and propaganda." In the same way, it was important to evaluate every single race crossing: if, in general, marriages with the "negro race," marked by the mental and physical stigmata of infantilism,[120] were to be prohibited by law, as the main cause of social and moral biological disaster for the white race,[121] "family telecrossing," that is, the "mesogamic crossing between the different stocks of the same race and between the ethnic families of the same people,"[122] had to be enhanced. It was toward such objectives that the Internal Migration Commissariat (*Commissariato della Migrazione Interna*) had to direct its political action:

> Internal migration encourages unions between individuals from different regions, favoring mesogamic crossings between stocks of the same race, and between ethnic families of the same people [...], so diluting the blood relations of the race. This will demonstrate to ourselves and the world that the best agent of happier chromosomal coupling to improve the Italian race and produce new universal genius will come from the renewal of the ancient Italian genetic material without the need to go outside borders.[123]

Even when deterministic hereditarianism was reaffirmed, it did not imply negative eugenics, but rather the exaltation of a policy of protection of the fascist woman-mother as a factor of "conservation of the race." This view was supported, for example, by Cesare Serono, director of the National Medical-pharmacological Institute, by recycling the eugenic and statistical model of the "average man" in a racist and sexist interpretation:

> Man, the energetic and directive element of reproductive function, is the average man, healthy and balanced; and if the woman who is his mate is an intelligent being, even in a latent state, this will result in offspring with superior qualities to the norm. In this way, therefore, as Carrel has said, we must clearly

120 Tallarico, "Il problema degli incroci," 496.
121 Tallarico, "Il problema degli incroci," 481.
122 Tallarico, "Il problema degli incroci," 476.
123 Tallarico, "Il problema degli incroci," 476.

> separate the tasks of the two sexes, forbidding to women excessive education, jobs and pastimes reserved for men, if we do not wish to create hybrids that are ill-adapted to social life.[124]

Beyond the rejection of the Mendelian–Weismannian model, environmentalist racial eugenics was also distinguished, in some of its expressions, by the development of a racial concept based not on references to genotypes as much as the evaluation of "ecological" dimensions of the habitat. This was the case, for example, of the "anthropo-geography" of Giovanni Marro, professor of anthropology at the University of Turin and senator since 1939, and Edoardo Zavattari, director of the Institute of Zoology at the University of Rome: both assiduous collaborators of *Razza e civiltà*, but also of *La Difesa della razza*. The early eugenicist Giovanni Marro, the son of anthropologist and psychiatrist Antonio Marro, wrote a series of articles between 1938 and 1943, published in *La Stampa*,[125] Preziosi's *La Vita Italiana* [Italian life], *Razza e civiltà* and *La Difesa della razza*, and summarized in the synthesis *Primato della razza italiana* [Supremacy of the Italian race].[126] In these contributions, Marro interpreted the racial and anti-Semitic fascist campaign as the ideal political fulfillment of his previous scientific research in two different fields: prehistoric archaeology, as founder and director of the Turin University Museum of Anthropology and Ethnography; and Egyptology, as a member of Senator Schiaparelli's Italian Archeology Mission (*Missione Archeologica Italiana*), and curator of Bernardino Drovetti's vast collection of Egyptian antiquities.

The "new racial science"—which Marro proposed to re-baptize the "history of the human race"—implied abandoning traditional "zoological" and "morphological" physical anthropology. In polemical discussion with the

124 Cesare Serono, "L'importanza del fattore femminile nella conservazione della razza," *Razza e civiltà* 2, no. 8–12 (October 1941–February 1942): 681.

125 In particular, see Giovanni Marro, "Il primato della razza italiana," *La Stampa* (30 July 1938); Giovanni Marro, "La razza italiana e l'ambiente," *La Stampa* (5 August 1938); Giovanni Marro, "Il problema delle origini della razza italiana studiato attraverso il materiale raccolto nel Museo di Antropologia," *La Stampa* (12 August 1938); Giovanni Marro, "La razza italiana e il suo linguaggio," *La Stampa* (24 August 1938); Giovanni Marro, "Egiziani, Fenici, Ebrei nella civiltà mediterranea," *La Stampa* (17 August 1939) (30 August 1939) and (22 October 1939); Giovanni Marro, "La razza italiana e il suo ambiente naturale," *La Stampa* (23 February 1940).

126 Not surprisingly, the volume was dedicated to the memory of his father, Antonio Marro, "pioneer of racial eugenics"; see Giovanni Marro, *Primato della razza italiana: Confronti di morfologia, biologia, antropologia e di civiltà* (Milan–Messina: Giuseppe Principato, 1940), 2.

approach of anthropologists Giuseppe Genna and Sergio Sergi,[127] Marro refused every "materialistic" interpretation of race:

> Many current anthropologists—under the influence of the positive or materialistic School (which also has great merits for its contribution to scientific progress) that is today in strong decline due to the press of different beliefs more suited to modern knowledge and aspirations—continue still to consider themselves simply as the zoologists of Man. [...] Exactly for the excessive restrictions that they themselves establish for their object of study, anthropologists often preclude the possibility of studying in depth, with appropriate guidelines and sure criteria, many questions and many problems that currently occur in the racial sciences.[128]

The most recent scientific classifications of the races—on which Marro often discoursed at length[129]—confirmed, in reality, the widespread tendency to enlarge the spectrum of taxonomic criteria from the exclusive consideration of "biological" factors toward the more "spiritual." It was therefore neither craniology nor anthropometry, but rather the synergy of man and environment that constituted the axis of Marro's racism.

In man, according to Marro, two elements could be recognized: one physical-somatic, the other spiritual. The first, common to all animal species, was "fatally subject to the modeling influences of the environment."[130] The capacity to oppose the environmental influence came, on the other hand, from "spiritual" elements, and varied according to the level of progress achieved and by the "particular characteristics of each ethnic group."[131] Therefore, while "primitive" man appeared strictly linked to the habitat in which he lived, "civilized" man was able to modify the natural environment that surrounded him. Consequently, it was this same human prog-

---

127 For Marro's polemic against Genna and Sergi, see Giovanni Marro, "Un allarme per il razzismo italiano," *La Vita Italiana*, 29, no. 336 (March 1941): 237–51. On Genna and Sergi, and the events of physical anthropology in Italy, see Pogliano, *L'ossessione della razza*, 369–440.

128 Giovanni Marro, "Nuovi orientamenti nella scienza razziale," *La Vita Italiana*, 29, no. 341 (August 1941): 139.

129 See Marro, *Primato della razza italiana*, 49–61; Giovanni. Marro, *Caratteri fisici e spirituali della razza italiana* (Rome: Istituto Nazionale di Cultura Fascista, 1939), 25–30.

130 Marro, "Nuovi orientamenti nella scienza razziale," 141–42.

131 Marro, "Nuovi orientamenti nella scienza razziale," 142. See also, Marro, *Caratteri fisici e spirituali della razza italiana*, 8.

ress, according to Marro, that made a purely "zoological" conception inadequate in considering race. In fact, "from a somatic point of view, human progress carries out—even indirectly—a leveling action on the human species, tending in reality to attenuate physical racial divergences, determined for the most part by the diversity of natural environments."[132] Secondly, in "superior races," somatic elements included the "exteriorisation" of the psychical, a sort of tool used by the psyche "not only for escaping the coercion of the environment, but also for dominating it, according with the grand design for human beings."[133] In the end, the esogenic environmental variation, as much as the frequency of racial crossings, contributed inevitably over the centuries to modifying the "purity" of the original races.[134] The same idea of man's adaptability to the surrounding habitat therefore opposed the notion of a race that remained morphologically and psychically pure through the ages.

In the face of such criticism against biological racism, it is not surprising that the definition of race proposed by Marro favored the "psychical" factor as a component of the environmentalist raciological paradigm:

> By "race," we intend a human grouping with a harmonious complex of endowments and spiritual tendencies constituting a specific mental entity; a grouping that has a historical basis as a formative substrate, represented as an uninterrupted patrimony transmitted from generation to generation [...]. The race can, therefore, also have more or less characteristic somatic elements, some of which are subject to a variation of place and time, and are generally directed to increasing efficiency and affirmation of the singularities of psychical personality. The race becomes better characterized, insofar as the mental complex is organic, harmonious and unmistakable.[135]

On the issue of the relationship between race and environment, Marro suggested a sort of philosophy of history, principally characterized by two aspects: on one side, the "anthropo-geographical" distinction between Mediterranean and Semitic civilizations; on the other, the description of a

132 Marro, "Nuovi orientamenti nella scienza razziale," 142. See also Marro, *Primato della razza italiana*, 62–63.

133 Marro, "Nuovi orientamenti nella scienza razziale," 142. See also Marro, *Primato della razza italiana*, 64.

134 Marro, *Primato della razza italiana*, 62–63.

135 Marro, *Primato della razza italiana*, 70–71. See also Marro, *Caratteri fisici e spirituali della razza italiana*, 31.

Mediterranean "historical destiny," articulated in three successive stages—Egypt, Greece and Rome—similar in the "climatic" and "anthropical" importance of their marine systems, but differentiated by their respective relationships between ethnic components and natural environment.

Regarding the first aspect, it was the lack of attachment to the soil—from which "real and proper sentiments of homeland" come—that, according to Marro, distinguished the basic racial structure of Semitism:

> The Semitic does not know how to establish an interdependent balanced relationship with the soil: he either remains foreign or he submits. Remaining foreign, as most often happens, he harbors mistrust, if not aversion, toward the ethnic group he lives among, effectively acknowledging him as master of the territory, and so constantly assuming [...] a cautious and defensive behavior. Instead, when, for whatever reason, the Semitic and "Semiticised" develop tight links with the natural environment, so as to be moved to no longer abandon it, they submit, continuing to be subjugated to the environment also in relatively advanced stages of their evolution.[136]

Even though they were well known in history as a dominating population of the Mediterranean, Marro declared that the Phoenicians had developed an "overly unilateral and egoistic activity," limiting themselves to the role of merchants of metals without scruples, ruled only by "primordial sentiments of aggression and abuse of power,"[137] and insensible to the "superior spiritual forms."[138] For all that they had in common (lacking links with the soil and desire for money), Marro believed that the Jews represented a further degeneration of Semitism as compared to the Phoenicians: while the latter had contributed to human progress through the development of mining, the Jews had instead used their money as the "most powerful factor in the fight against others."[139] Marro's anti-Semitism merged the usual stereotypes of the Jew as *heimatlos* and moneylender into a conspiratorial imagery that interpreted the traditional religious myth of the "Chosen People" as a disguise for "reverse racism" with the aim of dominating and exploiting other peoples. At the base of what he defined as the "degenerative men-

[136] Marro, *Primato della razza italiana*, 247.
[137] Marro, *Primato della razza italiana*, 250.
[138] Marro, *Primato della razza italiana*, 252.
[139] Marro, *Primato della razza italiana*, 249.

tal constitution of the Jew," Marro identified a "proud egoism, nurturing an exaggerated sense of self, together with a low estimation of others; always posing as victims of others' domineering ways."[140] The "bias of superiority"[141] clearly translated—according to a common motif of anti-Semitic literature—into Jewish duplicity: nationalism disguised universalism and the thirst for dominance; integration masked contagious infiltration. In the biological metaphor that Marro borrowed from Cesare Lombroso—also here adopting the usual anti-Semitic strategy of "letting the Jews talk against the Jews"—the Jewish race was comparable to an octopus:

> Not only because it mimetically takes on the most varied colorization and, disturbing the water around it through a black secretion from the glands, is able to render itself unidentifiable, but particularly because, slimy creature, it is almost symbolic of evasiveness, but it itself grasps everything, and everything sticks to the tentacles and suckers around its formidable masticatory apparatus.[142]

For Marro, only the category of "degeneration" could express the danger Jews posed. The "imprint of deformation" was manifest as much on a morpho-biological level—the consequence of racial crossing between Jews and "Negroes" or frequent interbreeding[143]—as on a moral and spiritual level. Through an explicit reference to Otto Weininger, Marro interpreted Judaism as a "specific form of moral deviation, ready to implant itself in any psychical structure where, for particular reasons, there is a lack of that reactive energy inherent in the personality, which normally acts to repel and preclude the entrance and settlement [of such a moral deviation]." In the face of a similar threat, "contacts of whatever nature with the Jewish race, particularly those which involve sentimental factors, could favor their infiltration into ours."[144]

In order to support an antithetical dichotomy between "Latins" and Jews, Marro elaborated a sort of tripartite "historical system" (Egypt–Greece–Rome) in which the results of his preceding anthropological and archaeological research converged. Not only Marro's definition of race, but also

---

140 Marro, *Primato della razza italiana*, 253.
141 Marro, *Primato della razza italiana*, 255.
142 Marro, *Primato della razza italiana*, 255.
143 Marro, *Primato della razza italiana*, 260–61.
144 Marro, *Primato della razza italiana*, 264.

his philosophy of history arose from the eugenic problem of the relationship between human characteristics and habitat. Egypt—the first stage in the process of "civilization" as delineated by Marro—appeared in fact to be characterized by total subordination of the race to the natural environment:

> The Egyptian civilization sprouted, grew, matured and fell [...] in a flat and monotone natural environment, every part of which presented the same simple constitutive elements of relation. The natural environment offered complex conditions singularly favorable for the development of human life, common to all the regions of the district; above all, the sweetness of the climate and the fertile excellence of the soil. Because of this enclosure within such an alluring vessel, their civilization carried the perennial stamp of being stationary, keeping separate from the fascination of the Mediterranean, unable, as it were, to acquire that space to breathe and achieve the force of expansion that constitutes a reason for life and for advanced levels of progress. And after a cycle, although it lasted millennia, the wave of civilization broke upon them.[145]

The greatness of the Egyptian civilization—its artistic monuments, hieroglyphics, cosmology, solar myths, the power of the pharaohs—was soaked in the influence of the surrounding natural environment, and could not help but reveal the negative impact the habitat exercised on the "psychical state" of the race, manifested by their "egocentrism," "cultural, religious and social" isolation, "inadaptability"[146] and a "naive and childish psyche."[147] Similarly, when regarding the ancient Greeks, Marro believed environmental influence was the vital starting point. The particular geographic conditions of Hellas sharpened the "conquering, inventive and speculative spirit," while on the other hand impeding "profound and intimate contact between the people,"[148] which might have allowed them to overcome local separatism in the name of unity for nation and state. Egypt and Greece therefore, "in their respective evolutions and declines," represented "two stages of a fatal historical system subordinate to natural laws."[149] In this anthropo-geo-

[145] Marro, *Primato della razza italiana*, 293–94.

[146] Marro, *Primato della razza italiana*, 267.

[147] Marro, *Primato della razza italiana*, 198. See also Giovanni Marro, "La razza e l'ambiente nella civiltà. I," *Razza e civiltà* 2, no. 2 (April 1941): 224–26.

[148] On the Greeks, see Giovanni Marro, "La razza e l'ambiente nella civiltà. II," *Razza e civiltà* 2, no. 3 (May–July 1941): 438–41.

[149] Marro, *Primato della razza italiana*, 267.

graphical philosophy of history, the third stage—the power of Rome—signaled therefore a "real and proper apogee" in which the "endogenic energy" of the race overcame the conditions of the habitat:

> As the ethnic element is able to tame the natural environment and develop itself, above all in the fight against adverse natural elements, this complex of endogenic energy allowed them to achieve an early civilized maturity and ensured the potential for a continued renewal.[150]

Italy's position at the center of the Mediterranean, and Rome's in the center of the peninsula,[151] together with the influences of the Alpine arc "on the physical, anthropic and economic conditions,"[152] were presented in Marro's works as a kind of geographical "predestination" to be "the hearth of the greatest and most enduring Mediterranean civilization."[153] The variables of the natural environment corresponded with regional differentiations of the Italian race's morphological characteristics. These had remained unchanged through the centuries, notwithstanding frequent foreign invasions, as demonstrated by the analysis of anthropological types immortalized in artistic iconography and archaeological investigations of sites such as Monticello d'Alba and the Susa Valley. It was the environment, Marro claimed, that exercised a selective action capable of maintaining the "native types" of the Italian race:

> This, in reality, seems to be a characteristic of our country: it exercises a selective action, promoting the disappearance of inferior morphological characteristics and determining the persistence, the assimilation and even the improvement of those of a superior order. This explains how, notwithstanding the influx of many ethnic elements [...], the native types have always overcome, albeit with some not so important variations that do not disturb the overall balance.[154]

As a last point, it was a "spiritual unity" that connected and melded the various "regional somatic types" into a "distinctly qualifiable ethnic group"

150 Marro, *Primato della razza italiana*, 268.
151 Marro, *Primato della razza italiana*, 268–69.
152 Marro, *Primato della razza italiana*, 291.
153 Marro, *Primato della razza italiana*, 269.
154 Marro, *Primato della razza italiana*, 47. See also Marro, *Caratteri fisici e spirituali della razza italiana*, 23–24, and on this same theme, Giovanni Marro, "Dell'armonia fra razza ed ambiente naturale in Italia," *Razza e civiltà* 1, no. 2 (April 1940): 165–82.

from ancient Rome to fascism.[155] The Italian and Roman mindset[156] could be expressed, according to Marro, in its "dynamism," "pragmatic sense," tendency to "universalism,"[157] and adaptability.[158] The faces of the Italian "spiritual personality" were multiple. In the language, above all, "one of the characteristic elements of a superior race"[159] could be recognized:

> Among the languages of the Latin branch, Italian essentially conforms not only to the laws of "minimum force" in Ribot's sense [...] but also to the satisfaction caused by the synchronization and therefore graceful play of musicality in the phonetics. For this reason, Italian language is also to be considered as subordinate to the fulfillment of the aesthetic sentiment already well developed in the Italian race.[160]

Another characteristic of the Italian race—the "endogenic tendency to movement"—showed, through the course of the centuries, in multiple forms, ranging from the celebrated Roman aqueducts and the railway lines of the era of Italian political unity to the fascist roads in Libya and Albania; from the great names of Italian navigators to, more recently, the invention of the radio.[161] Marro believed that the "juridical and political thoughts of the Italian race" represented an age-old supremacy, which from the Roman Law stretched to the new fascist code,[162] while the "adaptability," the "spirit of universality" and the "adherence to reality" of the Italian race was mirrored in emigration, the resistance to sanctions, the colonization of Africa, and religious missions.[163] Marro did not hesitate to identify, in Drovetti's epistolary archives,

> a high and realistic demonstration that the Italian race has continued to produce greatly, at home and abroad, with fervor and versatility. The Italian race mostly finds harmonious agreement in the excellent elements in every job, and in the desire - more, the greed - to learn, to prove itself, to produce, to be use-

155 Marro, *Primato della razza italiana*, 44.
156 Marro, *Primato della razza italiana*, 303.
157 Marro, *Primato della razza italiana*, 278.
158 Marro, *Primato della razza italiana*, 302–03.
159 Marro, *Primato della razza italiana*, 325.
160 Marro, *Primato della razza italiana*, 325. See also Marro, *Caratteri fisici e spirituali della razza italiana*, 35–37.
161 Marro, *Primato della razza italiana*, 345. See also Marro, *Caratteri fisici e spirituali della razza italiana*, 38–46.
162 Marro, *Caratteri fisici e spirituali della razza italiana*, 46–47.
163 Marro, *Caratteri fisici e spirituali della razza italiana*, 48–51.

> ful to any community, due not only to its comprehensive mental endowments, with particular accentuation on dynamism.[164]

The final component of the "Italian psychical orientation" was the "perpetuation of the love of agriculture and arms." Here, under the perspective of a single, undivided "racial heritage," Marro's discourse included Saint Benedict and Cincinnatus, Cavour, and the rock engravings of Valcamonica. The final paragraph was dedicated, in an excess of rhetoric, to the "Dux" and the "grandiose historical cycle of an ethnic group—firmly homogenous and compact, rich in endogenic energy, completely identifiable from ancient times—returned to the natural grandiose civil and social mission."[165]

Like Marro, Edoardo Zavattari, director of the University of Rome's Zoological Institute, also saw the relationship between humans and their habitat as the key to interpreting historical and social phenomena. Zavattari spoke, in that regard, of the "fauna element":

> If the fauna factor [...] is one of the major elements to have characterized, dominated, and modeled the most ancient human cultural phases; if the fauna factor has imprinted the activity of Paleolithic man and the most primitive populations to live until now, as paleontology and ethnology demonstrate, this same factor has not exhausted its capital function, but has continued to develop in the successive millennia, and continues still today, determining a complex of very important human and social phenomena.
>
> The great migrations of populations, the settlement of ethnic groups in determinate regions, the abandonment of certain districts previously densely populated, and the adoption of customs that have assumed the value of true racial characteristics, have often been caused by this factor of essential importance.[166]

Racial hierarchy depended on unbreakable and necessary bonds between the environment and organisms:

164 Marro, *Primato della razza italiana*, 367. See also Marro, *Caratteri fisici e spirituali della razza italiana*, 52–54.

165 Marro, *Primato della razza italiana*, 375. See also Marro, *Caratteri fisici e spirituali della razza italiana*, 58–62.

166 Edoardo Zavattari, "Fauna e fenomeni sociale. II," *Razza e civiltà* 2, no. 3 (May–July 1941): 463.

> There exists an authoritative and unbreakable necessity that ties the environment and its organisms; a fate dominating the world, implicit in the nature of the living being, that imposes a categorical and axiomatic relationship on beings and natural factors: creating a condition, an expression of that life's essence, that fixes in an unequivocal and absolute manner the relationship of subordinate dependence, tightly conjoining the organisms to each other and to the innumerable environmental components.[167]

From Zavattari's racial interpretation of the concept of habitat, the scientific legitimization of fascist expansionism proceeded:

> The problem of living spaces, interpreted from a strictly biological point of view, pivots on these essential principles: the necessity that every species, both vegetal and animal, possesses an area in which they find all that is needed for the life and perpetuation of the species and in which their struggle for existence with other species is not such that they feel any lack, but on the contrary permits them a full expansion.[168]

Beyond reasons of political or economic character, the doctrine of living space was founded on a "general principle of biology," or rather, on the "authoritative, categorical, and absolute necessity that every organism has at its disposition a space in which to live, to develop and to reproduce."[169]

As well as the biological justification of *Lebensraum,* a second consequence of Zavattari's rigid racial differentialism was the biological threat of racial crossing. The legislator and the colonist must not, in fact, ignore that the environmental "plasticity" of the colored man was by far inferior to that of the whites. The degree of "plasticity" that distinguished the link between genotype and habitat was here seen as a criterion of racial hierarchy, and therefore, of discrimination:

> The white race, and our Italian race in a much more obvious way, has become highly pliable over the course of several thousand years of civilization, making it capable of transfer to highly different environments without being strongly

[167] Edoardo Zavattari, "Ambiente naturale e caratteri razziali," *La Difesa della razza* 3, no. 8 (20 February 1940): 7.

[168] Edoardo Zavattari, "La dottrina degli spazi vitali dal punto di vista biologico," *Scientia* 71, no. 361–62 (May–June 1942): 175. See also Edoardo Zavattari, "Leggi biolgiche e spazi vitali," *Il Giornale d'Italia* (13 May 1943): 3; Edoardo Zavattari, "Le basi biologiche di fascismo," *Critica medico-sociale,* no. 6 (June 1937): 21–28.

[169] Zavattari, "La dottrina degli spazi vitali dal punto di vista biologico," 178.

> affected. [...] The colored populations are much less plastic and much less adaptable; the more we descend toward primitive races, the more this plasticity is reduced; the inferior races are destined to subservience; others do not have this sad destiny, but must not be pushed beyond their extreme limits. A nomadic population will never be transformed into a sedentary one; a population of the forest will never become inhabitants of the savannah; a seafaring people will not become shepherds; none of these can ever assume that social form that the whites often delude themselves into believing these inferior races can achieve.[170]

If environmental "plasticity" produced an "overwhelming chasm" between the races, for Zavattari hybrids could not appear other than a sort of biological error, their natural maladjustment quickly revealing them as a social and political menace:

> Out of their environment they either cannot live, or they live at a disadvantage, live as strangers, as intruders, like an encrustation that is clinging, but has no roots, like an encrustation from which that primordial origin will always surface, that legacy of inferior quality that makes the blacks sensitive to tuberculosis and alcohol, that leads the blacks to burst out in attacks, rebellion, in violent acts against the race, in the midst of which they conduct their lives, estranged and far from the soil on which they were born, and to which they should be returned.[171]

It was from this ecological-racist axiom that, in Zavattari's writings, the justification of anti-Semitic discrimination originated. The Jews always carried with them, in every place and every time, the stigma of their "desert and nomadic" environmental origins. The Jew was eternal, beyond any form of integration or assimilation, because his "racial patrimony" was irremediably shaped by habitat. So environmental eugenics, at a hereditary level, seemed to carry with it a form of particularly radical anti-Semitism that identified in the Jew the definition of the "anti-race," the absolute difference, the totally inassimilable Other:

> The Jews always remain the same. Just as they cannot strip away their cerebral character, they cannot strip away their structural characteristics, nor amalgam-

[170] Zavattari, "Ambiente naturale e caratteri razziali," 10.
[171] Zavattari, "Ambiente naturale e caratteri razziali," 11.

> ate them, nor melt them, because they will always be that people who were born and lived on the other side of the sea, between the scree of Transjordan and the depths of the Dead Sea; because they are the people that had their laws dictated on the top of a mountain between lightning and storms, by a severe God without pity or love; because they are the people that have the immutable aridity and a contempt for other peoples printed on their souls; because they are the people who have always attempted to conquer the promised land, but have never conquered it, and can never conquer it because they can never stop, but will always have to go in search of a new mirage, just as they are still awaiting the coming of a new Messiah; because the desert that is at the bottom of their souls will drive them to be enemies, will drive them to rebel, will drive them to be nomads.[172]

Based on this premise, Zavattari's agreement with the measures introduced by Fascism to eliminate the Jewish "pollution" was prompt:

> Naturally, from the problem of our race's relationship with other extra-European races comes logically and inescapably our position as regards the Jewish problem. The Jews are Asiatic, transplanted for centuries in other continents and therefore also in Italy; coming from a race that, simply through the course of events, has conserved their original characteristics, has conserved their ethnic uniqueness, their profoundly different spiritual uniqueness, which in several aspects is naturally antithetical to ours. In a process of renewal of the position of our race, the Jewish problem is necessarily comprised; otherwise this position would not be completely clear. The purity of the race presupposes the elimination of every pollutant, whatever the nature and provenance; it must be totally achieved, without concessions and without hesitation. The laws of heredity that underpin the major processes of life have the job, through a complex but categorical procedure, to eliminate all those elements that have polluted it.[173]

However, while the environmental influence had inevitably produced a negative genetic impact on the blacks, hybrids and Jews, the lot of the "Italian race" had been otherwise miraculously molded by the beauty of the Mediterranean:

---

[172] Edoardo Zavattari, "Ambiente naturale e caratteri razziali (continuazione)," *La Difesa della razza* 3, no. 9 (5 March 1940): 49.

[173] Edoardo Zavattari, "Politica ed etica razziale," *Vita Universitaria* (5 October 1938): 3.

> This is exactly the Italian, firm and fast as the mountains, strong-minded and daring like the peaks that stretch skywards, fearless and brave in seeking new paths like the courses of his rivers and the horizons of his sea, plastic in his intellectual and proactive capacities, as required by the natural aspects so mutable and different, pliable as called for by the necessities of his hard life, which must now be lived on the mountains, now on the plains, now in the snow and now by the sea.[174]

Beyond the somatic aspect, the harmony of light, sound and the actual form of the Italian landscape had forged in the Italian "the most perfect, most complete cerebral capacity":

> The cult of beauty, the joy of life, the search for harmony in form and acts, the profound devotion to nature, as the exaltation of the self, the profound sense of solidarity of the Italian finds its origin in this constitution of the natural environment.[175]

Therefore, environmental and anthropo-geographical eugenics, despite running against the hereditarian current, came to nourish the same racist discourse, based on anti-Semitism, condemnation of racial crossing and affirmation of the superiority of the Italian race.

3. Esoteric-traditionalist Racism and Eugenics: Julius Evola

Esoteric-traditionalist racism—particularly represented by its principal exponent, the philosopher Julius Evola—developed a hereditarian discourse in the field of eugenics, showing singular convergence with the position of biological racism.

Evola's "totalitarian" racism devoted a specific place to the "dynamic theory of heredity," as opposed to the "static" and "deterministic" theories of biological racism. Against the "fatalism of heredity" derived from a "scientific assumption of the laws of heredity" and from an "excessively unilateral and materialistic" interpretation,[176] Evola declared: "Race and hered-

[174] Zavattari, "Ambiente naturale e caratteri razziali (continuazione)," 51.

[175] Zavattari, "Ambiente naturale e caratteri razziali (continuazione)," 51.

[176] Julius Evola, *Sintesi di dottrina della razza* (Milan: Hoepli, 1941), 21. On the theory of heredity, see also Julius Evola, *Il mito del sangue* (1937; repr., Milan: Hoepli, 1942), 91–116.

ity must not only be understood in terms of naturalistic determinism, but, essentially, like forces, like creative energies from within and, to a certain extent, even from above."[177]

According to Evola, the determinism of Mendel's laws was lessened the moment they were applied to human beings. A complete explanation of human hereditarianism necessitated the presupposition of a "spiritual" element. The esoteric-traditionalist point of view hypothesized, in particular, the existence of a metaphysical natural "force," which organized the hereditary transmission of the several spiritual and physical elements constituting the racial types:

> How is it that in certain "pure" types we find exactly that quality of body, of character, of spirit, if you like, that exact group of genes, united and stable? It is evident that here we need to think of a force, of a unifying and organizing force [...]. It is at the heart of race, constituting [...] the ultimate essence. Now, nothing prevents us from thinking that such a force, linked to the specific bundle of qualities or genes of every type, is transmitted in an ethnic mixture, reacting with it, choosing, coordinating, and producing a type similar to greater or lesser degrees.[178]

In race-crossing, this "profound force" that synthesized the human genotype, could be "dominant" or "recessive":

> When one of the parents is a carrier of "dominant" qualities—that is, we would say, when his "type" wholly conserves the energy, as the giver of "form"—the qualities of the parent of a (relatively, not absolutely) different race can also be present in the product of the crossing, but stifled and latent. If we were to continuously unite these descendants with new races of superior origins we would have practically cancelled out the dishybridisation, that is, the recurrence of characteristics of the parent of the "recessive" race.[179]

Racial types were maintained as long as "internal lesions, cessation of that tension which creates the type's dominant value," could be verified: "only then can dishybridisation take over, that is, the disassociation and the re-emergence of the subjugated recessive elements."[180]

177 Evola, *Sintesi di dottrina della razza*, 22.
178 Evola, *Sintesi di dottrina della razza*, 93.
179 Evola, *Sintesi di dottrina della razza*, 93.
180 Evola, *Sintesi di dottrina della razza*, 94.

Spirit, therefore, preceded genes; and consequently, hereditary transmission followed not only Mendel's laws, but also the doctrine of *karma*: earthly birth, biological history, was nothing less than "the consequence of transcendental affinity," the "point of intersection" between a horizontal heredity (earth: race, blood, caste) and a vertical heredity (transcendence).[181]

Evola therefore identified the more or less "dominant" characteristics of a race from its degree of "spiritual tension." Even the evaluation of the effects of racial crossing was based on this esoteric-traditionalist view. First of all, masculine heredity was always "dominant," while the feminine could not be other than "recessive": consequently, the descendants of a racial crossing between the man of an inferior race with a woman of a superior race would result in the stifling and contamination of the latter's genes. In the opposite case, the woman of inferior race would, on the other hand, be "rectified and practically neutralized."[182] Secondly, the deleterious characteristics of the racial crossing did not consist so much in the "deformation of unnatural or deformed human types in respect to their original racial body" as much as in the creation of a hybrid, understood to be a "lacerated being," "semi-hysterical," in whom the "internal" (soul and spirit) and the "external" (body) no longer corresponded.[183] Beyond this, the hybrid, in Evola's view, was also a "transcendental hysterical," devastated by the interior disagreement between the "central will of the incarnation," which was realized in the body, and the "minor wills," which were realized in the character.[184] In the same way, a race's decadence was due, in the first place, to "internal extinction" of the spirit, while the mixing of blood was a secondary cause.

Spiritual decadence was essentially the entrance hall of genetic mutation. It was only at this point, when race had lost almost all contact with metaphysical forces, that Mendel's laws had value, because "then race, lowered to the plane of natural forces, submits—and cannot help but submit—to the laws and contingencies of that level." Given this premise, a "defense of the race" of a "totalitarian" type must put forward two objectives: to preserve biological heredity from one side, and from the other "preserve the

181 Julius Evola, "Razza e nascita, ovvero: gli isterici trascendentali," *Il Regime Fascista*, 14 (15 March 1939): 3. See also Julius Evola, "La razza, l'ideale classico e gli 'spostati spirituali,'" *Roma Fascista* 19, no. 9 (22 January 1942): 3.

182 Julius Evola, *Indirizzi per una educazione razziale* (Naples: Conte, 1941), 48ff.

183 Evola, *Sintesi di dottrina della razza*, 80.

184 Evola, "Razza e nascita," 3.

spiritual tension, the superior fires, the internal formative soul, which originally elevated that material to that determinate form."[185]

The criticism of deterministic hereditarianism did not, however, prevent Evola from underlining the validity of Mendel's laws in cases of "pathological heredity." In an article in December 1940, published in the newspaper *Corriere Padano*, the combination of genius and mental disorder, a bulwark of moderate Italian eugenicists, was strongly refuted. As well as being a theoretical product tendentiously coined by the "Jew" Lombroso, such a relationship could not be employed to confirm the objections against the "racist prophylaxis of inherited defects." Critics of negative eugenics exaggerated the disadvantages, Evola argued, obscuring healthy benefits derived from the elimination of defectives. From the traditionalist point of view, the cases in which "something truly spiritual" manifested "through disintegration, illness and psychical-physical disequilibrium" could be considered "exceptional and sporadic."[186] Secondly, the resultant loss from the "elimination of a defective and physically inferior descendent" could be "compensated for, because, bit by bit, the path would be rediscovered, along which the internal action of the spirit on the spirit" would favor the return of the model of "traditional ancient humanity."[187] Obviously, in Evola's view, the practice of "prophylactic" eugenics had an essentially "negative" value, namely "removal of obstacles":

> He who claims to realize the superior aims of racism and recall to life, in a certain sense, the superior pure racial type by means of purely biological and prophylactic procedures, would repeat attempts to create a *homunculus*, an artificial man: a vain and absurd undertaking. Prophylactic means already alluded to could serve only to remove obstacles, in such a way that faculties previously blocked, whose origins are super-biological, could manifest themselves again: but it is not possible to create, nor, by itself, to reawaken these faculties, because nothing comes from nothing. Prophylactic racial means of heredity and selection of heredity must therefore be considered as part of an action much vaster and more complex, and put into practice without ever losing sight of the whole picture.[188]

185 Evola, *Sintesi di dottrina della razza*, 82.

186 Julius Evola, "Problemi della razza. Lo spirito e gli epilettici," *Il Corriere Padano* 18 (27 December 1940), now in Julius Evola, *I testi del 'Corriere Padano'* (Padua: Edizioni di Ar, 2002), 386–87.

187 Evola, "Problemi della razza," 387.

188 Evola, *Sintesi di dottrina della razza*, 99.

In essence, what was a point of arrival for biological racism was only a starting point for esoteric-traditionalist racism, a basis for successive and deeper selection of psychological and spiritual characteristics. Notwithstanding the diversity of the two orientations, the space of convergence was represented by a common adherence to negative eugenic solutions. This was clearly demonstrated in the debate that occurred, between January and June 1940, on the pages of Giovanni Preziosi's *La Vita Italiana*, between Guido Landra, advocate of biological racism, and Julius Evola. The strongly hereditarian position of Landra, rich with references to the German and American eugenic literature,[189] stirred up criticism from Evola who repeated the need to consider the "spiritual" dimension of human heredity. This did not signify, however, that he rejected the validity of Mendelian laws. On the contrary, they remained a reference point for negative eugenics, the importance of which was clearly declared.[190] Landra's response to this was conciliatory, sharing "the opinion of Arthos [Evola's pseudonym] of the need for a clear stand and to avoid the unilateralism and materialism which some environmental scientists fall into."[191] He even agreed that, at the extremes, a "hereditary doctrine" could lead to a "biological determinism that absolutely disgusts our mentality." On the other hand—Landra continued—the risk at the door was that of falling once more into environmentalism, which was dear "to the defunct demo-liberal mentality." In the nature/nurture debate, the truth was perhaps "in the middle," since "the hereditary factors must be considered as much as the environmental ones." The problem was basically that of "establishing the reciprocal limits of influence of the two factors with exactness."[192] Not surprisingly, in the next article published in *La Vita Italiana,* Landra once more held out his hand to esoteric-traditionalist racism, hypothesizing the existence of a "general law of variability," on the basis of which "psychological and physical hereditary characteristics" could mutate, depending on environmental influences or the force of will. For biological racism, race remained "an objective reality

189 See Guido Landra, "L'eredità dei caratteri razziali," *La Vita Italiana* 28, no. 322 (January 1940): 29–31; Guido Landra, "L'eredità delle qualità psicologiche," *La Vita Italiana* 28, no. 324 (March 1940): 286–90.

190 Arthos [J. Evola], "Sui limiti del razzismo: il problema dell'eredità," *La Vita Italiana*, 28, no. 323 (February 1940): 178–79.

191 Guido Landra, "L'eredità delle qualità psicologiche," 290.

192 Landra, "L'eredità delle qualità psicologiche," 290.

of Nature," but this did not exclude that "we, with our will, could influence race, optimizing those things that can be changed."[193]

4. Assortative Mating and Racism

In the different directions of eugenic research, fed by the constitutionalist medical school, statistical studies on matrimonial attraction flowed particularly well into the racist ideology of fascism.

After 1897, the statistician Rodolfo Benini began a series of studies dedicated to the modality of choice and selection of conjugal couples, inventing a new method of measurement (the index of attraction) and coming up with an analysis of the laws of matrimonial attraction between similar individuals: "Whatever character is at the base of a group, whatever city, whatever state the spouses come from, a clear empathy between the individuals with identical characteristics is always visible."[194] Alfredo Niceforo's work on the rigid matrimonial segregation between social classes,[195] Federico Chessa on the "hereditary transmission of profession"[196] and Franco Savorgnan on the matrimonial choices between groups of different nationalities,[197] all followed from this innovative frontier of research.

Eugenics was grafted onto such statistical studies in the moment in which the object of analysis changed from the study of social conditioning to that of physical and psychical characteristics. Francis Galton, with his family records, had been the first to attract his attention on the combination of marriage according to stature, the color of the eyes and artis-

---

193 Guido Landra, "Ereditarietà e ambiente," *La Vita Italiana* 28, no. 327 (June 1940): 651.

194 Roberto Bachi, "Gli indici della attrazione matrimoniale," *Il Giornale degli Economisti* 69 (November 1929): 895. Benini's studies on the topic are: "Probabilità statistica e probabilità matematica (prolusione al corso di statistica letta nella R. Università di Pavia il 1° dicembre 1897)," partially published in *Rivista italiana di sociologia* 2, no. 2 (March 1898) :152–71, under the title "Le combinazioni simpatiche in demografia"; *Principii di demografia* (Florence: Barbera, 1901), 129–58; "Sulla rappresentazione in diagramma cartesiano di fenomeni classificati secondo caratteri qualitativi," *Rendiconti della R. Accademia dei Lincei—Classe di scienze morali, storiche e filologiche. Serie V*, 24, no. 12 (December 1915): 21; "Gruppi chiusi e gruppi aperti in alcuni fatti collettivi di combinazioni," *Bulletin de l'Institut International de Statistique* (Le Caire: Imprimerie Nationale, 1928): 362–83.

195 Alfredo Niceforo, *Antropologia delle classi povere* (Milan: Vallardi, 1910), 94.

196 Federico Chessa, *La trasmissione ereditaria delle professioni* (Turin: Fratelli Bocca, 1912), 92.

197 Franco Savorgnan, *La scelta matrimoniale. Studi statistici* (Ferrara: Casa Editrice Taddei, 1924).

tic tendencies.[198] Further contributions were then made by Karl Pearson, who broadened the range of considered characteristics, identifying a selective mechanism, parallel to Darwinian preferential mating,[199] in assortative mating. In the 1920s, the Galton-Pearson line of research was further developed by the analysis of Harris and Govaerts (1922),[200] Rosiński (1923),[201] Kretschmer (1926) and Nicolaeff (1929).[202] Meanwhile in Italy, it was above all Franco Savorgnan, in line with Ludwig Gumplowicz's concept of "syngenism," who identified in "racial sameness" one of the factors of cohesion between individuals:

> The white man chooses by preference a white bride, the Italian man an Italian woman and so on in this way, because the women of one's own race and nationality respond synergistically, awakening one's empathy and offering a certain guarantee of conjugal happiness. [...] In judging the global indices, the syngenistic factor of racial communality have a much more intense influence than nationality upon homogamy, and therefore the *fusion of heterogeneous races appears generally more difficult than that of nationalities.*[203]

For Savorgnan therefore, the "hatred of race" had a precise biological origin, recognizable in the "visual sensation produced by the color" of the skin: racism was "continuously kept aroused by visual sensations, produced by inalterable somatic characteristics."[204] In Italian social sciences, however, the study of eugenic assortative mating only established itself at the start of the 1930s, in close connection with statistical-biometric material furnished by already cited studies on large families commissioned by CISP, under the direction of Corrado Gini. At the 1931 International Congress for Studies on Population, Giuseppe Genna expounded, in fact, the results

198 Francis Galton, *Natural Inheritance* (London: Macmillan, 1889).

199 Karl Pearson and Alice Lee, "On the Laws of Inheritance in Man: Inheritance of Physical Characters," *Biometrika* 2, no. 3 (1903): 357–462; Karl Pearson and Alice Lee, "Assortative Mating in Man: A Cooperative Study," *Biometrika*, 2, no. 4 (1903): 481–98.

200 James A. Harris and Albert Govaerts, "Note on Assortative Mating in Man with Respect to Head Size and Head Form," *American Naturalist* 56, no. 645 (July 1922): 381–83.

201 Bolesław Rosiński, "Charakterystyka antropologiczna ludności pow. Pultuskiego," *Kosmos*, 48 (1923); Bolesław Rosiński, "Antropogenetische Auslese," *Antropologischer Anzeiger* 6, no. 1 (1929): 49–64.

202 Léon Nicolaeff, "Les corrélations entre les caractères morphologiques des époux," *L'Anthropologie* 41, no. 1–2 (1931): 75–93.

203 Savorgnan, *La scelta matrimoniale*, 63–64; italics added.

204 Savorgnan, *La scelta matrimoniale*, 63–64.

of his inquiry on nearly three hundred couples from the city of Trapani—an "ethnic group determined to be homogenous"[205]—and analyzed the correlations between the "morphological characteristics of the spouses," in particular those "most expressive from an anthropological point of view," including stature, and cephalic, facial and nasal indices. Genna's conclusions seemed to confirm the hypothesis that matrimonial attraction was determined by purely anthropological causes:

> Summarizing the results, we can confirm that the stature and cephalic, facial and nasal indices of around 270 Trapani couples with seven children or more showed a positive correlation between each spouse, a stronger correlation in stature, and much less for the various indices of the head, which could explain not only aesthetic, and in certain cases, social (this last indirectly) causes, but also perhaps a purely anthropological cause, able to influence human sexual choices.[206]

Drawing always from the pool of large family studies, one year later, the biologist Carlo Jucci and his assistant T. Amendola, during the course of the 21st Congress of SIPS in Rome (9–15 October, 1932), confirmed Genna's view, expounding the results of their anthropometric measurements and identifying the facial index as an "element of matrimonial attraction."[207]

In 1934, Albino Uggé, a student of Boldrini at the Laboratory of Statistics of the Catholic University in Milan, reanalyzed Franz Boas' reports on emigrant families to New York, determining the level of resemblance between spouses from seven different groups based on coefficients of correlation: Sicilian, Central Italian, Bohemian, Hungarian and Slavic, Polish, Scottish, and Jewish. The resemblance was relative to stature, cephalic index, bizygomatic diameter, the relationship between the largeness of the face and the largeness of the head, and the color of the hair and eyes. According to Uggé, sexual choice was determined not only by the correlation between elements of somatic character (such as stature), but also "constitutional type," that is,

205 Giuseppe Genna, "Correlazione fra i caratteri morfologici degli sposi, " in Corrado Gini, ed., *Atti del Congresso internazionale per gli Studi sulla Popolazione (Roma, 7–10 settembre 1931)* (Rome: Istituto Poligrafico dello Stato, 1934), vol. 4, 796.

206 Giuseppe Genna, "Correlazione fra i caratteri morfologici degli sposi," 803.

207 Carlo Jucci and T. Amendola, "L'indice facciale come elemento di attrazione matrimoniale," in Lucio Silla, ed., *Atti della SIPS. XXI riunione (Roma, 9–15 Ottobre 1932)* (Rome: SIPS, 1933), vol. 3, 318–19.

the links between temperament (or neuropsychical constitution) and morphological aspects, theorized from the constitutional medical school:

> Next to direct selection operating by several physical traces that are striking and easily appreciable, is the influence of indirect selection, dependant on assortative mating according to constitutional type, which provides a reason for several results otherwise unexplainable.[208]

With the launch of the state racial campaign, the welding of statistical investigations of assortative mating to fascist racism was almost immediate. The studies on conjugal selection quickly developed the role of scientific legitimization of discriminatory policies. In the reassuring light of figures and percentages, the racist legislation could more easily be presented not as an expression of a radical and violent political measure, but as the confirmation of *natural* data, demonstrated by the homogamic tendencies inherent in matrimonial selection. Fascism, to put it shortly, had not invented anything: it was the citizens, with their matrimonial choices, who behaved as *naturally* racist. Giuseppe Genna again, in an article published in *Razza e civiltà* [Race and civilization], elaborated Rosiński's data into a racist framework:

> The homogamic racial tendency, operating against racial crossings, tends to maintain the racial composition of the population unaltered, generation after generation, with all its physical and psychical attributes. And, as things stand, we can say that the racial policies of the Regime, with the prevention of marriages between Italians and people of other racial descent, even if from within the same national area, is inspired by a biological inclination generally inherent in the masses. The conservation of racial purity appears to be a spontaneous natural tendency, before being a codified will of the state.[209]

If then, Genna concluded, the research demonstrated a connection between "racial homogamy" and "fertility," the fascist measures against racial crossing would be doubly valid, as they would "not only tend to maintain the racial structure of the population unaltered, but also augment its consistency; in a word, optimizing the race not only in quality but also in quantity."[210]

208 Albino Uggé, "Sulla rassomiglianza fra coniugi per alcuni caratteri somatici," in *Contributi del Laboratorio di Statistica. Serie III* (Milan: Vita e Pensiero, 1934), 168.

209 Giuseppe Genna, "Razza e sangue," *Razza e civiltà* 1, no. 5–6–7 (July–September 1940): 461.

210 Genna, "Razza e sangue," 462.

Similarly, the statistical and eugenic studies on matrimonial attraction were used in the journal *La Difesa della razza* as justification for the racial discrimination and mixophobia of fascism. Restating Genna's arguments, Vincenzo De Agazio formulated a precise link between "matrimonial attraction and physical characteristics" and racial laws:

> The racial policies of the Regime, by preventing marriages of Italians to people of other racial descent are in accordance with this natural biological tendency inherent in the masses, and so, while racially different individuals are repelled, those racially similar attract, in this way perpetuating original racial types through the generations.[211]

The "purity of our race"—exulted De Agazio—clearly corresponds to a "spontaneous aspiration of the masses." The Regime had done well, therefore, by using the law to forbid marriages between "individuals of the Italian race and the Semitic race" who "altered the biological tendency," threatening to "pollute the purity of the race."

5. Toward a National Genetic Center

Between 1936 and 1943, the dream of fascist eugenics to achieve a "national genetic program" seemed to be one step from realization.

In May 1936, at the Clinic for Nervous and Mental Diseases of the University of Florence, directed by Mario Zalla, the First Genealogical Day (*Giornata Genealogica*), was organized by the LIPIM. Around the theme of the day—*Does a similar and dominant heredity really exist in the manic-depressive phrenosis?*—twelve reports were presented.[212] This confirmed the growing

---

211 Vincenzo De Agazio, "Attrazione matrimoniale," *La Difesa della razza* 2, no. 20 (20 August 1939): 10.

212 The reports, collected in *Atti della Lega italiana di igiene e profilassi mentale*, 1936, were the following: C. E. Roberti, "Il concetto nosografico e clinico di frenosi maniaco-depressiva in rapporto al problema dell'ereditarietà," 23–26; Luisa Gianferrari and Giuseppe Cantoni, "Ricerche in una popolazione endogama circa l'epoca d'origine di idiovariazioni," 27–32; Giovanni Fattovich and N. Nicolai, "Considerazioni sull'ereditarietà in alcuni casi di psicosi maniaco-depressiva," 33–39; Emilio Rizzatti and Vittorio Martinengo, "L'ereditarietà nella psicosi maniaco-depressiva," 40–47; Antonio D'Ormea, "Considerazioni clinico-genealogiche sulla frenosi maniaco-depressiva," 48–51; Gino Volpi-Ghirardini, "Su l'ereditarietà nella psicosi maniaco-depressiva," 52–58; Annibale Puca, "Ricerche eredo-biologiche nei psicoastenici e nei maniaco-depressivi," 59–63; Alberto Rostan, "Sulla ereditarietà delle psicosi maniaco-depressive," 64–78; Gino Calzavara, "Qualche rilievo statistico sulla ereditarietà generica nelle psicosi circolari," 79–82; A. Coen, "L'eredità

interest in the heredity of nervous pathologies, but was also evidence of the uncertainty still diffused around "heredity measuring techniques." From this contradictory situation, through LIPIM, the project to construct a "genealogical archive of mental illnesses" grew, based on the model offered by German psychiatry. In the discussion that accompanied the First Genealogical Day, Giuseppe Pintus,[213] the young assistant of Lionello De Lisi, director of the Clinic of Nervous and Mental Illnesses (*Clinica delle Malattie Nervose e Mentali*) of the University of Genoa, was charged by the League with the task of deepening the issue of "the methods of genealogical research," through apposite sojourns at specialized German psychiatry centers in Munich, headed by Rüdin, and Berlin, headed by Richard B. Goldschmidt. In 1937, during the 5th General Meeting of LIPIM, De Lisi stressed the importance of creating a national "genealogical archive" of mental illnesses, following the example of "these kinds of institutes, such as that of Goldschmidt and that in Berlin [sic]."[214]

In the same year, on 18 June, the Lombard Society of Medicine (*Società Lombarda di Medicina*) approved zoologist Luisa Gianferrari's proposal regarding the constitution of a national genetic center particularly intended for research on the endogamic populations in the Central Alps of the Italian peninsula. The latter, in fact, for Gianferrari, represented the "spring from which common undesirable hereditary factors descend, which once introduced are then dispersed, their origins unrecognizable, in the chaotic sea of the city."[215] The investigation of endogamic nuclei could therefore be a starting point to deepen the knowledge of the hereditary transmission of characteristics, "especially recessive." Gianferrari included among these the "socially important" hereditary illnesses—schizophrenia, deaf-muteness, and epilepsy—and the "identification of defective branches" neces-

delle malattie mentali studiata in un gruppo etnico poco inquinato (Ebrei di Mantova)," 83–85; Giuseppe Pellacani, "Considerazioni sulla ereditarietà nella psicosi maniaco-depressiva," 86–108; Giuseppe Antonini, "Contributo statistico sulla ereditarietà nella psicosi maniaco-depressiva," 109–13.

213 Born in Iglesias in 1902, student of Carlo Ceni at the Cagliari Neurological Clinic, Pintus, from his very first period as assistant in Sardegna directed his scientific orientation toward genetic psychiatry, producing studies on the hereditary transmission of essential tremor (1932), progressive muscular atrophy, such as Charcot-Marie (1934) and Unverricht's myoclonic epilepsy (1937).

214 See "V Assemblea Generale della Lega Italiana di igiene e profilassi mentale," *Atti della Lega italiana di igiene e profilassi mentale* (1937): 114.

215 Luisa Gianferrari, "Importanza, urgenza di ricerche genetiche in popolazioni endogame," *Atti e memorie della Società Lombarda di Medicina* 5, no. 8 (1937): 582.

sary for "intimately penetrating into the genetic constitution of the population, identifying the distribution of pathological factors." Only in this way, Gianferrari continued, could an "unassailable base of effective prophylactic eugenics" be founded. The demographic, racial and eugenic framework of the project was explicit: the "reclamation of the alpine and rural populations" was presented as "the most solid fundamental of a population policy aimed at the increase and improvement of the stock."[216] Concretely, Gianferrari outlined the necessity of a "Center for the collection of material and of consultancy for human genetic research," constructed on the model of the London Bureau of Human Heredity. This center would aim, on one hand, to favor the diffusion of genetics in the hospitals and university classrooms and, on the other, to reinforce the links between geneticists, clinicians and hygienists, so that "profitable deductions" could occur "for prophylactics and hygienic reclamation" even in the field of hereditary illnesses. Beyond the approval of Luigi Zoja, president of the Lombard Society of Medicine, the proposal gathered, in the course of the session, the support of Eugenio Medea, president of LIPIM, ready in that role to guarantee the "enthusiastic compliance of the Italian Society of Mental Hygiene and Prophylaxis."[217]

A year later, LIPIM's psychiatric project and the Lombard Society of Medicine's biological project came together at the 2nd Genealogical Day, held in Genoa in May 1938 at De Lisi's neuropsychiatric clinic. Returned from study trips to the genetic institutes of Munich and Berlin, Giuseppe Pintus was the only speaker, on the theme of *Methodology of genetic research in psychiatry*. The conference's objective was to develop the proposal for a national neuropsychiatric genetic center to present to the authority of the fascist regime. De Lisi's general introduction to the sessions left no doubts regarding this: "The League will expound its opinions and proposals, which will then be assessed and made achievable, naturally, by the organs of the Regime interested in such questions of high racial importance, or, in terms more adapted to the thoughts and political practices of Fascism, of high national importance."[218] It is not surprising, therefore, that the long, in-depth examination of Pintus on the statistical and genealogical methods of

216 Gianferrari, "Importanza, urgenza di ricerche genetiche," 581.
217 Gianferrari, "Importanza, urgenza di ricerche genetiche," 584.
218 "II Giornata Genealogica (Genova, 21 maggio 1938)," *Atti della Lega italiana di igiene e profilassi mentale* (1938): 50.

research in psychiatric genetics, rich in references to German eugenic literature (Lenz, Rüdin, Kallmann, etc.), culminated in a pronunciation of the "indispensable norms for the functioning of a center for the study of genetics applied to mental illnesses." The genealogical archive, according to Pintus, had to count, above all, on the collaboration of the Statistical Center for Mental Diseases (*Centro Statistico per le Malattie Mentali*), directed by Gustavo Modena in Rome. On the other hand, the construction of a genealogical tree required the genetic center to adopt a precise methodology:

> For every mentally ill person, we must collect his exact name, surname, sex, paternity, maternity, place of birth; number, sex and health condition of different members of his family, particularly those of possible deceased family members, premature births and miscarriages.
> This data will be noted on the clinic card of the patient, and a copy of this card will be sent to the Genetic Center. The card will be used to note, in addition to the anamnestic information (starting period of the illness!), examination objective, etc., etc., also some diary notes. The diagnosis will therefore be well documented from the notes on the card.[219]

The genetic center did not need to contain all the data of all the mentally ill, but only "those of most stabilized and pathogenic forms and of sure diagnosis":

> More than a vast number of genealogical trees, the archive needs trees gathered with care and diagnostic precision. It will not be necessary for all institutes, even private ones, to send cards to the center. It is enough that all the provincial psychiatric hospitals do so, and that every province uses a special system for handling these practices (copies of the cards, etc.)[220]

Finally, the staff had to be composed of geneticists and psychiatrists: "the first must concern themselves with the genetic interpretation of the material received, and the second with the assessment of the clinical cards." In conclusion, Pintus particularly insisted on the necessity of sorting the most recent genealogical data, going back at the most to the limit of two generations, in

219 Giuseppe Pintus, "Metodologia delle ricerche di Genetica in Psichiatria," *Atti della Lega italiana di igiene e profilassi mentale* (1938): 100.
220 Pintus, "Metodologia delle ricerche di Genetica in Psichiatria," 100.

this way enabling correct diagnosis: "in matters of mental illness it is damaging and hazardous to make retrospective diagnoses on the basis of indications offered by relatives. It is far better not to consider those distant sick people, listed with dubious, and often, incorrect, diagnoses."[221] This point came under particular scrutiny in the debate that followed. Luisa Gianferrari proposed that the future genetic center should utilize the information kept in the archives of the mental hospitals, which went back many centuries, while the psychiatrists Francesco Bonfiglio (Rome) and Giuseppe Corberi (Milan) strongly opposed this approach, together with De Lisi:

> I believe it would be difficult enough to accept a psychiatric diagnosis of only 30–40 years ago or to value the descriptions of mental illness from a psychiatrist from that era, given the imperfections and incompetence of the diverse standards of past semeiotic psychiatry.[222]

However, the scientific refusal of the German method of sterilization appeared unanimous, energetically restated on this occasion by Arturo Donaggio and Giuseppe Corberi.[223]

The same critical stance against Nazi policies was repeated, again in 1938, at the annual congress of SIPS. At this event, Lionello De Lisi, who was an attentive critic of German sterilization laws, proposed a synthesis of the positions of Italian neuropsychiatrists, centered partly on the condemnation—essentially scientific, not moral—of negative eugenics, and partly on the deepening of genetic studies, aimed at "preventing dangerous unions" and favoring "natural processes of elimination of abnormalities and defectives":

> In the face of the prophylaxis of hereditary illnesses of the nervous system, particularly mental ones, the program of Italian neuropsychiatrists is the following:
> 1) Optimize the therapy of the hereditary illnesses, offering hope of curing practices and the predictability of their attenuation.
> 2) Develop all the forms of prevention that come under the heading of constitutional and social medicine [...].

221 Pintus, "Metodologia delle ricerche di Genetica in Psichiatria," 101.
222 *Atti della Lega italiana di igiene e profilassi mentale* (1938), 109.
223 *Atti della Lega italiana di igiene e profilassi mentale* (1938), 106 and 108.

> 3) Develop studies on heredity, that today are not nearly complete or brought to definite conclusions; dissect the many aspects of neuropathological and psychopathological heredity with the method and technical coordination that will make a center of genetic neuropsychiatry possible.
> 4) Study and enforce with the maximum care, those means already existent, or possible (individual advice, forms of propaganda and assistance, proposals for laws), which being based on genetic studies are those most adapted to preventing dangerous unions for the mental health of the stock and of favoring the natural processes of the elimination of abnormals and defectives.
> Particular attention should be given to the prevention of dangerous blood-related marriages and to those spontaneous processes of mixing of the populations coming from different regions of the same nation. In single populations, this could avoid the accumulation of recessive heterozygotes, carriers of diseased genes, produced by blood-related marriages.[224]

Not sterilization, therefore, but the monitoring of statistical and genealogical data of mental illnesses on a national scale would be set up, at the end of the 1930s, as the principal objective of Italian psychiatric eugenics.

The project developed at the 1938 conference in Genoa was nevertheless rejected two years later, meeting with the complete refusal of the Department of Public Health (*Direzione Generale della Sanità*). The problem was placed in the order of the day at the meeting of the directive board of LIPIM, held on 28 January 1940: the aim was to decide upon the foundation in Rome of a center for psychiatric genetics, hosted by the Statistical Center for Mental Diseases that operated in Rome psychiatric hospital. The president Medea quickly took a distant stance from the racial line of Gianferrari's proposal, so enthusiastically welcomed in 1937:

> As I have indicated not long ago, the proposal to institute an Italian center of psychiatric genetics was formed after Pintus's presentation at the meeting in Genoa of our League. A similar project was advanced in Milan, by Professor Gianferrari, and was discussed widely in a conference called by the Chancellor of the Milan University. I was invited to this conference in my role as president of LIPIM, but did not see it as opportune for the League to become part of the

[224] Lionello De Lisi, "Profilassi delle malattie ereditarie in Psichiatria," in Lucio Silla, ed., *Atti della SIPS. XXVII riunione (Bologna, 4–11 Settembre 1938)* (Rome: SIPS, 1939), vol. 6, 138.

> Institute project given that the direction of this initiative was prevailingly biological and racial rather than psychiatric.[225]

The intention of the League, continued Medea, was "purely scientific, and strictly psychiatric":

> It is that of instituting a center that would coordinate and regulate the research on heredity in mental illnesses with rigorous methods, similar to those used for statistical data by the office in Ancona, now involving the psychiatric hospital in Rome.[226]

Ugo Cerletti intervened in the debate, quickly posing the question of financing and suggesting the adoption of the Nazi model of a genetic center closely dependent on the state, as recorded in the minutes:

> *Prof. Cerletti*: Apprised us that the first question to resolve was that of financing because conducting research of this type requires numerous personnel, frequent travel, archives, etc. needing a lot of money. Who could give it to them? In Germany, the research is conducted by the State, which did not begrudge means, given its particular aims.[227]

Germany was also the point of reference for Lionello De Lisi, who, confronting this economic problem, repeated the opportunity to identify competent professional figures, particularly psychiatrists with statistical and genetic skills:

> *Prof. De Lisi*: Certainly an economic question exists, but the most important and difficult thing seems to be the personnel suited to manage such a center. It would need one psychiatrist with statistical and genetic skills. It is a complex function for which not only medical competence, but also mathematical skill is needed. This illustrates what is being done in Germany.[228]

While Cerletti and De Lisi looked with favor at the Nazi psychiatric model, the director of the Department of Public Health, Giovanni Petragnani, lined up on the opposite side:

---

[225] "Riunione del Consiglio Direttivo e dei Delegati regionali," *Atti della Lega italiana di igiene e profilassi mentale*, (1940), 120; italics added.

[226] "Riunione del Consiglio Direttivo," 120.

[227] "Riunione del Consiglio Direttivo," 120.

[228] "Riunione del Consiglio Direttivo," 120–21.

> *Excellency Petragnani*: Keep in mind that in this field we cannot take models that come from other countries, given that the means adopted there might have been suggested by political and racist reasons rather than strictly scientific and health ones. The research that they [the Germans] are doing is executed by force of law, and can not be officially allowed in Italy, where such laws have no currency. In this way the planned center would lack that richness and precision of data, without which its actions would be nothing. [...]
> The modest economic help that the Province of Rome could give would be unequal to the task and nothing could be demanded from the State for the reasons already mentioned. Without these means, the anamnestic research would be limited to information provided by the relatives that accompany the sick people to hospital, information of little value. Having completed its purely scientific work, this center could not carry out its prophylactic scope without imposing a repressive order on the reproduction of the hereditarily mentally ill. This would, for us, be absolutely inadmissible.[229]

In the face of Petragnani's immediate and opposing reaction, Eugenio Medea attempted a difficult mediation that focused exclusively on the center's "scientific" character:

> *Prof. Medea*: Wished to assure his Excellency Petragnani that the desire to imitate or copy just any foreign initiative aimed at forced prevention of hereditary mental illnesses was far from his or his colleagues' ideas. [...] The scope of the center should be, as mentioned previously, solely scientific. This, in the intentions of the proponents, would function as a collection in Rome of all the research, statistical data, etc. in the psychiatric institutes of the kingdom, a point of reference and consultation for all the problems regarding hereditary psychopathology. The center, rigorously organized, would unite all the studies on the subject.[230]

De Lisi also pinpointed his orientation, declaring his opposition to the adoption of forms of negative eugenics:

> *Prof. De Lisi*: Wished to better clarify his point of view that Italian science could not accept, without controls, the results of research conducted in other

229 "Riunione del Consiglio Direttivo," 121; italics added. On Petragnani's nearing to the position of Pende's constitutionalism, see Maiocchi, *Scienza italiana e razzismo fascista*, 240.

230 "Riunione del Consiglio Direttivo," 121.

> nations. The center would have the exact task of systematically and rigorously controlling methods, data and conclusions, still susceptible to criticism. All this would occur independently from those eugenic applications to which Italian psychiatrists have always been openly contrary.[231]

But the debate was completely closed by the response of Petragnani, confirming, should need arise, the unavailability of the state in terms of any legislative and financial support for the Center:

> *Excellency Petragnani*: Continued to state that the works of a Center would be ineffective if it was not supported by laws that authorized the research of the families of the mentally ill. Such laws do not exist in Italy. No one would forbid the free research of scholars, but the State could not recognize the legitimacy and authority of a Center of such a kind. Scholars could continue for their part the methods adopted in other countries consulting the works of those scientists.[232]

For Medea there was nothing left to do, therefore, but register the political veto, and postpone the discussion to 1942.

However, while the psychiatric project of LIPIM was put aside in January 1940, December of the same year saw the inauguration of a Study Center in Human Genetics (*Centro Studi di Genetica Umana*), in Milan at the Institute of Biology and Zoology of the Faculty of Medicine, as hoped by Gianferrari in 1937. Supported by the majority of the local political and academic authorities, the Center was financed by both private and public entities, including the provincial administration of the Savings Bank (*Cassa di Risparmio*) of Lombardy Province, Edison, Italviscosa, Montecatini, Pirelli, Marelli and Olivetti. The Council Presidency consisted of the president Luigi Zoja, and advisors Antonio Cazzaniga, dean of the Faculty of Medicine and Surgery at the University of Milan; Pietro Rondoni, director of the Institute of General Pathology; Marcello Boldrini and Luisa Gianferrari. Gianferrari was also director of the Center, while his assistant, Giuseppe Cantoni, was vice-director.[233] In the same year—and with a second edition

---

231 "Riunione del Consiglio Direttivo," 121; italics added.

232 "Riunione del Consiglio Direttivo," 121.

233 "Centro di studi di genetica umana," *Gli Annali della Università d'Italia*, 2, no. 4 (28 April 1941): 374.

in 1942—Gianferrari and Cantoni published a genetics handbook, with a preface by Luigi Zoja. In a brief concluding chapter entitled *L'eugenetica od eugenica* [Eugenetics or eugenics], a significant passage appeared that described the German eugenic legislation, first Weimarian and then Nazi, in laudatory terms: "Germany, with the two laws of 8 October 1925 (laws of matrimonial restriction) and 14 July 1933 (sterilization law), is—among all the nations—the one which has confronted the problem of eugenic reclamation with the most thoroughness and energy."[234]

The aim of the Study Center, according to article 1 of the regulations of the Institute, was to "collect data on the physiological and pathological characteristics of man, with the aim of carrying out genetic studies, also with a focus on health and demographic problems."[235] From an organizational point of view, the Study Center was established as a link in a chain of collaboration between the University of Milan, public administrations and health structures.[236]

The first relevant initiative of the Milan Study Center was the elaboration of a sort of national genetic index of the transmission of hereditary characteristics. The index drew on the archives of the hospitals, surgeries, special schools, mental hospitals and many other institutes. The collection of information was entrusted to volunteers, recruited from the students of the Faculty of Biology and Surgery, "who have passed the Genetics and Biology of Races exams," especially prepared with an exam on human genetics. Each volunteer was provided with a form by the Center. Luisa Gianferrari mentioned 510 "student field-researchers":[237]

> They collect the data from clinic cards and surgery registers kept in institute archives, supported by the advice of medical personnel. Some of them, admitted to the surgeries, collect directly from the live voices of the ill or their families. As a guide, they use an apposite report card that the management of the Center has edited after examining and discussing the systems used for the col-

234 Luisa Gianferrari and Giuseppe Cantoni, *Manuale di Genetica con particolare riguardo all'Eredità nell'Uomo* (Milan: Vallardi, 1942), 451.

235 Luigi Zoja and Luisa Gianferrari to Giuseppe Montalenti, October 30, 1948, Montalenti papers (hereafter AM), b. 24.

236 Luisa Gianferrari, "Il contributo dell'Università al Centro di studi di genetica umana," *Gli Annali della Università d'Italia* 3, no. 1 (29 October 1941): 25.

237 Luisa Gianferrari, "Sull'organizzazione e sull'attività svolta dal Centro di studi di genetica umana nel primo quadriennio dalla sua fondazione," *Natura* 35 (1944): 114.

> lection of data in the major institutions of human genetics in other countries, and a list of the diseased forms of deformity in which the study of heredity is particularly interested.[238]

Already in 1944, by sifting through a number of relevant clinics, hospitals and institutes in Milan and, more generally, Lombardy, the Study Center had gathered a good one hundred thousand index cards. Of these, over one thousand regarded twins—"object of an inquiry about their concordance and discordance for diverse hereditary forms"—and about a thousand documented genealogical pedigrees.

Zoja and Gianferrari's ambition however was to transform the Milan Study Center into a broader National Center for Human Genetic Studies (*Centro Nazionale per Studi di Genetica Umana*), based on the coordination of a network of regional seats spread across the national territory:

> The results we could achieve would be much broader and of more national interest if our organization, limited to Lombardy, was extended to other regions. If a center for the genetic study of the population under its sphere of influence was created, with a unity of direction, in every university, and every regional center sent and requested the genetic index cards from the other centers, for those who changed residency, moving to regions far from their origins, an uninterrupted network of genetic investigation would be spread across our country and could, in such a way, lead us to a genetic census of the Italian population.[239]

This would mean the construction in every region of "center sections," based on the Milanese model, which would use "the same index cards, the same categorization and organization." Every section would send "a copy of their collected index cards" to Milan "in such a way that they could be inserted in the general index":

> The advantages that a national center could offer to the sections are as follows:
> 1) the possibility to use, for scientific research, the material already gathered in the general index;
> 2) the possibility to collaborate with other sections on studies that regard materials collected in respective regions;

[238] Gianferrari, "Sull'organizzazione e sull'attività svolta dal Centro di studi di genetica umana," 113.
[239] Gianferrari, "Il contributo dell'Università al Centro di studi di genetica umana," 28–29.

> 3) the possibility to use the specialized library and eventually the collaboration of the national center personnel;
> 4) the possibility to avail themselves of the "Rh" section of the Milan center, which arranges the hemodiagnostic serums necessary for the determination of the type and subtype of systems ABo, MN, Pp and Rh.[240]

The general index, which identified "the branches that could be useful to study from the point of view of the hereditary transmission of several traits,"[241] formed the basis for the principal studies of the Milan Study Center during the war. A first field of research regarded the identification and localization of "defective branches." In different zones of Lombardy the field-researchers discovered "original foci of various pathological hereditary forms."[242] Numerous studies were also carried out on hereditary pathologies:

> Several studies have also been carried out on the heredity of dental deformities (students Pazardjklian and Gilioli); fistula auris, cerebral ptosis, atresia ani, and plicated tongue (prof. Luisa Gianferrari); tumours (prof. L. Gianferrari and prof. G. Cantoni); schizophrenia and manic-depressiveness, considering the population of an entire village dating back to the beginning of the 1600s to identify probable local variants (Gianferrari L. and Cantoni G.); and Laurence-Moon-Bardet-Biedl syndrome (Dr. R. Oldrini, center assistant).[243]

As well as the heredity of pathological traits, another line of eugenic research was undertaken by Luisa Gianferrari and his collaborators, regarding the hereditary transmission of talents, starting with "pictorial":

> Many studies have already been carried out in Val Vigezzo, well-known for the great frequency of talented painters, designers, and sculptors, many of whom have achieved well-deserved fame; particularly gifted branches have been identified, going back many generations. The study of material gathered in Val Vigezzo would make a notable contribution to the knowledge of hereditary transmission of talented painters.[244]

240 Zoja and Gianferrari to Montalenti, 30 October 1948, AM, b. 24.
241 Gianferrari, "Sull'organizzazione e sull'attività svolta dal Centro di studi di genetica umana," 113.
242 Gianferrari, "Sull'organizzazione e sull'attività svolta dal Centro di studi di genetica umana," 115.
243 Gianferrari, "Sull'organizzazione e sull'attività svolta dal Centro di studi di genetica umana," 116.
244 Gianferrari, "Sull'organizzazione e sull'attività svolta dal Centro di studi di genetica umana," 116.

Gianferrari's project of a national genetic index collapsed with the fascist regime, leaving only the slim heritage of the research activities of the Milan Study Center in Human Genetics, at least in the first half of the 1940s. Yet Gianferrari's institute did not completely disappear with the demise of the fascist regime, but underwent a second birth from 1948, in the name of a *new* eugenics, that renounced the project of demographic and racial improvement of the stock and instead turned its attention to "eugenic" counseling for couples.

CHAPTER VI

TOWARD A NEW EUGENICS

From 24 to 31 August 1953, the 9th International Congress of Genetics was held in Bellagio, on the banks of Lake Como. Some of the most important names of the discipline were present among the 863 participants, including Haldane, Penrose, Dobzhansky and Darlington. At the end of the Congress, two excursions offered participants the chance for an Italian summer trip: the first comprised visits to the scientific institutes and "main monuments" of Pavia, Milan, Bologna, Arezzo, Rome and Naples; the other, shorter trip visited the Gran Paradiso National Park, as well as Pavia, Milan and Turin.[1]

Following this, a national symposium of genetics applied to zootechnics was held in Turin on 3 September 1953. This congress was organized by the Observatory of animal genetics founded three years previously by the Turin Chamber of Commerce, the Province and the Valle d'Aosta Region.[2] Prior to this, Adriano Buzzati-Traverso, taking advantage of the presence of so many illustrious colleagues, on behalf of the *Union internationale des sciences biologiques,* organized another symposium on population genetics, with the participation of, among others, Dobzhansky, Fisher, Haldane, Mather, Mayr and Waddington.[3]

In many ways the Bellagio Congress represented the expression and the product of the development which had occurred in Italian genetics from the second half of the 1940s. In fact, in March 1947, a Study Center for

[1] Giuseppe Montalenti and Alberto Chiarugi, eds., *Atti del IX Congresso internazionale di genetica. Bellagio (Como, 24–31 agosto 1953)* (Florence: Florentiae, 1954), 1, 16.

[2] Montalenti and Chiarugi, eds., *Atti del IX Congresso internazionale di genetica* 1, 1265–98

[3] Adriano Buzzati-Traverso, ed., *Symposium on Genetics of Population Structure. Pavia, Italy, August 20–23, 1953* (Pavia: Tip. succ. Fusi, 1954).

Genetic Cytology was established at the Italian National Research Council (CNR), under the direction of Giuseppe Montalenti, member of CNR's National Consultancy Committee for Biology and Medicine (*Comitato nazionale di consulenza per la biologia e la medicina*), and the first professor of genetics in Italy.[4]

A few days later, on 25 March, a convention between CNR and the University of Pavia christened the birth of a Study Center for Genetics at the University Institute of Zoology and Genetics, directed by Carlo Jucci.[5] Finally, in July 1947, at the Botanical Institute of the University of Pisa, a Study Center for Plant Cytogenetics was inaugurated, presided over by Alberto Chiarugi, while in December 1948 a Study Center for Biophysics began activities, at the Institute of Hydrobiology "Marco De Marchi," in Verbania-Pallanza. 1948 was also the year in which two professorships of genetics were created, won by Claudio Barigozzi in Milan,[6] and by Adriano Buzzati-Traverso in Pavia.[7]

4 Giuseppe Montalenti (1904–1990) studied with Grassi in Rome, as an internal student in the Laboratory of comparative anatomy. He graduated in natural sciences in 1926 and was appointed as assistant at the Institute of zoology in the University of Rome, directed by Federico Raffaele. In 1937 he obtained the position of *aiuto* at the Institute of zoology at the University of Bologna, directed by Alessandro Ghigi, and stayed until 1939. Between 1933 and 1937 he taught courses of genetics in Rome. In 1939, Montalenti became head of the department of zoology at the Naples Zoological Station. The following year he was appointed to hold the first professorship of genetics in Italy, instituted at the Faculty of Science of the University of Naples, a chair that he held until 1960, at the same time keeping his position as department head of the Station until 1944. See Stefano Canali, "La Biologia," in Raffaella Simili and Giovanni Paoloni, eds. *Per una Storia del Consiglio Nazionale delle Ricerche* (Bari: Laterza, 2001) 1, 534–35; Alessandro Volpone, *Gli inizi della genetica in Italia* (Bari: Cacucci Editore, 2008), 124–25; Fabio de Sio and Mauro Capocci, "Southern Genes: Genetics and its Institutions in the Italian South, 1930s–1970s," *Medicina nei Secoli* 20, 3 (2008): 791–826.

5 Carlo Jucci (1897–1962) graduated in natural sciences in Rome in 1920, spending time in Giovan Battista Grassi's laboratory, before transferring to the Bacological Institute of the High School of Agriculture in Portici and graduating in medicine in Naples in 1925; he also worked as an assistant to the chair of physiology, under Filippo Bottazzi. Thanks to a Rockefeller Fellowship, he spent a year in Plymouth (Massachusetts, USA) before receiving, in 1930, a position teaching in zoology and anatomy in Sassari, from where he transferred to Modena (1932), and finally to Pavia (1934). For a biographical profile, see Paola Bernardini Mosconi, ed., *Carlo Jucci nel centenario della sua nascita. Testimonianze e documenti*, (Milan: Cisalpino, 2000); Maurizia Alippi Cappelletti, "Jucci Carlo," in *Dizionario Biografico degli Italiani* (Rome: Istituto della Enciclopedia Italiana, 2004), vol. 62, 687–90; Volpone, *Gli inizi della genetica in Italia* (Bari: Cacucci Editore, 2008), 128–33.

6 Student of Cesare Artom in Pavia, Claudio Barigozzi (1909–1996) from the start of the thirties studied the chromosomes of the mole cricket and the crustacean *Artemia salina*. In 1937, he worked as a non-staff lecturer in genetics, and in 1939 became assistant of Silvio Ranzi at the institute of zoology at the University of Milan. In the 1940s, he began to research the drosophila and, in particular, the genetic basis for its diverse reactions to light, and genetic control of the dimensions of the cells. For an autobiographical profile, see Claudio Barigozzi, *La stanza di genetica* (Luino: Francesco Nastro, 1981); see also on this topic, Mauro Capocci and Gilberto Corbellini, "Il contesto culturale della ricerca biomedica in Italia nel secondo dopoguerra," *Nuova Civiltà delle Macchine*, 19, (2001): 29–41.

7 Adriano Buzzati-Traverso was born in Milan, the younger brother of the writer Dino Buzzati. In 1934 he spent one year in the US studying population genetics at Iowa University with Ernest W. Lindstrom.

In this climate of rapid development of Italian genetics, eugenics went through a sort of *no-man's-land*, particularly in the 1950s, in which tensions and oppositions were articulated on different levels. These various conflictual dynamics could be summarized as follows:

1) institutional and academic conflict: between SIGE's statisticians-demographers and the geneticists, who formed a new association (AGI) in 1953; and between the latter and physicians, who in their turn formed the Italian Society of Medical Genetics (*Società italiana di genetica medica*) in 1951;
2) political conflict: between mainline, neo-fascist and racist eugenics (SIGE) and reform/new anti-fascist and anti-racist eugenics;
3) ideological conflict: between catholic, familist and natalist eugenics, and secular, birth control-oriented eugenics.

In such a gladatorial context, in the 1950s and 1960s, the debate over so-called genetic counseling seemed to play a unifying role. In fact, applied medical genetics was generally presented as a sort of extension of "eugenics." Genetic counseling was conceived as a worthy and modern form of eugenics, "even if its aim was relief of individual suffering rather than changes in differential birthrate or improvements in the genetic pool, and its means—provision of information to those who asked for it—were wholly voluntary."[8]

In 1938, Buzzati-Traverso went to Berlin where he began a five-year collaboration with Timoffieff-Ressovsky, with whom he developed the theories and methods of radiogenetics. That same year Buzzati-Traverso introduced radiogenetics to his Italian audience, and the views of Timoffieff-Ressovsky and Delbrück concerning the physical dimension of the gene, in so doing developing the concept of an experimental approach to evolutionary mechanisms at the University of Pavia. Professor of genetics in Pavia from 1948, between 1944 and 1948 he directed the Italian Institute of Hydrobiology in Pallanza, while from 1947 he was the director of the CNR Study Center for Biophysics. Between 1953 and 1956 he worked at the University of California, where he founded and directed the genetics division of the Scripps Institution of Oceanography in La Jolla. In 1962, he founded in Naples the International Laboratory for Genetics and Biophysics. For a biographical profile, see Bernardino Fantini, "Buzzati-Traverso Adriano," in *Dizionario Biografico degli Italiani* (Rome: Istituto della Enciclopedia Italiana, 1988), vol. 34, 563–67. See also Mauro Capocci and Gilberto Corbellini, "Adriano Buzzati-Traverso and the foundation of the International Laboratory of Genetics and Biophysics in Naples (1962–1969)," *Studies in History and Philosophy of Science Part C: Biological and Biomedical Sciences*, 33, 3 (2002): 489–513; Francesco Cassata, *Le due scienze. Il "caso Lysenko" in Italia* (Turin: Bollati Boringhieri, 2008).

8 Diane B. Paul, *The Politics of Heredity. Essays on Eugenics, Biomedicine, and the Nature–Nurture Debate* (Albany: State University of New York, 1998), 134.

1. SIGE Schisms: Genetics against Eugenics

In 1938, the Third congress of SIGE, presided over by Corrado Gini since 1924, included the participation—in the section dedicated to human genetics—of biologists Montalenti, Barigozzi and Buzzati-Traverso. Ten years later, in 1948, it was again Gini who led the Italian delegation at the 8th International Congress of Genetics in Stockholm, and who read, as representative, the communication that accepted the invite to host the next congress in Italy.[9]

It is not surprising therefore, that in January 1949, it was the General assembly of SIGE who nominated the Provisory committee[10] to organize the Italian congress. A few months after the Swedish congress, Corrado Gini had resumed his role as Dean of the Faculty of Statistical, Demographic and Actuarial Sciences at the University of Rome, after having risked suspension from service during the post-fascist purging.[11] Having regained his academic power, Gini began once again to draw together the threads of SIGE, which had almost vanished after the end of the war. On 31 December 1948, Gini sent a letter with five attachments to all the members of SIGE. At the heart of the document was the intention to reactivate the organization, recognizing the increasing specialization of genetics with respect to eugenics:

> The President of the Italian Society of Genetics and Eugenics, now that the conditions of Italian academic and scientific life have assumed approximate normality, is about to reanimate the Society, which in the inauspicious wartime and post-war period was forcedly inactive.
> His first act has been to contact the previous members and to find new supporters. To that end, he is approaching people who seem particularly suited to be part of the Society. [...] As the number of new and old members is by now around a hundred, it seems opportune to proceed immediately to the reorganization of the Society, making it more fitted to the times and responding to

9 Giuseppe Montalenti, "L'VIII Congresso internazionale di Genetica (Stoccolma, 7–14 luglio 1948)," *La Ricerca Scientifica*, 19 (1949): 130–31.

10 In the documents, the Provisory Committee was also defined as a Provisory Commission.

11 On this trial, see: Francesco Cassata, "Cronaca di un'epurazione mancata (luglio 1944–dicembre 1945)," *Popolazione e Storia* no. 2 (2004): 89–119.

> the growing number of members, as well as to the specializations of the discipline of genetics on one hand, and eugenics on the other.[12]

In view of an assembly of members, to be held on 15 January 1949, the attachments to the president's letter aimed at the rapid resolution of several organizational questions that were still unclear. First of all, "members" would be defined as all those who, upon the invitation of SIGE, paid the annual society fee of 500 lire before 10 January 1949. Second, the members were asked to approve a new statute, with essentially two characteristics. Article 2 sanctioned the constitution of two "special sections" to distinguish between the spheres of genetics and eugenics within SIGE. The general frame of reference however remained that of racial eugenics, as seen in article I:

> The aim of the Italian Society of Genetics and Eugenics (SIGE) is to promote and support the studies, research and initiatives that seek to grow and perfect the knowledge of the laws of heredity and the improvement of the races, with particular attention to the human races.[13]

The modified statute also consolidated the markedly presidential structure, above all concerning the positions of leadership and the organizational activities. This can be seen in the following articles:

> V. The president administers the society and provides for the inscription of new members. He has the capacity to constitute the committees of article II; and, on standard request by at least 10 members, the special sections of the same article. Every section or committee will have its own president and can have its own secretary;
> VI. [...] The vice secretary general and the treasurer are nominated by the president of the society, assisted by the office of the president.
> The secretaries of the sections are nominated by the president of the society in accordance with the presidents of the respective sections [...].
> VII. The president calls the ordinary and special meetings of the society, seeking to schedule them concordantly with the meetings of the Italian Society for the Progress of Sciences. [...]

12 Corrado Gini to members, 31 December 1948, Montalenti Papers (hereafter AM), b. 24, f. 2, sf. 8.
13 Gini to members, 31 December 1948, AM, b. 24, f. 2, sf. 8.

VIII. The president organizes the national congresses called by the society, presides over them, and oversees the publication of the minutes.
IX. In governing the Society, the president is assisted by the office of the president.
Ex-presidents and ex-vice-presidents of the society, the presidents of the committees, the vice-secretary general, the treasurer and the secretaries of sections can be invited to the office of the president, in an advisory capacity.[14]

The revised statute was accompanied by a questionnaire, which represented the basis of an internal referendum by SIGE on several aspects of general relevance. In particular: the approval of two distinct sections of genetics and eugenics; the declaration of membership of one or the other, or both, sections; the assignation of a secretary to both sections; eventual useful proposals for the organization of the 9th International Congress of Genetics in Italy (partners, contributions to expenses, etc.). A voting card followed, for the election of the president, vice-president and secretary general (already indicated in the respective persons of Corrado Gini, Ottavio Munerati—director of the sugar-beet Experimental Station in Rovigo—and Carlo Jucci) and for the nomination of three proposals for the presidency of the genetic and eugenics sections. A final attachment contained the list, in alphabetical order, of SIGE members as of 31 December 1948: a total of 99 names, of whom 52 were in the eugenics section[15] and 47 in the genetics section.[16]

Gini's convocation, in strict continuity with the past, immediately aroused the opposition of the principal exponents of budding Italian genetics, particularly Adriano Buzzati-Traverso and Claudio Barigozzi. The nature of the clash was clearly expressed in a letter of 1 January 1949 sent by Buz-

14 Gini to members, 31 December 1948, AM, b. 24, f. 2, sf. 8.

15 Alfieri, Argenti, Armanini, Barberi, Barison, Benini, Bisceglie, Buonomini, Cattaneo, Caranti, Castellano, Castrilli, Canella, Costanzo, Dechigi, Eugeni, Federici, Fiore, Floris, Forlini, Fortunati, Gatto, Gemelli, Gini, Giovanardi, Giudici, Imbasciati, Laurincich, L'Eltore, Maggio, Malcovati, Margaria, Martinolli, Maroi, Moracci, Paolinelli, Petrini, Quinto, Revoltella, Robaud, Romaniello, Satta, Savorgnan, Scaglione, Scopelliti, Seppilli, Severi, Sfameni, Sofia, Tesauro, Tripi, Tortora.

16 Baldi, Bambacioni Mezzetti, Barajon, Barigozzi, Baschini Salvadori, Battaglia, Battistin, Beer, Benazzi, Blanc, Bonarelli, Bonvicini, Bronzini, Buzzati-Traverso, Cavalli, Chiappi, Chiarugi, D'Ancona, Dionigi, Draghetti, Dulzetto, Galeotti, Granderi, Guareschi, Jucci, Marcheson, Marcozzi, Maymone, Melis, Montalenti, Monterosso, Morselli, Mosti, Munerati, Pasquini, Piacco, Pirovano, Pompilj, Ranzi, Reverberi, Scossiroli, Taibel, Tallarico, Tria, Valle, Vezzani, Zannone.

zati-Traverso to Montalenti and also signed by Barigozzi.[17] By hand, above the date, Buzzati-Traverso added these few, ironic accompanying lines:

> Dear Monti [Montalenti], before receiving your letter of the 30th, following that circular from the unmentionable one [Gini], I wrote this epic, with the intention of sending it to you, Barigozzi and Jucci. Bari [Barigozzi], as you see, has approved it. For the "strange man" [Jucci] it is difficult to make predictions.
> Read it and think about it. If you share the proposals, or if you have some modifications, let me know urgently, so that we can communicate the proposals and the lack of intention to participate in the statistician's assembly, and inform those members who are friends of genetics before 12 January.[18]

The main point of the document, underlined by Buzzati-Traverso and Barigozzi, was represented, in first place, by the necessity of abandoning any reference to eugenics:

> In its title, the society contains the two expressions of Genetics and Eugenics; this has a historical justification, insofar as the foundation of the society dates back to times in which eugenics was the more widely used term and was appreciated in a way it is not today, while genetics—at least in Italy—had not yet reached the same broad significance with which it is used in various languages. It is highly doubtful that today the two expressions can be used side by side. It is above all certain that, while the term eugenics is falling into disuse, the term genetics corresponds, with unanimous consensus, to a dominion of experimental and exact research that is identified with the most vital and functional part of current biological thinking.
> There is little relevance in conserving a title for traditional reasons, if the structure and the style of the society becomes shaped by this situation. But, in the communication that we have received, there are several points which lead to the conclusion that new conditions have not been considered in the form planned for the functioning of the re-established society.[19]

Another critical remark concerned the form adopted by Gini for the reactivation of SIGE: a hasty assembly, in which the members had only twelve

17 Adriano Buzzati-Traverso and Claudio Barigozzi to Giuseppe Montalenti, 1 January 1949, AM, b. 24, f. 2, sf. 8.
18 Buzzati-Traverso and Barigozzi to Montalenti, 1 January 1949, AM, b. 24, f. 2, sf. 8.
19 Buzzati-Traverso and Barigozzi to Montalenti, 1 January 1949, AM, b. 24, f. 2, sf. 8.

days at their disposal for deciding on fundamental questions regarding the nature and scope of the association. This inexplicable haste risked excluding new members from SIGE who would better represent Italian genetics.

Regarding this, Buzzati-Traverso listed several "facts," dating from 1947 and 1948, which the new post-war SIGE could not ignore:

> 1) The creation and coverage of two new university professorships for Genetics: Milan and Pavia, added to the existing chair in Naples; 2) the development of four Centers of Research of the National Council of Research for experimental activities in the field of genetics. These two facts demonstrate also in an official form that today in Italy an active nucleus of experimental geneticists exists, which can worthily represent our nation on an international level and which must be congruently represented in the heart of a society of genetics, and cooperate and guide the activities; 3) our Delegation to Stockholm has proposed that the next International Congress be held in Italy.[20]

The designation of Italy as the seat of the next International Congress of Genetics decided in Stockholm in 1948, placed SIGE in a position of responsibility to the international scientific community and therefore did not allow a simple maintenance of the status quo:

> A very serious responsibility hangs over Italian geneticists and the institution that has assumed the role of representing and coordinating them, for the obvious reasons of prestige and to demonstrate the level and dignity that these studies and their environment have reached among us. This role must not be underestimated: transactions, compromises and accommodations that might be accepted—for lack of anything better—in our own home, could be severely judged on an international level, and must therefore be avoided.[21]

In addition to these general considerations, Buzzati-Traverso added some accurate observations regarding the new SIGE statute proposed by Gini. Essentially he remarked on four defects: the draft, subjected to the vote of the members, gave excessive power to the president, conceding him the faculty of organizing congresses and deciding the admission of new members; it "armor-plated" the role of president, vice-president, section presidents

20 Buzzati-Traverso and Barigozzi to Montalenti, 1 January 1949, AM, b. 24, f. 2, sf. 8.
21 Buzzati-Traverso and Barigozzi to Montalenti, 1 January 1949, AM, b. 24, f. 2, sf. 8.

and secretary general, by presenting just one name for each; it limited the elections to the meager number of old members, automatically excluding "a quite broad crowd of young experimental geneticists who certainly have the right to have their say"; and finally, it centralized the organization in Rome, without taking into account the "geographic distribution of genetic activities in Italy" concentrated prevalently in the north of the country.

On the basis of the fundamental and formal problems pointed out, Buzzati-Traverso and Barigozzi proposed, first, that the convocation of the SIGE assembly be delayed to a commonly agreed upon date. Secondly, they suggested radical reforms to the SIGE statute: admission of new members before the voting; inclusion in the office of the president of the presidents of committees, the vice-secretary general, the treasurer and secretaries of the sections, with deliberative vote; ordinary and special meetings of the society and the sections; proposals from the outgoing office of the president of three names for each leadership position. The evident key to the revision of the statute was the strong restructuring of the role of the president in the name of a greater "democratization" of the society:

> The modification of the statute and an eventual internal regulation should occur in one of the following ways: a) the character of the society could be transformed from "presidential" to "parliamentary," so that the president has a less prevalent function in the activities of the society, favoring the presidents of the sections and relative secretaries; in particular, the organizational activities of the International Congress of Genetics would be devolved to the president of the genetic section; or b) the "presidential" character could be maintained, but, in view of the Congress, the role of president must be given to a professional geneticist, who, above all on an international level, can more specifically represent Italian genetics.[22]

Therefore, just a few days from the general assembly that was to have signaled the return of SIGE to the public scene, an internal fracture had occurred, as much scientific as it was ideological-political. On one side, the statisticians and demographers gathered around the figure of Corrado Gini and the University of Rome, compromised by their past commitment to fascist eugenics and supporters of a line of substantial continuity; on the

22 Buzzati-Traverso and Barigozzi to Montalenti, 1 January 1949, AM, b. 24, f. 2, sf. 8.

other side was the "Lombardy" group, guided principally by Buzzati-Traverso and Barigozzi, expression of emerging Italian genetics, wanting to eliminate the eugenic past.

The scission, which seemed imminent, was avoided due to mediation by Giuseppe Montalenti, whose strategy was founded on the following objectives: maintain the unity of SIGE under Gini's presidency; give internal autonomy to the genetics section; and remove the organization of the future International Congress of Genetics from Gini's control.

Although Adriano Buzzati-Traverso refused to recognize the validity of the voting, nevertheless the general assembly on 15 January 1949 represented the success of Montalenti's moderate line. Approving the draft proposed by Gini, the assembly elected the president (Gini), vice-president (Munerati), secretary general (Jucci), president of the genetics section (Montalenti) and eugenics (L'Eltore).[23] As for the organization of the International Congress of Genetics, Montalenti successfully promoted the constitution of a provisory committee, presided over not by Gini, but by Alessandro Ghigi.

It was Montalenti who informed Ghigi, clearly disclosing the meaning of his own mediation:

> After many discussions and objections on the part of the Lombardy geneticists, my criteria has prevailed, which was, to *not schism and create another society on our account, but to group ourselves within the existing one, and bite the bullet of Gini's presidency, at least for three years.*
> You have received the relative documents. We then reserve for the genetics section, which has been entrusted to me as president, the right to move with a certain autonomy.
> [...] To avoid individual uncoordinated actions (such as have already been done, for example, by Jucci) regarding the international congress, it seemed necessary and urgent to me that the society nominate a provisory committee to take care of this important problem. The recent general assembly of members held in Rome on 15 January have accepted my proposal for the committee, and that is, to offer the presidency to you. I feel that this is a deserved homage to you on the part of Italian geneticists, and I am also certain that you are the most appropriate and able person for this important undertaking. [...]

23 Gini to members, 23 February 1949, AM, b. 24, f. 2, sf. 8.

I warmly urge you to accept this title. *We did not propose you for the presidency of SIGE, as we had wished, for diplomatic reasons... In this moment it seemed important to revive the society as soon as possible, without creating a fracture, saving different options for a future election.*[24]

But the clash was merely delayed by a few months. A new *casus belli* occurred in April–May 1949, apparently deriving from a banal misunderstanding—an overlap between the date of the general meeting of SIGE and that of the genetics section. Between the lines however, it was possible to clearly read the incompatibility between Gini's centralizing strategy and the system of autonomy pursued by Montalenti for his genetics section.

At the beginning of April, a circular from Gini to the members announced the first scientific meeting of SIGE, to be held in June in Milan, on the occasion of the Congress of Experimental Biology.[25] In the meantime, Montalenti and Buzzati-Traverso were organizing a meeting of the genetics section. On 21 April, in a letter to Gini, Claudio Barigozzi, charged with organizing SIGE's scientific meeting, fixed the date of the genetics section's meeting for 9 June. On 23 April, in a letter to Buzzati-Traverso, Montalenti already suggested the danger of an overlap, although he wasn't concerned:

> It would be good to issue the invitations for the meeting in Milan, in agreement with Bari [Barigozzi], with whom I had a brief telephone conversation in Milan, so that people can prepare their presentations. With Bari, I've agreed that it would be good to open the meeting with a presentation, and I entrust this work to you, on a theme of your choice. I hope you accept. The moment you let me know, I will inform the president. Meanwhile, you tell Jucci, who I suppose will not object.
> *The problem is that, as I see from the circular from our president, the meeting in Milan will not be only of our section, but all of SIGE. However, given that we are more numerous, I don't think it is worth opposing this.*[26]

On the same day, Montalenti officially named Buzzati-Traverso the secretary of the genetics section:

---

24 Giuseppe Montalenti to Alessandro Ghigi, 19 January 1949, AM, b. 24, f. 2, sf. 8; italics added.

25 Gini to members, 4 April 1949, AM, b. 24, f. 2, sf. 8.

26 Montalenti to Buzzati-Traverso, 23 April 1949, AM, b. 24, f. 2, sf. 8; italics added.

> I would be grateful to you if you would take on the job of secretary for our section. I am sure of your acceptance, and ask you to communicate the composition of the advisory board of the society to our sister societies in other countries, and in particular to the English Genetical [sic] Society, telling them to address correspondence to you or me, so that we can stay in contact.[27]

Several days later, Montalenti as president and Buzzati-Traverso as secretary sent a circular that called a meeting of the genetics section in Milan, on 9 June, and invited the members to send the title of their presentations to Barigozzi. On 5 May 1949, Gini wrote to Barigozzi fixing the meeting of SIGE for 7 June, and trusted in the "active participation of the northern geneticists."[28] The fuse was lit. Barigozzi, alarmed, called on Montalenti:

> I have received this amazing letter from Gini: I send it to you urgently, in order for you to respond. I will not respond to Gini, because I would more or less say: take it up with Montalenti.
> It is obvious that a response of this type could create an unpleasant situation between the two of you. You probably have the possibility of diplomatically sorting things out.[29]

Giovanni L'Eltore, president of the eugenics section, and Giuseppe Pompilj sent two incendiary letters to Barigozzi (with a copy to Gini, Jucci and Montalenti), crying conspiracy and sabotage. L'Eltore, on 16 May, declared:

> In consequence of this complex of facts that, I confess, I find very unpleasant, I must categorically protest against the methods followed, reserving the right to take this question to the office of the president, and eventually to the assembly, or, should it be the case, some other forum; *I personally recognize, beyond the lack of regard for the president and the members, the manifest purpose of sabotaging the functioning of the society and damaging the good relationships between its members.*[30]

Pompilj reacted, on 19 May, in an even more dramatic tone:

---

27 Montalenti to Buzzati-Traverso, 23 April 1949, AM, b. 24, f. 2, sf. 8.

28 Gini to Barigozzi, 5 May 1949, AM, b. 24, f. 2, sf. 8.

29 Barigozzi to Montalenti, 18 May 1949, AM, b. 24, f. 2, sf. 8.

30 Giovanni L'Eltore to Claudio Barigozzi, with a copy to Giuseppe Montalenti, Corrado Gini and Carlo Jucci, 16 May 1949, AM, b. 24, f. 2, sf. 8; italics added.

> I discovered in passing that the next 10 June [sic] in Milan a scientific meeting of only the genetics section of SIGE will be held, which, it seems, will substitute that of the whole society, decided on in our last general assembly and pre-announced for early June in a circular from the president.
> It was my intention to take part in this meeting [...]. Naturally now I no longer see fit to participate given that, by mysterious mutation, the scientific meeting of SIGE has been transformed into a meeting of only the genetics section, and not even all of this section, seeing that I have not received any notice, despite having requested a membership of both the genetics and eugenics sections.
> I write to you, distinguished Professor, because I see *something very serious in these facts, that so strongly involve the general interests of science, and therefore, of our society.*[31]

In commenting on the last phrase, underlined, Montalenti added in pencil, ironically, "BAM!!" In the following lines, Pompilj interpreted the entire occurrence as the fruit of a clash between the geneticists, on one hand, and the statisticians and mathematicians, on the other:

> It is with painful surprise that I have to recognize in this small episode an evident attitude of hostility, if not even of *provocation, of the biological geneticists toward the statisticians and mathematicians, whose work deals with the analysis and interpretation of data.*
> As collaboration between different categories of scholars is always fertile, with substantial results for Science, it is therefore desirable, and in the case of genetics such collaboration is indispensable, as the modern development of this science has proved. Why then do you wish to refuse such collaboration? [...] *To this can be added that, on principle, I see this accentuation of the distinction between the two sections of genetics and eugenics as inopportune, all the more because, as things stand today, everything seems to me reduced to the distinction between genetics of the Drosophila and human genetics!*[32]

Montalenti replied to L'Eltore on 20 May 1949, rejecting the accusation of sabotage and emphasizing, on the contrary, his extensive mediation:

31 Giuseppe Pompilj to Claudio Barigozzi, with a copy to Giuseppe Montalenti, Giovanni L'Eltore, Corrado Gini and Carlo Jucci, 19 May 1949, AM, b. 24, f. 2, sf. 8; italics added.

32 Pompilj to Barigozzi, 19 May 1949, AM, b. 24, f. 2, sf. 8; italics added.

> In judging the functioning of the society in the years between Liberation and the meeting of 15 January 1949, which I personally insisted on with Prof. Gini, you could not say that this functioning that we intended to sabotage has been very active.
> Prof. Gini can bear witness to all my actions and efforts to reconstruct SIGE, so that it works, *and to conciliate the opposing currents represented by those who wanted the society governed by professional geneticists and those that wished to continue the direction prevalently by statisticians, demographers and eugenicists.* I worked in this way both out of deference to Prof. Gini, and in order to not divide our forces.[33]

According to Montalenti, Gini was informed of the date of the meeting of the genetics section as early as February. It was therefore up to the president to fix the date of the scientific meeting of SIGE, pre-announced in the 4 April circular.

Montalenti evidently had only just been informed about this development. On 8 June, he in fact wrote to Gini:

> I received, from Barigozzi, in a letter dated 18 May, your letter to him dated 5 May, in which you propose a general meeting of SIGE the 7th of June.
> This date is notably inconvenient because the Society for Experimental Biology will begin its meeting on the 8th and the major part of the members can not arrive in Milan a day earlier, without adding that the Milanese will be very busy with the preparations for the assembly of the next day.
> Therefore, I believe it most opportune that the SIGE general meeting be held on the 8th, or the 10th or 11th, with the meeting of the genetics section on the 9th. Eventually, if the 9th is better, we could also hold the general meeting on the 9th and delay the genetics section's to the 10th.[34]

On 28 May, Corrado Gini answered Montalenti, in a lengthy letter that assumed the shape of a declaration of war. The attempt to find an agreement on the date was radically rejected:

> As for the date of the meeting (and I speak of the general meeting) I do not understand how you can propose the 10th or 11th when you already know that many from here will be busy in Rome on those days, or even the 8th, on which

[33] Montalenti to L'Eltore, with a copy to Gini, Jucci and Barigozzi, 20 May 1949, AM, b. 24, f. 2, sf. 8; italics added.
[34] Montalenti to Gini, 21 May 1949, AM, b. 24, f. 2, sf. 8.

> date the biology colleagues are evidently attending their congress, and even if they were interested in our meeting, could not participate, whilst many of our members may desire to assist in the work of the biology congress.[35]

Leaving aside the calendar, the heart of the question was another entirely, and regarded the possible autonomy of the genetics section from the rest of SIGE:

> Above all, I do not understand how you feel that there could be two distinct meetings, one of the entire society, and one of the genetics section. Evidently, if we hold a general meeting of the society, this must comprise both sections. *It would be truly new if one section was in competition with the society it was a part of!*[36]

Gini's argument culminated in a personal attack on Montalenti, in his role as president of the genetics section:

> Various members [...] have told me, now and on other occasions, the impression that you are not doing for the society that which would be hoped of a member of the office of the president. In particular: the lack of all collaboration with personnel during the period of the reconstruction of the society, the slowness of which you later complained; the lack of contribution from the National Research Council (CNR) at the Center of Theoretical Genetics requested by my Faculty; the lack of inclusion of the president of this society as a CNR delegate at the Congress in Stockholm; the lack of a meeting, already agreed upon at Stockholm, of Italian delegates on occasion of the 1948 Italian competition for a chair of genetics, which delayed the reorganization of SIGE, of which you then complained; the development of every activity concerning the next congress outside of the Society and not in your role as section president, sending, if anything, communication to the society only of things already done; of giving support, if only partial, to the objections recently raised by Prof. Buzzati-Traverso, both about the assembly of 15 January (that you urged) and the modifications of the statute (that were submitted to you as to every other member, without any observations from you) and the leadership positions (the nomination of which was completed after consulting you and without observations on your part); for not allowing the participation in the meetings, not even

35 Gini to Montalenti, 28 May 1949, AM, b. 24, f. 2, sf. 8.

36 Gini to Montalenti, 28 May 1949, AM, b. 24, f. 2, sf. 8 [italics in the original].

> inviting or informing the society, the members or the president of the meetings pertaining more or less strictly to genetics, which will be held in Italy.[37]

In this situation, any meeting of the genetics section in Milan would have been interpreted by the members "no long as a negative attitude, but as positively contrary to the interests of the society." Further, there were the irritated sensibilities of the Roman SIGE members toward the "northern geneticists." Gini continued:

> Keep in mind, regarding this, that the members resident in Rome intended to give proof of sympathy to their geneticist colleagues from the north by accepting their invitation to Milan and by putting themselves to the expense of the travel and stay, and that those that have been informed of this hitch have been strongly irritated by seeing the general meeting already announced in the circular of 4 April [...] substituted, in a unilateral initiative, by a meeting of only one section of the society.[38]

In the face of these "irritated souls," the only solution to "placate the discontent" appeared to Gini to be that of postponing the date and place of the meeting of the genetics section to one "possibly contemporaneous with a meeting of the eugenics section."

As for the violations of the statute, Gini substantially attributed two to Montalenti: first, the rules did not allow unilateral convocations of sections; secondly, the secretaries of the sections were nominated by the president of the society in accordance with the president of the sections, and consequently, Buzzati-Traverso's appointment as secretary could not be considered valid.

Giuseppe Pompilj also asked for clarification, writing an indignant letter to Barigozzi on 31 May (with a copy to Montalenti and Buzzati-Traverso). Pompilj's letter faithfully followed Gini's line:

> This unilateral action has not only undermined the foundations of our society, violating the statute, has not only created a situation of grave embarrassment between the members, certainly damaging a profitable collaboration, but it has also accentuated the separation between the sections of genetics and eugenics, with procedures that we absolutely can not permit.[39]

---

37 Gini to Montalenti, 28 May 1949, AM, b. 24, f. 2, sf. 8.

38 Gini to Montalenti, 28 May 1949, AM, b. 24, f. 2, sf. 8.

39 Pompilj to Barigozzi, 31 May 1949, AM, b. 24, f. 2, sf. 8.

On 2 June, Montalenti sent a technical communication, in which he postponed the meeting of the genetics section to an indefinite date, but possibly coinciding with the international conference on Rh groups of the Milan Serotherapeutic Institute. In the same letter he requested that Adriano Buzzati-Traverso be nominated secretary of the Genetics section.[40] Montalenti's personal response to Gini came several days later, on 6 June. Montalenti did not enter into the merits of the accusations, which he considered "miserable," "inappropriate" and completely "alien to the activities of the society," instead emphasizing his role in helping to reconstruct SIGE. However, the position in defense of Buzzati-Traverso was clear, almost a lesson of democracy inflicted on Gini's autocratic methods:

> Particular attention however must be paid to one of your points, which has truly surprised me. You have accused me "of giving support, if only partial, to the objections recently raised by Prof. Buzzati-Traverso" etc. It would be truly strange if I could not, or rather must not, take into account the objections raised by a member, above all when I am persuaded of their complete or partial justice. I feel that in doing so, I would act against the interests of the society and against every principle of liberty.
> I am sure that you agree with me that the office of the president must serve the society, and that the authority with which it is invested must come from the members, not be imposed on them from above.[41]

As for the presumed difficulty of movement deriving from the choice of Milan (and not Rome) as the location of the meeting, Montalenti declared: "There is no reason in the world that those in Rome should consider it a great condescension to move to Milan, and that it is a natural and dutiful thing that the Milan colleagues come to Rome. If things are put in these terms it will be difficult to agree."[42] To definitively close an "overly long, unpleasant and tiresome correspondence," Montalenti confirmed the convocation in Milan of the genetics section, requesting the complete list of members in both sections and strongly claiming the need for autonomy of young Italian genetics:

40 Montalenti to Gini, 2 June 1949, AM, b. 24, f. 2, sf. 8; see also the circular from Montalenti to the members of the genetics section, 31 May 1949 and from Gini to members, 3 June 1949, both in AM, b. 24, f. 2, sf. 8.
41 Montalenti to Gini, 6 June 1949, AM, b. 24, f. 2, sf. 8.
42 Montalenti to Gini, 6 June 1949, AM, b. 24, f. 2, sf. 8.

> The situation in Italy has changed greatly from before the war: there are now many professors of genetics, each one with a group of young scholars. They are prevalently situated in the north. It is necessary that their opinions are heard and that they are left with a certain liberty of movement.[43]

The best comment on the entire affair was, however, contained in the stinging letter sent by Adriano Buzzati-Traverso to Pompilj. With disarming lucidity and irony, Buzzati-Traverso collapsed Gini's accusatory house of cards, retracing the stages of the clash:

> It seems to me that things went as follows: due to a series of circumstances [...] the president was informed too late that a meeting of the genetics section had been organized; he was not able to make the announcement himself, nor did he want to extend the meeting to the entire society. And so what did he do? He simply waited for several members to directly manifest their amazement and "righteous indignation" at Barigozzi and Montalenti, and then asked at the last moment that Prof. Montalenti postpone the meeting. It would have been greatly preferable if, due to these very exigencies of collaboration of which he speaks, the president, the moment he was informed of the affair on his return from Spain, had written to Montalenti something like this: Dear Montalenti, I have heard what has been done, and am very sorry that you have followed unorthodox practices, because you have erred in a, b, c; let us use every means to remedy this, immediately announcing a meeting also of the eugenics section, so that SIGE meets all together and in this way we have the chance of discussing together the problems of life and relations between the two sections. [...] Instead, we are still here, writing each other recriminating letters of various types truly constructive and essential for the future development of Italian genetics and SIGE in particular! That's what I have to say regarding the past.[44]

As for the future, Buzzati-Traverso underlined the necessity of separating the two sections of SIGE and making them independent. If this did not happen, it was not so bad. Italian genetics could even survive outside of SIGE:

43 Montalenti to Gini, 6 June 1949, AM, b. 24, f. 2, sf. 8.

44 Buzzati-Traverso to Pompilj, 16 June 1949, AM, b. 24, f. 2, sf. 8.

> If this is not possible, I will console myself with the thought that fortunately, Italian genetics has came a long way and has also managed to gain a certain consideration abroad, thanks to the activities of several colleagues, even in the absence of a functioning society that unites all the experts in this subject; if this has happened in the recent past it could even happen in the future.[45]

From these few lines the game in play is neatly visible: while, on one side, Buzzati-Traverso and Barigozzi advocated the necessity of creating an alternative association to SIGE, Montalenti still believed in the possibility of mediation. A solution was represented by the acceptance of Buzzati-Traverso as secretary of the genetics section. Montalenti clearly confirmed the relevance of this candidature, in a letter sent to Barigozzi, and copied to Alessandro Ghigi, on 29 October 1949:

> My intentions are by now sufficiently clear to me: if Buzzati accepts and Gini nominates him secretary of my section, I will remain president. Otherwise, I will stand down (from presidency). I will wait and see what you do about the institution of a dissenting association. I confess that I do not much like the idea. I reserve every decision regarding my eventual participation in your movement: personally I would prefer to stay here tranquilly.[46]

Already in the meeting of the office of the president on 25 July, Gini had declared himself against the admittance of Buzzati-Traverso, judging his position in January 1949 in contrast to the interests and aims of SIGE. Montalenti, on that occasion, had proposed delaying every decision to a general assembly, threatening to stand down if Buzzati-Traverso wasn't nominated.[47] In September, at a convention on *Recent contributions of human genetics to medicine*, organized by the Milan Serotherapeutic Institute "S. Belfanti," Montalenti's introduction, which reprised a 1939 article,[48] implicitly contained a sense of distance on the part of Italian genetics from an embarrassing past, incarnated by the uncomfortable figure of Gini. In front of foreign guests of the caliber of Haldane and Fisher, the condemnation of racism could not have been more complete:

45 Buzzati-Traverso to Pompilj, 16 June 1949, AM, b. 24, f. 2, sf. 8.
46 Montalenti to Barigozzi, with a copy to Ghigi, 29 October 1949, AM, b. 24, f. 2, sf. 8.
47 Montalenti to Buzzati-Traverso, 30 July 1949, AM, b. 24, f. 2, sf. 8.
48 Giuseppe Montalenti, "Utopie," *Rivista di psicologia*, 35 (1939):197–99.

> All of us have noticed the enthusiasm that has accompanied the development of the first studies of human genetics, thanks to Galton in England, and which has produced the birth of eugenics. This discipline, of an eminently applicable character, drawing inspiration from the principles of genetics, was to have rendered man complete master of his destiny, allowing him to improve the species, which, among all the animal species, greatly needs it.
>
> Following this early enthusiasm came a sense of discouragement and skepticism. I do not wish to analyze the causes of these two attitudes, which would carry me too far away. I will cite two recent works: one that mirrors hopeful enthusiasm, the other, jeering skepticism.
>
> The first is a book by Hermann J. Muller, *Out of the Night*, the second, the novel by Aldous Huxley, *Brave New World*.
>
> If we wish to start to consider the negative side of eugenics—which suits my slightly pessimistic temperament—we cannot stay silent on a sad argument, the development of which has recently carried terror throughout the world: the question of race.
>
> When lunatic legislators believe they can seize possession of the destiny of humanity for the advantage of a race that they consider superior, or for an idea that—in good or bad faith—is considered just, the consequences can be terrifying. It is not necessary to remind you of this, as all our hearts are still full of dismay.[49]

The specter of Nazism did not seem, however, to limit the possibility of a reform eugenics, based not on prejudice of race or class, but on irrefutable scientific knowledge, and above all, conducted with liberal, non coercive methods. In the section of the Milan Congress dedicated to the issue *Hereditary illnesses and defects*, the Danish eugenics model was illustrated by Tage Kemp, director of Copenhagen's Institute of Human Genetics, founded in 1938 with a relevant contribution from the Rockefeller Foundation.[50] Since 1938, Denmark had become the major human genetics laboratory in

49 Giuseppe Montalenti, "Genetica umana ed eugenica," in *Atti del convegno dedicato a "I recenti contributi della genetica umana alla medicina"* (Milan: Istituto Sieroterapico Milanese S. Belfanti, 1949), 5.

50 Denmark was the second European country (after the Swiss Canton of Vaud in 1928) to adopt a eugenic legislation in 1929, with the introduction of voluntary medical sterilization, to which was added, in 1934 and 1935, decidedly coercive measure in dealing with the mentally ill and sexual criminals. The application of the law was distinguished by a relatively moderate attitude: from 1935 to 1939, 1380 people were sterilised in Denmark, of whom 1200 were in the category judged "mentally retarded." For a detailed discussion, see Bent Sigurd Hansen, "Something Rotten in the State of Denmark: Eugenics and the Ascent of the Welfare State," in Broberg and Roll-Hansen, eds., *Eugenics and the Welfare State*, 9–76.

the world: several factors (complete civil records, stable and homogeneous population, small distances, advanced state of social health programs) had fostered the development of a national registration of hereditary diseases. After having presented the characteristics of this genetic registration project, and summarized the Danish eugenic legislation, Kemp defended the necessity of basing the negative eugenics method of sterilization on the consent of patients and on scientific prudence:

> Experience demonstrates that the patients themselves, like their relatives, are almost always able to understand the value of the eugenic operations or precautions, and therefore do not object. Notwithstanding this, it is obvious that measures that interfere so radically with the destiny and the intimate life of man could cause friction or differences of opinion. Physicians and other authorities that have to do with eugenics cases are always very prudent and delicate in their research; the guiding principle must always be that too few eugenic operations are preferable to too many.[51]

To avoid the risk of an uncontrollable excess of sterilizations, Kemp hoped that other nations would follow the Danish example, putting into place "eugenic registration based on records concerning all the patients in the country affected by any important hereditary disease, and also their families."[52] In Kemp's view, only "intensive and close international scientific collaboration between medicine and genetics" could make eugenics effective: it would also be necessary that preventive or prophylactic medicine controlled the most important hereditary diseases in the same way in which it monitored and controlled epidemic diseases.[53]

The Marxist biologist John B. S. Haldane was asked by Piero Malcovati, director of the Provincial Maternity Institute, to explain his views on the eugenic effectiveness of sterilization according to the "criteria explained by Prof. Kemp."[54] His response was to praise the Danish model, while specifically rejecting coercive methods:

51 Tage Kemp, "Malattie e difetti ereditari," in *Atti del convegno dedicato a "I recenti contributi della genetica umana alla medicina,"* 17.

52 Kemp, "Malattie e difetti ereditari," 17

53 Kemp, "Malattie e difetti ereditari," 17

54 Piero Malcovati, "Discussione," in *Atti del convegno dedicato a "I recenti contributi della genetica umana alla medicina,"* 69.

> I believe that sterilization could be recommended without reserve only if we could have the security that it would be applied in every country with the same humanity used in Denmark. Although we cannot eliminate the occurrence of hereditary diseases, we can greatly diminish their incidences. Disgracefully, sterilization can and has been used as a weapon of tyranny and, in the current state of human civilization, tyranny is certainly a greater danger than hereditary disease.
> For this reason, I believe that before we make sterilization obligatory, we must use every way to attempt to convince the carriers of serious and dominant abnormalities to abstain from procreation.[55]

Buzzati-Traverso agreed that the research of Tage Kemp reinforced the importance of reconstructing the *human pedigree*. Referring to the Congress of the Serotherapeutic Institute, in the pages of *L'Europeo*, the geneticist overturned the traditional resistance of Italian eugenics in applying the selective practices in use with other animal species to human beings:

> A Danish man, Tage Kemp, had the idea of considering his country of only four million inhabitants as a huge experimental breeding ground, and to gather data on all the families that present any elements of interest. And like the breeder who knows that the offspring of the mare "Tromba" have the defect of biting, Professor Kemp knows that if Signorina Anderson marries, half her children may be deficient.[56]

Buzzati-Traverso believed that the fact that the study of hereditary illness in Denmark was accompanied by the possibility of voluntary sterilization was reasonably positive:

> The number of individuals who request sterilization is gradually increasing year by year. It is calculated that today around half of the mentally defective are sterilized and that day in which the major part of hereditary defects have little probability of being diffused in the population is not far off.[57]

Nevertheless, this did not mean that "horrible malformations and illnesses" could be considered eradicated forever. The process of genetic

55 John B. S. Haldane, "La selezione naturale nell'uomo: Discussione," in *Atti del convegno dedicato a "I recenti contributi della genetica umana alla medicina,"* 69.
56 Adriano Buzzati-Traverso, "Il pedigree umano," *L'Europeo* 5, no. 41 (9 October 1949).
57 Buzzati-Traverso, "Il pedigree umano."

mutation was always possible and according to Buzzati-Traverso, "one of the major dangers of the atomic bomb and nuclear energy is the fact that the radiation emitted in the process of nuclear disintegration greatly increases the normal rates of mutation, so increasing the probability that individuals with hereditary defects will be born."[58] Instead of giving in to fear, it would be better to convert to the progressive confirmation of a "new hygiene": while in the past, bacteriology and pharmacology had won over a large series of infectious diseases, in the future, due to the development of genetics, "we will develop the ability to control hereditary diseases and deformations, habituating men to value their pedigree."[59] In this way it would be possible to avoid dangerous unions, render some marriages infertile, and cure carriers of hereditary defects with new medical procedures. Obviously the "non-worsening of the human type" should be based on "free choice" and on the "development of a biological responsibility of the citizens":

> Some will object that the sanctity of the family and the mystery of procreation confer intrinsic value on the genetic phenomena of human beings, of a moral and religious order that cannot be cancelled by some scientific discovery. Even admitting this criticism, it is worth observing that the diffusion of practices for the improvement, or rather the non-worsening of the human race, must be achieved through free choice and not imposition. With the development of biological responsibility of the citizens we will entrench new persuasions similar to that of not marrying between siblings.[60]

While, therefore, the Congress of the Serotherapeutic Institute was characterized, on one side, by the condemnation of racist eugenics, and on the other, by the sympathetic presentation of Danish reform eugenics, it is not surprising that the SIGE general assembly, following immediately, confirmed the fracture between Corrado Gini and Giuseppe Montalenti, which had by now become unavoidable. On 29 October 1949, Montalenti wrote to Gini:

58 Buzzati-Traverso, "Il pedigree umano."
59 Buzzati-Traverso, "Il pedigree umano."
60 Buzzati-Traverso, "Il pedigree umano."

> As you were in a position to ascertain in Milan, a fracture can be seen in our society that I have always tried to avoid. If this happens, it is clear that I must stand down as president of the genetics section, because this signifies that my policy has completely failed.[61]

Despite the attempts at mediation at the last moment by Carlo Jucci between December 1949 and January 1950, Montalenti's resignation was irrevocably presented to Gini on 30 March 1950:

> Mister President,
> You are in full knowledge of the situation that has occurred in SIGE and it has been the subject of several discussions between yourself, me and other colleagues last September in Milan.
> In particular, you did not wish to accept my proposal to nominate Prof. Adriano Buzzati-Traverso as secretary of the genetics section presided over by myself, making my position difficult and giving me no guarantees that I would be able to represent in good faith all the different currents of Italian genetics.
> As you know, I continued attempts, after September, to allay the disagreement, with no success.
> In these conditions, I do not feel I can continue as the genetics section president, and therefore pass into your hands, Mister President, my resignation.[62]

In an internal referendum on 15 April 1950, Carlo Jucci and the statistician Gaetano Pietra were elected as, respectively, president of the SIGE section of applied genetics and president of the section of mathematical genetics. Even the four honorary foreign members nominated for the occasion reflected Gini's personal scientific relationships: Felix Bernstein, Gunnar Dahlberg, Tage Kemp and René Sand.[63] Ten days later, SIGE's office of the president accepted Montalenti's resignation, attempting to conceal the complete internal division behind the formal quibble of the impossibility of section secretaries—as in the case of Adriano Buzzati-Traverso—residing outside of Rome:

---

61 Montalenti to Gini, 29 October 1949, AM, b. 24, f. 2, sf. 8.

62 Montalenti to Gini, 30 March 1950, AM, b. 24, f. 2, sf. 8.

63 Gini to members, 31 May 1950, AM, b. 24, f. 2, sf. 8.

> All the members of the office of the president have agreed that it is indispensable for the section secretaries to reside in Rome. Regarding this, Nora Federici has revealed the already onerous nature of the work of the central secretary, which would be made unsustainable if correspondence with section secretaries in other places was added.
> As for me personally, you will recall that I clearly wrote in the circular of February 1948 issued to old and new members interested in reviving SIGE, that it was more convenient for the secretaries to reside in the region of the society, a convenience that I felt was agreed with, given that no one raised objections regarding this.[64]

Montalenti's mediation strategy had therefore failed, and Buzzati-Traverso's and Barigozzi's line had prevailed: to constitute an anti-SIGE association of genetics.

And it was this precise intention to definitively distinguish genetics from Gini's eugenics, heavily involved in the fascist past, which became the reason to form the Italian Genetics Association (*Associazione Genetica Italiana*), or AGI, founded in 1953.

2. From Premarital Examination to Genetic Counseling

After the Second World War, Milan became the new Italian capital of eugenics. In fact, in 1946, the first Italian genetic counseling center was based here, part of the Milan State University, as a direct emanation of the Study Center in Human Genetics, directed by Luisa Gianferrari. A few years later, in 1948, the first "municipal eugenic counseling" was founded, at the Milan Policlinic, also entrusted to Gianferrari's Study Center. Individuals and organizations were eligible to approach the counseling center upon presentation of a medical certificate that "clearly specified the diagnosis of the form of illness of the proband, and as many members of the family as possible."[65] The activities of the two consultancy centers were principally

64 Gini to Montalenti, 22 July 1950, AM, b. 24, f. 2, sf. 8.

65 Luisa Gianferrari, "Il Centro di Studi di Genetica umana dell'Università di Milano ed i Consultori di genetica umana dell'Università e del Comune di Milano," *Natura* 41 (1950): 76.

concerned with premarital counseling for betrothed couples and counseling for maternal-foetal haematic group or transfusional incompatibility. The first kind of counseling included mental (psychosis, manic-depressiveness, paranoia, oligophrenia) and nervous diseases (spastic spinal paralysis, progressive muscular atrophy, Little's disease, Huntington's chorea); malformations (cleft lip, congenital dislocation of the hips, metatarsus varus); eye diseases (congenital glaucoma, congential cateracts, Retinitis pigmentosa, juvenile glaucoma, blepharoptosis), haemopathy (hemophilia). The second one was generic and was almost always requested for blood related marriages.[66]

In addition to the Study Center in Human Genetics, there were, in Milan, the premarital prophylactic counseling center of the Red Cross, opened in 1946 on the initiative of Giuseppe Leone Ronzoni, Piero Malcovati and Emilio Alfieri, and the Catholic counseling center *Istituto La Casa* (also called *Opera Cardinal Ferrari*), inaugurated in 1948 and presided over by Antonio Cazzaniga, dean of the Faculty of Medicine at the University of Milan.[67]

It is not surprising therefore, that it was in Lombardy that the discussion of premarital eugenic examinations once again started up in 1946. The occasion was the Congress for Social Assistance and Welfare Studies (*Convegno per gli studi di Assistenza Sociale*), organized in Tremezzo (Como) from 16 September to 6 October 1946, by Michael Schapiro, director of the UNRRA Welfare Division for Lombardy, and by Francesco Vito, professor of political economics at the Catholic University of Milan. Just like thirty years earlier, the debate on the eugenic control of marriages was stimulated by the process of modernization of the welfare system, motivated by the dramatic consequences of the Second World War. In the section *Social welfare and the legislation of work* of the Tremezzo Congress, the presentation of Sergio Mantovani, director of the journal *I problemi dell'assistenza sociale* [The problems of social welfare], dealt directly with this question. In his contribution, Mantovani declared himself in favor of the introduction of compulsory premarital examinations, possibly with a prohibitory character:

66 Gianferrari, "Il Centro di Studi di Genetica umana dell'Università di Milano," 76.

67 On the Istituto La Casa, see Don Paolo Liggeri, "A proposito di consultori prematrimoniali," *Riflessi* 2 (1950): 6.

> It would be easy to conclude the sanitary examination with the exhibition of a certificate of fitness to a public official or a priest, containing at worst, prudential advice. It would be more difficult to conclude with a prohibition, which implies inquiry, control and security measures. I believe that society has the right to take these preventive measures in its defense, even if this leads us to damage some of its members.[68]

Eugenics was invoked in the name of civic education, hygienic awareness and the secular affirmation of the preservation of public health:

> I believe that the introduction of civil habits of control for those who are united in marriage could signal the start of a true, if still uncertain, hygienic conscience, very necessary for the moral and material well-being of all. If it is Christian to bear pain, I do not believe that ignorance or brutalization are Christian.[69]

A year later, on 20 and 21 September 1947, the International Congress for the Treatment of Medical and Social Problems of Premarital Prophylaxis (*Convegno internazionale per la trattazione dei problemi medico-sociali di profilassi pre-matrimoniale*) was held in Milan at the obstetrics and gynecology clinic of the State University, directed by Emilio Alfieri. The positions of the various participants—prevalently syphilographers and gynecologists—reflected the plurality of opinions which was typical of Italian eugenic debate. Piero Malcovati, director of the Provincial Maternity Institute and manager of the premarital prophylactic counseling service of the Red Cross, declared himself in favor of "optional premarital prophylactic consultancies, equipped for clinical and genealogical research and confidential individual counseling of an educational and informative character on the problems of eugenics and familial orthogenics."[70] If Italy was to introduce the principle of a sanitary premarital control into legislation, the ideal solution, according to Malcovati, could be that already adopted, for example, in the Soviet Union, based on the reciprocal exchange of information between the betrothed:

68 *Atti del Convegno per studi di assistenza sociale* (Milan: Marzorati, 1947), 169.

69 *Atti del Convegno per studi di assistenza sociale*, 170.

70 *Profilassi pre-matrimoniale (problemi medico-sociali). Atti ufficiali del Convegno internazionale per la trattazione dei problemi medico-sociali di profilassi pre-matrimoniale* (Bologna: Cappelli, 1949), 52–53.

> I believe that the State must, through the municipal hygiene office:
> 1) quickly inform the future spouses in a simple and persuasive manner, of the main dangers that could beset the spouses and descendants (venereal infection, tuberculosis, hereditary illnesses), appealing to their senses of responsibility;
> 2) then demand the declaration that the future spouse had reciprocally exchanged an explanatory medical certificate, which could eventually be filled in based on a specifically designed form or questionnaire, so that the physician (or consultant) must necessarily direct their attention to the individual fundamental points.[71]

For Giuseppe Morganti, researcher at Gianferrari's Study Center in Human Genetics, the preservation of public health and the reduction of the costs of the welfare system were more than sufficient reasons to justify the necessity of an effective prevention of hereditary diseases:

> Each year considerable sums are spent with the intent, often, unfortunately, in vain, of bringing a physically or psychically abnormal person back to social life, when many times, the birth of this person could have been avoided if only we had informed the parents of the impending danger. Without counting that, if the most caring assistance is an unquestionable duty to this unhappy individual, reinserting him into the social life is often a biological absurdity, because every rehabilitated individual could represent the uncontrollable possibility of procreation of other unhappy people.[72]

In particular, Morganti hoped for the development of a kind of genetic counseling service, with a purely informative character, and the constitution of a national genetic index, "in which, without exception, all the cases of illness of interest to genetics and eugenics should be obligatorily recorded, by all the organizations and people designated by appropriate laws (clinics, hospitals, special schools, physicians)."[73] Carlo Armanini, head physician in obstetrics and gynecology of the Hospital Maggiore in Milan, was also completely opposed to any coercive measures: the prohibition of marriage could in fact carry with it "appalling consequences, such

71 *Profilassi pre-matrimoniale*, 53.
72 *Profilassi pre-matrimoniale*, 186.
73 *Profilassi pre-matrimoniale*, 185.

as the accentuation of Malthusian practices, the spread of illegitimate pregnancies, abortive practices and perhaps also sterilization."[74] Far from being compulsory, premarital examination had to be contained within "the limits of a strictly confidential counseling service of a prophylactic and hygienic kind, that allowed the future spouses for whom it is necessary to understand their situation and eventually put themselves in conditions in which they can spontaneously and freely postpone, or even definitively renounce, their marriage."[75]

Although foreign participants at the Milan congress described medical experiences, such as the premarital counseling in Switzerland[76] or Britain's Marriage Council,[77] in which the "optional" character seemed dominant, voices in favor of eugenic coercive measures were not completely absent. For Sergio Mantovani, for example, premarital prophylaxis, in a country like Italy, "overpopulated, with eight million illiterate, with chronic alcoholism spread throughout the poorest classes, with two million unemployed,"[78] represented an indispensable need for "hygiene" and "social education." Mantovani's sympathies lay in particular with the French legislative model of a compulsory premarital certificate, approved in 1942.[79] Rosario Ruggeri, head physician in infant neuropsychiatry department of the Milan Psychiatric Hospital, also advocated the prohibition by law of marriages between "defective subjects":

> If I were to present those entire families in the psychiatric hospital, father, mother and many children, I am sure that even the fanatical defenders of liberty would be perplexed.
> In these subjects, mental deficiency is clearly imprinted on their faces and it is certainly not necessary to be a physician to see it. Nevertheless, neither the municipal official, nor the parish priest has had the good sense to refuse to unite such people in marriage. [...]
> I am convinced that in some cases coercive measures are necessary to prevent marriages from which we can presume defective subjects will issue.[80]

74 *Profilassi pre-matrimoniale*, 226.
75 *Profilassi pre-matrimoniale*, 227.
76 *Profilassi pre-matrimoniale*, 35–36.
77 *Profilassi pre-matrimoniale*, 123–60.
78 *Profilassi pre-matrimoniale*, 199.
79 *Profilassi pre-matrimoniale*, 199.
80 *Profilassi pre-matrimoniale*, 216.

Equally rigid and intransigent was the position of Cesare Ducrey, professor in Clinical Dermatology at the University of Milan and president of the Italian Society of Dermatology and Syphilography. His point of reference was the legislation of some States of the American confederation, with respect to the introduction of a compulsory premarital examination, with a prohibitory character.[81] The congress closed by accepting Ducrey's invitation to send the papers from the conference to the Parliamentary Medical Group "so that they can utilize them in the next reorganization of the services of hygienic-sanitary defense of our population."[82]

In September–October 1949, the 4th International Congress of Catholic Physicians dedicated a specific session to "premarital eugenics."[83] In a context of complete refusal of any practice considered damaging to Christian morals, from sterilization to birth control, most of the speakers, predominantly Spanish, Portuguese and Latin-Americans, hoped for some form of non-obligatory, non-prohibitory eugenic counseling, accompanied by adequate hygienic education.[84] This was the position of the two principal speakers, João Maria Porto, professor of therapeutic clinical medicine at the University of Coimbra, and Antonio Castillo de Lucas, professor of medical hydrology at the University of Madrid. The latter, in particular, believed that "eugamy"—that is, the biotypological selection of the betrothed—had to complement premarital eugenics. Next to organic treatments, a spiritual preparation for the spouses was also necessary, a sort of "eugenics of the soul." In this sense, the premarital certificate could not help but be spontaneous, dictated by Christian medical conscience, while chastity remained the only permissible solution for the prevention of venereal illnesses.[85]

At almost the same time as the Congress of Catholic Physicians, on 28 September 1949, the Christian Democrat senator Monaldi presented a bill, with the precise intention of providing some suitable prophylactic mea-

81 *Profilassi pre-matrimoniale*, 211.

82 *Profilassi pre-matrimoniale*, 238.

83 *Atti del IV Congresso internazionale dei medici cattolici (Roma, 24 settembre – 2 ottobre 1949)* (Rome: Orizzonte Medico, 1950), 75–158.

84 Amadeo José Cicchitti (Cuyo, Argentina) was in favour of a obligatory premarital certificate but without a punitive character; José Malaret Vilar (Barcelona) condemned sterilization and therapeutic abortion; Antonio M. de Figuereido Meyrelles do Souto (Lisbon) supported a premarital certificate only with an informative nature and exchange of information between betrothed; Victor Manuel Santana Carlos (Lisbon) desired premarital medical counselling; Giacomo Santori (Rome) proposed obligatory syphilis cures, accompanied by a hard fight against prostitution and the introduction of an informative premarital certificate.

85 *Atti del IV Congresso internazionale dei medici cattolici*, 103–04.

sures to combat the menace of venereal infection that would be the consequence of the approval and application of the Merlin law (this law, which in 1958 abolished the Italian system of legal brothels or "closed houses," was already in discussion in 1949). Monaldi's bill, in article seven, insisted on the mandatory nature of premarital visits and on a certificate that simply attested that an examination had occurred.

The Monaldi bill vividly interested the National Center for Prevention and Social Defense (*Centro Nazionale di Prevenzione e Difesa Sociale*, CNPDS), the prestigious Milan cultural institution founded in 1948 by the magistrate Adolfo Beria di Argentine to study the social effects of the process of modernization in post-war Italy.[86] The CNPDS appointed a "Commission for legislation of matrimonial prevention" (*Commissione per una legislazione di prevenzione matrimoniale*), with the aim of deepening the study of article seven of Monaldi's bill. At the end of the work, the commission published a document that briefly summarized the critical considerations of the experts who had participated.

Many members of the medical section of the commission declared themselves in favor of examinations being non-obligatory, as an obligatory exam not only partly corroded individual freedom, but, in many cases, did not facilitate the individual's collaboration in the genetic research, therefore compromising the eugenic prognosis. Luisa Gianferrari, in particular, pronounced herself in favor of non-obligatory visits that regarded not only venereal illnesses, but all hereditary diseases.[87] According to Gianferrari, given the diagnostic difficulties, the coercive and unilateral charac-

86 Mimmo Franzinelli and Pier Paolo Poggio, *Storia di un giudice italiano. Vita di Adolfo Beria di Argentine* (Milan: Rizzoli, 2004), 45. The presence of several people in particular at the constitutional meeting of the CNDPS, in July 1948, gives the idea of its cultural and political relevance: Antonio Banfi (philosopher and communist senator), Riccardo Bauer (president of the Humanitarian Society), Alessandro Casati (War minister of the first post-war government), Ettore Conti (financier and president of the National Development Society for Industrial Entities), Giovanni Demaria (rector of the Bocconi University), Antonio Greppi (socialist mayor of Milan), Achille Marazza (then Christian democrat delegate for CLNAI and later undersecretary in De Gasperi's government), Ferruccio Parri (Prime minister in the first post-war government), Alfredo Pizzoni (then president of CLNAI), Umberto Terracini (president of the Costitution assembly). On CNPDS, see Franzinelli and Poggio, *Storia di un giudice italiano*, 37–138; Vincenzo Tomeo, *Il Centro nazionale di prevenzione e difesa sociale. Un caso di ricerca sociale e di azione sui centri di decisione politica* (Milan: Giuffrè, 1961); Mirella Larizza Lolli, *Le scienze politiche e sociali*, in *Storia di Milano. Il Novecento*, vol. 18 (Rome: Istituto della Enciclopedia Italiana, 1995), 854–58.

87 CNPDS, *Prevenzione matrimoniale e introduzione di un certificato prematrimoniale obbligatorio nella legislazione italiana. Relazione della Commissione di studio – art. 7 del progetto di legge del sen. Monaldi* (Milan: CNPDS, 1951), 23.

ter of the examination would certainly be counter-productive. Instead of enforced methods, a diffuse activity of education and "eugenic propaganda" was without any doubt preferable, in order to familiarize citizens with the existence and services of the counseling centers: "Therefore, only well-intended propaganda, that stimulates those interested in knowing their own specific risks and their own eventual descendency, together with an adequate genetic counseling, could achieve the aims."[88]

Agostino Crosti, director of the Dermo-Syphilopathic Clinic at the University of Milan, was also in favor of a purely consultative counseling function,[89] as was Piero Malcovati, who particularly insisted on the importance of cultural propaganda:

> The public appreciate the concept and the initiative and understand the problem; but due to a singular form of inertia, they need to be pushed by propaganda to go to the counseling. When the political papers or the weeklies speak of the possible dangers of marriage and the necessity to prevent them with a premarital examination, the counseling center has many patients for some months; no sooner has the propaganda slowed than the public also thins out.[90]

Carlo Alberto Ragazzi, head of the medical staff of the municipality of Milan and responsible for hygiene at the Polytechnic of Milan, also considered "hygienic and moral propaganda" more effective for a "reawakening of awareness" than a legislative measure, which he judged "insufficient in its structure and social effects."[91] Instead, Ducrey found himself in an isolated position, advocating the adoption of a mandatory certificate with serological exams for syphilis for men only, and the extension to both sexes of radiological exams for tuberculosis.[92]

In conclusion, the medical section of the CNPDS commission, presided over by Eugenio Medea, declared itself in favor of a premarital examination of a consultative-educational character—"it is a question of comprehension, of civil education, of sense of responsibility"—hoping for the broadening to all hereditary diseases, "above all mental and nervous."[93]

88 CNPDS, *Prevenzione matrimoniale*, 24.
89 CNPDS, *Prevenzione matrimoniale*, 11–12.
90 CNPDS, *Prevenzione matrimoniale*, 17.
91 CNPDS, *Prevenzione matrimoniale*, 22.
92 CNPDS, *Prevenzione matrimoniale*, 13–16.
93 CNPDS, *Prevenzione matrimoniale*, 25.

The juridical section of the commission however, faced a crucial argument: what would happen if the betrothed did not exhibit the certificate? The president of the section, Gaetano Scherillo, declared that introducing an obligatory sanitary measure without any kind of punitive mechanism ran "the risk of proclaiming a principle, but without any practical effect, not even that of [...] creating a new custom."[94] To resolve the problem, Domenico Medugno, president of the Milan Juvenile Court, and Mario Dondina, university lecturer on penal law and penal procedure at the Faculty of Law of the State University of Milan, supported the impeditive effectiveness of premarital examinations.

However, Domenico Barbero, professor at the Faculty of Law of the Catholic University of Milan, and Antonio Donati, magistrate and judge for the Milan Civil Court, proposed that the lack of presentation of the premarital examination certificate be elevated to the level of prohibitive impediment, or in other terms, that the presentation of this certificate be necessary not for the validity, but for the regularity of marriage. This would necessitate the introduction therefore of a fine for any municipal official who celebrated a marriage without the registration of the premarital visit. This moderate line came to be the general position of the juridical section of the Commission, which, in its final resolution, interpreted an eventual compulsory premarital examination as the first step along a dangerous path that necessarily led to eugenic sterilization. Gaetano Scherillo concluded:

> The concerns for sanitary protection and social defense are sacrosanct, but we must be careful, as it is the start of a path that leads to a consequence that no one wishes to see arrive in Italy: eugenic sterilization. If we commence with prohibiting marriages, step by step, that is where we will end. And perhaps we should not forget that man is man, and not a bovine or equine race to improve with progressive selection.[95]

The moderate positions of the medical and juridical sections of the CNPDS commission were contrasted by the more radical one of the sociological section. The speakers of this section—Eugenio Pennati, professor of political sociology at the University of Pavia, and Mario Dal Prà, professor of

94 CNPDS, *Prevenzione matrimoniale*, 39.
95 CNPDS, *Prevenzione matrimoniale*, 40.

history of ancient philosophy at Milan State University—supported the obligatoriness of the premarital visit, even if it would initially be without "impeditive effectiveness." This was the case, for example, of the evolutionary and illuminist view of Mario Dal Prà:

> Obstacles should be placed before a physically defective person who could compromise, through marriage, the possibility of physical and spiritual life of the children, impoverishing or compromising at the same time the general equilibrium of the life of society. [...] A sign of the moral poverty of a society is its arrest of its acquired forms of behavior, even when the critical and social senses have shown the need for change and progress toward new experiences. We can not claim that the old forms will mature by themselves into new experiences. Suitable legislative acts are needed to break the crust of tradition, and open it toward always deeper integration.[96]

Finally, the sociological section, presided over by Antonio Banfi, senator of the Popular Democratic Front and historian of philosophy, proposed these conclusive resolutions:

> 1) the necessity of developing sexual education;
> 2) the necessity of the diffusion and facilitation of syphilis diagnosis and cures with institutions adapted to the social environment;
> 3) the necessity of a gradual development of legislation, in the sense that the legislator must not intervene only to sanction an ethical custom, but to provoke and confirm an ethical conscience, taking into account the conditions in which the action takes place;
> 4) the recognition of the social problem that underlies all these particular problems of defense and prophylaxis, of ethical, juridical and sanitary education.[97]

Considering the criticisms of the three different sections, the CNPDS commission agreed to propose to legislators the removal of article seven from Monaldi's bill and suggested the formulation of another bill on premarital prophylaxis, which could acknowledge the conclusions reached by the commission.

In December 1949, a bill written by Mary Tibaldi Chiesa referred to the analytical report of the CNPDS. Tibaldi Chiesa had been delegated by the

[96] CNPDS, *Prevenzione matrimoniale*, 43.
[97] CNPDS, *Prevenzione matrimoniale*, 44.

Italian Republican Party (PRI) to study the problem of the institution of premarital consultancies. There were three "points of view" expressed in this bill:

> 1) recognition of the need of premarital examinations and consequent determination on the part of the legislators to make them obligatory with appropriate measures; without however, the result of the visit constituting a possible obstacle to marriage;
> 2) recognition of the need not only for the visit, but for a medical premarital certificate, and consequent determination to make both obligatory by law, avoiding marriage in cases where the results of the visit are unfavourable;
> 3) recognition of the need of premarital examinations as a guarantee for the protection of the spouses and the offspring, not obligatory by law, but rather as an opportunity to create an awareness of the problems of marriage and offspring, and to exercise, with appropriate means, solutions and measures that promote the knowledge of the danger constituted by infective and hereditary illnesses, the maximum propaganda and works of persuasion and conviction around the efficiency and utility of centers of premarital counseling, with a free and confidential examination.[98]

Specifically, the creation of counseling services should be as broad as possible: every hospital in every capital had the obligation to institute a premarital counseling service, and those towns that, although not capitals, "were relatively important," had the ability to institute one, first asking the advice of the High Commission of Hygiene and Health (*Alto Commissariato d'Igiene e Sanità*), or ACIS. The counseling service would be directed by "the head physician of the hospital, with the advice of the head surgeon, a gynecologist, a neuropsychiatrist and a social assistant." Contrary to Monaldi's bill, the counseling was voluntary, free and secret:

> Whoever comes to the counseling service is not under the obligation to give their details, but only all the data useful to the consultant, and has the right to receive, at the end of the consultation and the tests, a written declaration justifying the advice for the better fulfillment of marriage.[99]

98 *Atti Parlamentari, Camera dei Deputati,* Bill proposed by deputies Tibaldi Chiesa Mary, Chiostergi, Targetti, Capua, Ceravolo, Cornia, De Maria, Perrotti, Riva, Migliori, Giannini Olga, Zerbi, Cucchi, announced 19 December 1949, n. 1000, entitled: *Istituzione di Consultorii prematrimoniali*: 1.

99 *Atti Parlamentari, Camera dei Deputati,* Bill no. 1000: 5.

To favor the "necessary hygienic matrimonial awareness," premarital counseling centers must carry out "adequate information campaigns, and upon publication of the marriage banns, the municipalities must distribute to the future spouses a booklet that clearly illustrates the principles of premarital prophylaxis, and the aims and functions of the genetic advisory services."[100] The functional costs of the counseling services would be charged to the hospitals where they were based, but the State, through the ACIS, could contribute to the financing with subsidies proportional to the activities of single counseling services, "to a sum of 40 million annually for the entire national territory."

In the wake of the Tibaldi Chiesa project, in March 1950 the *Istituto La Casa* in Milan also developed a draft bill, signed by Giuseppe Canino and Luigi Migliori: every capital of the provinces would be obliged to institute a premarital counseling center, under the control of the provincial administration and the ACIS. The response would be verbal and free, without obligation to provide details. All those who requested a visit or simply a verbal consultation would be given a free "prophylactic booklet." All the genetic advisory services would then send useful information to the Milan Study Center in Human Genetics, as a contribution to the "national genetic index."[101]

Luisa Gianferrari[102] also intervened several times, in the first half of the 1950s, to support the introduction of "eugenic premarital prevention in the Italian sanitary organization." The years in which Gianferrari had praised the effectiveness of Nazi legislation seemed far away. Now the condemnation of "compulsory eugenics" was nearly obligatory:

> We believe that any compulsory eugenics is unacceptable and we are contrary to every measure damaging to the moral and juridical rights of man, even if it is limited to the obligation to present a premarital certificate or attend a consultation. Our experience of over a decade of eugenic counseling has demonstrated to us that even from a technical point of view, eugenic counseling must necessarily be based on the collaborative activity of those interested,

100 *Atti Parlamentari, Camera dei Deputati,* Bill no. 1000: 5.

101 "Inquadramento della istituzione dei consultori prematrimoniali nella legislazione italiana," *Riflessi* 2 (1950): 6–7.

102 On the role of Luisa Gianferrari, see Giovanni Widmann, "Pionieri della medicina genetica preventiva in Italia. Luisa Gianferrari e l'esperienza dei consultori genetici prematrimoniali," in *Atti della Accademia Roveretana degli Agiati,* 3, no. B (2003): 35–66.

> which must derive from a voluntary act, determined by the conscience to fulfill the moral obligation that marriage carries in regards to the health of the unborn children.[103]

Eugenic counseling had to be, therefore, the result of free individual choice:

> The only means available to prevent the diffusion of pathological factors [...] in our current state of knowledge, is the selection of coupling. We must be aware that such a measure, in eugenics, can be distinguished into compulsory and non-compulsory. The first possibility includes premarital certificates and sterilization, carried out through surgery or radiology. The second option includes preventive birth control, based on limitation of births, sexual education and eugenic counseling.
>
> We declare ourselves completely contrary to every compulsory measure—and therefore also to premarital certificates, even if they are purely "informative"—because they contravene the moral and juridical rights of man. We believe moreover that preventive birth control in practice fails eugenic aims, due to selfishness and hedonism. What remains is education and eugenic counseling.[104]

Gianferrari identified two forms of "prophylaxis of hereditary diseases." The first, "idiotypic" prophylaxis, comprised the classic forms of negative and positive eugenics:

> Idiotypic prophylaxis comprises both classical eugenics, which aims for the improvement of the stock through the selection of spouses, favoring the reproduction of individuals particularly endowed, and impeding as much as possible that of defective individuals, and idiotypic therapy. This can be practiced through amphimixis, that is, the insertion in defective plasm of factors that act to correct or block pathological factors, or by favoring a return mutation, if the pathological form is influenced by mutational factors.[105]

"Phenotypic" prophylaxis, on the other hand, worked on environmental conditions, inhibiting the manifestation of idiotypic defects: "Phenotypic

103 Luisa Gianferrari and Giuseppe Morganti, "Appunti per una organizzazione eugenica in Italia," *Acta geneticae medicae et gemellologiae* 1, no. 2 (May 1952): 214.

104 Luisa Gianferrari, "Proposte per l'inquadramento della prevenzione eugenica prematrimoniale nell'organizzazione sanitaria italiana," *La settimana medica* 37, no. 21 (1949): 4–5.

105 Luisa Gianferrari, "Introduzione alla profilassi delle malattie ereditarie," *Acta geneticae medicae et gemellologiae* 1, no. 2 (May 1952): 116; see also Luisa Gianferrari, "Genetica e matrimonio," *Riflessi* 1 (March 1959): 1–11.

prophylaxis aims to impede and attenuate the manifestation of hereditary illnesses by modifying the environmental conditions necessary for phenotypic realization."[106] Therefore, in Gianferrari's view, preventive measures of hereditary diseases were matrimonial selection, voluntary control of reproduction and, last but not least, direct action on environmental variables:

> From a theoretical point of view, we are therefore authorized to declare that if we are able to understand the environmental components necessary for the manifestation of hereditary pathological characteristics and their active momentum, there will be only one limitation to possible intervention, that of law, omnipresent.[107]

As with infective illnesses, when considering hereditary illnesses, geneticists had to operate in strict contact with clinicians and hygienists, while "eugenic awareness" could be developed by adequate education and information. To this end, Gianferrari proposed the distribution by the municipality of a "sanitary booklet to inform those affected by morbose hereditary forms or who come from defective pedigrees of the serious responsibility toward the offspring that marriage carries with it,"[108] to every youth who came of age—and not only to engaged couples in the act of publishing their marriage banns.

Starting from this theoretical position and from the activity of the Milan Study Center, in 1952 Gianferrari and Morganti, partially integrating the Tibaldi Chiesa proposal, listed several "points for eugenic organization in Italy," which consisted of the development of state and private structures for a campaign of sensitization of "all strata of the population," for the training of specialists in "eugenic counseling" and for the old proposal of a national genetic index:

> Even if we limit eugenics to free counseling, State measures are necessary to diffuse and control it, possibly improving current private initiatives.
> In our opinion, the State must:
> carry out efficient campaigns that reach all strata of the population in every region;

[106] Gianferrari, "Introduzione alla profilassi delle malattie ereditarie," 116.
[107] Gianferrari, "Introduzione alla profilassi delle malattie ereditarie," 117.
[108] Luisa Gianferrari, "Genetica umana," in *Atti del IV Congresso internazionale dei medici cattolici*, 129.

> create apposite courses to offer eugenic counselors the possibility to adequately prepare themselves for their difficult work;
> institute a qualification exam for eugenic counselors;
> exercise vigilance and control over the eugenic counseling;
> oblige eugenic counselors active in a premarital counseling center, or free professionals, to always provide a certificate with conclusions clearly justified and to keep a copy for the sanitary authority; and favor the gathering and the analysis of statistical data for hereditary illnesses with eugenic relevance in our population.[109]

It was necessary however to wait until 1956 to see the approval of a law that seemed to reconcile the principles that inspired the Monaldi and Tibaldi Chiesa bills: on one side, in fact, the 25 July 1956 law, no. 837 (the so-called "Monaldi law") again dealt with the measures for "the control of venereal illnesses"; on the other, it provided for non-mandatory premarital examinations. Article seven, in particular, read as follows:

> Whoever intends to contract a marriage can ask a provincial physician or a municipal sanitary official to arrange, through a recommended sanitary institute, the ascertainment of their current state of health, comprising a serological blood test for syphilis [...]. The results of the examination should not be indicated on the certificate.[110]

The Italian legislation implicitly confirmed the principle of positive premarital eugenics and recognized the appropriateness of a premarital medical-prophylactic examination. In practice, the issue was resolved by confirming the voluntary nature of the act: with Monaldi's law, the Italian State invited the citizens to accept such a principle voluntarily, offering them the possibility to freely obtain the examination and medical certificate upon request.

Premarital counseling, in the 1950s, underwent significant development: in 1951 a counseling center was founded in Trieste, as part of the municipal hygiene and health office. In 1956, another opened in Florence, at the university's Institute of Medical Semeiotics; another in 1957 in

109 Gianferrari and Morganti, *Appunti per una organizzazione eugenica in Italia,* 214. See also Luisa Gianferrari, "Piano per un'organizzazione eugenetica in Italia," *L'economia umana* 2 (1952): 5–7.

110 For a copy of the text of the law in Italian, see Giovanni Davicini, *Lex-Legislazione italiana* 42, July–December (Turin: UTET, 1956): 1254–59.

Rome, in the offices of ONMI, under the direction of Aldo Marcozzi, central dermo-syphilographic inspector.[111]

During this decade, the problem of the "eugenic" prophylaxis of genetic diseases was particularly connected in Italy with the implementation of the anti-thalassemia campaign. It was, in fact, in the last half of the 1950s that the Italian public health system finally recognized the relevance of the studies and sanitary program that Ezio Silvestroni and Ida Bianco, at the time pathologists at the Medical Clinic of the University of Rome, had formulated since 1943.[112] Thanks to the mediation of the Institute of Hygiene at the University of Rome and to financing from the ACIS, in 1954 the first Microcythemia Study Center was founded in Rome, followed in 1956 and 1961 by another seven regional sections.[113] In 1954, the Rockefeller Foundation, on the basis of a research project coordinated by Giuseppe Montalenti, decided to finance the research of Silvestroni and Bianco and the Roman newborn center, repeating the necessity of confronting "the eugenic aspect of the microcythemic problem, the establishment of official registers of persons carrying this gene, marriage counseling in some form."[114] In 1961 the network of centers, directed by Silvestroni and Bianco, was offi-

111 See Giacomo Perico, "Visita e certificato prematrimoniali," *Aggiornamenti sociali* 12, no. 1 (January 1961): 13. On the Rome counseling center, see the testimony of Aldo Marcozzi in "Voci diverse," *Riflessi* 3 (September 1960): 71.

112 Ezio Silvestroni (1905–1990) graduated magna cum laude in medicine and chirurgy from the University of Padua in 1934. From 1936 to 1939 he worked at the Cancer Institute of Milan, directed by Pietro Rondoni. From 1939 to 1956 he was an assistant in the Medical Clinic at the University of Rome, where he developed his scientific activities with the collaboration of Ida Bianco. He lectured in general pathology, medical pathology, clinical medicine and hematology. From 1947 to 1953 he participated in four competitions for the chair of medical pathology, but did not win the chair despite having a scientific curriculum vitae already well-known and appreciated on an international level. From 1957 to 1975 he was the head haematologist at the Sant'Eugenio Hospital in Rome. In 1943, at the Medical Academy of Rome, Silvestroni and Bianco described the existence of healthy subjects who were carriers of a haematological framework both characteristic and hereditary, which they named microcythemia (today, thalassemia minima). Soon afterwards, they studied a vast group of microcythemic families, collected in various regions of Italy with great difficulty, given that it was during the war. This led to the discovery of the etiological link between microcythemia and Rietti-Greppi-Micheli illness (today, thalassemia intermedia). It demonstrated, completely independently from the analogue research of American scientists, which was then unknown in Italy due to the war, that Cooley's anaemia (today, thalassemia major or mediterranean anaemia) was the expression of the homozygotic condition for microcythemia.
In 1949, presenting the results of their studies at the 50th Congress of the Italian Internal Medicine Society), Silvestroni and Bianco proposed the introduction of "eugenic" measures, specifically a premarital control and an obligatory blood test for students. See Ida Bianco Silvestroni, *Storia della microcitemia in Italia* (Rome: Giovanni Fioriti editore, 2002).

113 In order: Ferrara (1956); Cosenza (1957); Palermo and Cagliari (1958); Naples, Reggio Calabria and Lecce (1960). For a study of greater depth on the entire affair, see Stefano Canali and Gilberto Corbellini, "Lessons from Anti-Thalassemia Campaigns in Italy, before Prenatal Diagnosis," *Medicina nei secoli* 14, no. 3 (2002): 739–71.

114 R. R. Struthers, director of the European Office, to Montalenti, 22 January 1954, AM, b. 125.

cially included in the special projects, financed by the Ministry of Health, and assumed a juridical character under the name of National Association for the Fight against Microcythemia in Italy (*Associazione Nazionale per la Lotta contro le Microcitemie in Italia,* or ANLMI). This national association provided the first example worldwide of a prophylactic campaign against thalassemia, and not surprisingly, was successively adopted in Greece and Cyprus in initiatives based on the same model. ANLMI's activities, between 1954 and 1971, were founded essentially on the preventive "eugenic" model conceived by Silvestroni and Bianco, and characterized by mass screening of the school population and a vast and simultaneous campaign of information and premarital prophylaxis. In 1963 in Ferrara—one of the zones most hit by microcythemia and in which the activities of ANLMI were particularly intense and effective—the entire school population in the provincial territory was screened. The identification of carriers of microcythemia led to a successive investigation on family nuclei, and the parallel development of a provincial haematological register, accompanied by an intense campaign of information and genetic counseling, in order to favor the development of a "premarital eugenic mentality."[115]

At the end of the 1950s, the problem of the prevention of thalassemia was the starting point for a timely and direct intervention by the Pope, Pius XII, on the issue of marital morality. On 5 September 1958, in a special audience to the participants of the 7th Congress of the International Society of Blood Transfusion, the Pope cited the example of the Dight Institute at the University of Minnesota as a model to imitate for eugenic counseling in Italy, in order not to damage individual freedom:

> In a general sense, we must, first of all, underline the necessity of providing the public with the indispensable information on blood and its heredity, so as to permit individuals and families to be on their guard against this terrible eventuality. With such an aim, we can organize, in the manner of the American "Dight Institute," services of information and counseling, where the betrothed and spouses can examine the questions of heredity in good faith, with an aim of better ensuring the happiness and security of their union. These services will not just give information, but help those interested to carry out the appropriate measures.[116]

[115] Canali and Corbellini, "Lessons from Anti-Thalassemia Campaigns," 752.

[116] Pius XII, *Discorsi ai medici* (Rome: Orizzonte Medico, 1959), 680–81.

The would be parents, therefore, were also able to eventually choose the "dysgenic" option:

> Informed of the danger and its effects, the parents can then take a decision that will be "eugenic" or "dysgenic" regarding the hereditary characteristic taken into consideration. If they decide not to have children, their decision is "eugenic," which means that they will not propagate the defective gene, generating both ill babies and normal carriers. If, as usually happens, the probability of producing a child who is a carrier of this defect is less than was feared, they may decide to have other children. This decision is "dysgenic" because it propagates the defective "gene" instead of arresting its diffusion.[117]

In sum, the result of the "genetic counseling," according to Pius XII, should be that of "encouraging the parents to have more children than they would have had without it, as the probability of an unfortunate case is less than was thought."[118] In a clinic such as the Dight Institute, counseling would not, however, involve the problem of number of children and would not aim to "repress fertility." The Pope emphasized: "You would not give information on the way to 'plan' families, because such a question does not enter your objectives."[119] It is interesting to note here how the Pope reproduced, in an almost literal way, several passages from the essay *Counseling in Medical Genetics*, by Sheldon C. Reed, director of the Dight Institute from 1947 to 1977:[120] this classic essay was translated in Italian in 1959 and published in the series *Analecta Genetica*, edited by Luigi Gedda.[121] In Reed's essay, the typical topics of American eugenics abounded, such as the proposal of the segregation of children with low IQs in special institutes,[122] the statement of the "dysgenic" nature of insulin[123] and the identification of "diagnostic criteria" useful in adoption for estimating whether a child "of mixed racial ancestry" could "pass for white" and therefore enjoy better socio-economic conditions in life.[124] Despite these ambiguous refer-

117 Pius XII, *Discorsi ai medici*, 681.
118 Pius XII, *Discorsi ai medici*, 682.
119 Pius XII, *Discorsi ai medici*, 682.
120 On Reed and the Dight Institute, see Kevles, *In the Name of Eugenics*, 253.
121 See, for comparison, Sheldon C. Reed, *Consulenza in Genetica medica* (Rome: Edizioni dell'Istituto Gregorio Mendel, 1959), 12–13.
122 Reed, *Consulenza in Genetica medica*, 77–86.
123 Reed, *Consulenza in Genetica medica*, 160.
124 Reed, *Consulenza in Genetica medica*, 130.

ences, the Pope's discourse nevertheless explicitly condemned racism and negative eugenics. In the face of the progress of genetics, men—the Pope claimed—must "themselves avoid, and help others to avoid, the numerous difficulties of physical and moral character," in this way respecting the "community of blood" that represented the material basis of human nature:

> They must be alert to all that could cause permanent damage to their descendants and lead to an interminable length of disgrace. In regards to this, let us remember that the community of blood between people, whether in the family or the community, imposes certain duties. Although the formal elements of every human community are of a psychological and moral order, the descendants constitute the material basis that must be respected and not damaged.[125]

Applied to the "human stock," this same principle required great prudence, given the "exaggerated insistence of the significance and value of racial factors":

> Those excesses that can lead to racial pride and hatred are unfortunately overly marked. The Church has always been energetically opposed to this, both in cases of attempts at genocide, and in those that have been called a "color-bar" (color barrier). It also disapproves of any genetic experience that takes the spiritual nature of man too lightly and treats it as an example of any animal species.[126]

A few days later, on 12 September 1958, Pius XII received the participants of the 7th International Congress of Hematology to Castel Gandolfo, and on this occasion, responded directly to several questions posed by physicians on the issue of "defective heredity" and genetic counseling. Four questions specifically addressed the problem of Mediterranean anemia. The first was: "In general, and especially in Italy and the Mediterranean basin, are premarital examinations and in particular blood exams, advisable?" The Pope's answer was affirmative, even going so far as to hypothesize, in particularly serious localized situations, an obligatory character:

[125] Pius XII, *Discorsi ai medici*, 683.
[126] Pius XII, *Discorsi ai medici*, 683–84.

> This examination is advisable, just as, if the danger is truly serious, it could be imposed in certain provinces or localities. In Italy, in the entire Mediterranean basin, and where groups of emigrants from this country are gathered, we must keep special track of this Mediterranean hematological disorder. The moralist will avoid apodictic "yes" or "no" pronouncements about particular cases; only the observation of the data will allow us to determine if we find ourselves in front of a serious obligation.[127]

Marriage could be advised against, but not prohibited: this was the response of the Catholic Church to the second question. Pius XII referred to the encyclical *Casti Connubii*, highlighting the difficulty of "reconciling the two points of view, the eugenic and the moral."[128] The third question—"For existing marriages in which 'Mediterranean hematological disorders' are ascertained, is it permissible to advise against offspring?"—was satisfied with similar arguments. It was permissible to advise against, but not prohibit, and the Church proposed, as acceptable contraceptive methods from a Catholic moral point of view, abstinence and the Ogino-Knaus method, and also approved of adoption of children.[129] As for a question regarding the validity of a marriage contracted by carriers of "Mediterranean hematological illness"—"If the spouses are ignorant of their condition at the moment of marriage, could this be grounds for an annulment of marriage?"—the Pope responded in the negative:

> Neither simple ignorance, nor fraudulent concealment of a hereditary defect, nor moreover the positive error that would have impeded the marriage if the defect had been discovered, are sufficient to cast doubt on its validity. The object of the matrimonial contract is too simple and too clear to be able to plead ignorance.[130]

In the photographs that accompanied the publication of Pius XII's two discourses, the figure constantly at the Pope's side was that of Luigi Gedda, president of the Catholic Action, as well as director since 1953 of the "Gregorio

---

[127] Pius XII, *Discorsi ai medici*, 710–11.

[128] Pius XII, *Discorsi ai medici*, 712.

[129] Pius XII, *Discorsi ai medici*, 712. On the Catholic Church's acceptance of the Ogino-Knaus method, see Anna Treves, *Le nascite e la politica*, 372.

[130] Pius XII, *Discorsi ai medici*, 713.

Mendel" Institute in Rome, and authoritative voice of medical genetics, close to the orientation of the Vatican. In Gedda's interpretation, eugenics was one of the "knots" that characterized the links between medicine, on one side, and on the other, the family, in the Catholic sense. On 7 June 1958, at the 5th Health Congress (*Convegno della salute*) in Ferrara, Gedda confirmed:

> Eugenics is today rapidly earning public notice, so that Renzo and Lucia would be more likely to consult the physician before going to Don Abbondio or to Azzeccagarbugli. [...] A family rationally oriented by their physician must tighten a eugenic knot between the spouses, that is, a rapport which, in the probabilistic approach of genetics, is destined to produce healthy children.[131]

In this view, "eugenic counseling" was a "delicate but necessary service, worthy of science and modern civilization"[132] and its development had to be based on respect for the sacredness of life and individual liberty. From this, Gedda explicitly condemned any form of mandatory premarital certificate:

> We are against that exaggeration called mandatory premarital medical certification, clearly being of the view that the free consultation of a physician on the part of the betrothed is at least as important as the consultation of a lawyer. Eugenic counseling by a physician revolves around two poles: knowledge that every man carries morbid defects; and discretion regarding the freedom of man, requiring the physician to give, with professional confidentiality, advice and not anathemas.[133]

Sterilization and the "systematic registration of defectives"—a technique supported by Scandinavian eugenicists at the World Population Conference in Rome in 1955—also provoked Gedda's net condemnation:

> However hereditarily defective he might be, man is endowed with values that are truly human, which cannot be deliberately ignored, or reviled in anyone. Registration, as discreet as the proposal suggests it would be, would never be secret, and would therefore classify, in front of public opinion, in a seriously

---

[131] Luigi Gedda, *Problemi di frontiera della medicina* (Turin: Borla, 1963), 164. Renzo, Lucia, Don Abbondio and Azzeccagarbugli are all characters of Alessandro Manzoni's novel *I promessi sposi* (in the English translation, *The Betrothed).*

[132] Luigi Gedda, "Eugenetica e profilassi mentale," in *Sanità mentale ed assistenza psichiatrica. Atti del II Congresso italiano di Medicina forense* (Rome: Homo, 1962), 84.

[133] Gedda, *Problemi di frontiera della medicina*, 172.

> damaging way, a category of people who, beneath other aspects, may be worthwhile, and who are not morally at fault in any way for having received a certain inheritance.[134]

In the same context, Gedda compared eugenic birth control to a sort of sterilization:

> The same moral principles just enunciated have weight for the birth control of defectives, with one addition. Birth control, including certain methods that have been publicly proclaimed, is not so different from the sterilization of defectives pursued by racism, and we cannot understand how those who are justly opposed to that procedure can consider themselves satisfied by birth control. [...] Also for the procreation of defectives the recourse to high prestige eugenic counseling is preferable, which, within the boundaries of moral laws, can create an imperative of conscience: this represents a strong impediment, but it respects the moral freedom of mankind.[135]

Rather than birth control, eugenic diagnosis had a precise function of supporting the birthrate, as Gedda declared in a seminar, in January 1969, on the theme of premarital counseling, at the Italian Institute of Social Medicine: "To summarize, [the aims of premarital counseling are] exclusion of sterility, exclusion of infertility, [...] and prevention of illness in those who could be the children of the couple."[136] According to Gedda, rather than a compulsory measure, a constant eugenic monitoring of the family was necessary, not limited to the premarital phase, but extended also to the postnatal and adolescent ages of the children.[137] Gedda proposed, in particular, the institution of an "individual sanitary identity card" that followed the person in all his relationships with the medical sphere.[138]

---

134 Gedda, *I problemi della popolazione* (Rome: Staderini, 1955), 21–22. The speech was made in Rome, on 14 January 1955, at the Bank of Rome, under the auspices of the Italian Center for International Reconciliation Studies.

135 Gedda, *I problemi della popolazione*, 22–23.

136 Istituto Italiano di Medicina Sociale, *La consultazione prematrimoniale (Roma, 24 gennaio 1969)* (Rome: Tip. Loffari, 1969), 8. Presenters at the seminar were Umberto Chiappelli, Giuseppe Del Porto, Dante Primo Pace, Cesare Chiarotti, Giovanni Villani, Ezio Borgognoni Castiglioni, Giorgio Alberto Chiurco, Tommaso Paladino, Francesco Di Raimondo, Adalberto Galante, Giuseppe Cardinali and Mino Bolognesi.

137 Gedda, *I problemi della popolazione*, 24.

138 Gedda, *Problemi di frontiera della medicina*, 167. Gedda was opposed to the idea of a general index of the population, as desired by Giorgio Alberto Chiurco: see Istituto Italiano di Medicina Sociale, *La consultazione prematrimoniale*, 15–16 and 23.

Therefore, from the end of the 1950s and for a good part of the following decade, the secular and the Catholic fields of Italian eugenics seemed to share, for different motivations, the approval of premarital eugenic counseling, based not on imposed and compulsory measures, but on the respect of individual freedom and the "construction of a hygienic and sanitary mentality."[139]

Nevertheless, the problem of mandatory premarital visits reappeared in the Italian legislative debate in 1969. Curiously, it was actually the Catholic battle against the divorce laws approved in 1970 that fed this *rentrée*. Explicitly recalling the Tibaldi Chiesa bill, the new proposal presented in July 1969 by the Christian Democrat deputy Beniamino De Maria[140] identified the premarital certificate as one of the indispensable sanitary instruments for defending the solidity of the family structure, which was increasingly under threat:

> The socio-economic and above all, moral, progress of our country has by now matured the principal problems that surround the institute of marriage and the formation of an increasingly advanced and civilized society. The dangers of such progress are well known and undermine the roots of matrimony as an indissoluble bind on which the family should be founded. In the face of these attempts at disintegration and annulment of family life, the necessity to identify instruments and institutions—*in the deplorable hypothesis of an opening of a "breach" in the connective tissue of the indissoluble link that unites two spouses*—which allow, on the contrary, the reinforcement and restoration of the institute of marriage, has come to the attention of public opinion, the Parliament and the country. These measures must work in such a way that the youth, who intend to unite themselves for all their lives, will be more responsible and

139 Giacomo Perico, "Visita e certificato prematrimoniali (continuazione)," *Aggiornamenti sociali* 12, no. 2 (February 1961): 82. The author specifically supported the Catholic position favouring an obligatory certificate without punitive character. On the Catholic position, see also Alfredo Boschi, "Visita e certificato medico prematrimoniale," *Palestra del Clero* 3 (1 March 1952): 193–204 and *Palestra del Clero* 11 (1 June 1952): 489–500; Bonaventura D'Arenzano, "La visita prematrimoniale," *Orientamenti pastorali* 3 (March 1960): 44–46; P. P., "La visita prematrimoniale," *Studi cattolici* 10 (January 1959): 61–63. For a summary of the debate on premarital visits in 1960, see also the symposium titled "Introduzione del certificato prematrimoniale obbligatorio in Italia," *Riflessi* 3 (September 1960): 51–71.

140 Lecturer of social medicine at the University of Rome and of hygiene at the University of Lecce, De Maria was president of the Parliamentary Commission for Public Hygiene and Health (*Commissione parlamentare Igiene e Sanità Pubblica*), manager of the Italian Catholic Physicians Association (*Associazione Medici Cattolici Italiani*) and on the board of administration of the Italian Institute of Social Medicine (*Istituto Italiano di Medicina Sociale*).

> knowledgeable of the act that they are about to undertake, and of the perspectives, rights and duties that attend the founders of a new family, from a juridical, moral, and in particular, hygienic-sanitary point of view.[141]

According to the introductory section of the bill, the wide-spread opposition to any form of mandatory premarital examination derived from the "hygienic-social immaturity (and in some cases, absolute ignorance) of vast sectors of the Italian population" and the "almost total inadequacy of the current sanitary and advisory structures in our country, in which sanitary centers of primary importance are scarce."[142] Referring to Pius XII's declaration at the Congress of Hematology in 1958, De Maria's bill proposed mandatory premarital examinations and certificates, with a simply informative character, because "society has the right to defend itself against the dangers that could strike its collective health."[143] The articles of De Maria's bill reproduced the contents of Tibaldi Chiesa's proposal, at least as far as concerned the constitution of counseling centers in provincial hospitals, the composition of the specialized medical staff, the characteristics of the examination and certificate, and the mode of financing. Added to such indications however, were the authorization of the release of the certificate to individuals for both public and private counseling centers, and the introduction of monetary fines, according to criteria already outlined by the CNPDS Commission.[144]

While Monaldi's and Tibaldi Chiesa's bills had notably exploited the technical advice of the CNPDS, De Maria's bill on the "mandatory nature of premarital visits and the institution of matrimonial counseling centers" was studied and discussed by a specific commission nominated by the members of the Italian Genetics Association (AGI), during the meeting in Erice on 16 October 1970.[145] The members of this special commission included Giuseppe Montalenti, Luigi Luca Cavalli-Sforza, Luigi Gedda, Franco Conterio, and Antonio Moroni. The results of the analysis, fruit of a first draft

141 *Atti Parlamentari, Camera dei Deputati,* Bill proposed by deputies De Maria, Anselmi Tina, Martini Maria Eletta, Micheli Pietro, Castelli, Pennacchini, Rausa, Barberi, presented 3 July 1969, n. 1656, entitled: *Obbligatorietà della visita prematrimoniale e istituzione di consultori matrimoniali*: 1; italics added.

142 *Atti Parlamentari, Camera dei Deputati,* Bill no. 1656: 2–3.

143 *Atti Parlamentari, Camera dei Deputati,* Bill no. 1656: 3.

144 *Atti Parlamentari, Camera dei Deputati,* Bill no. 1656: 5.

145 Angelo Bianchi to Montalenti, 27 October 1970, AM, b. 76, f. 6.

written on the 9 November 1970 and successively completed with the observations of Cavalli-Sforza and Italo Barrai, were available until May 1971.[146] The commission and the executive committee of the AGI declared themselves generally in favor of De Maria's proposal, but expressed many reserves on the formulation of the bill. First of all, the geneticists rejected the idea of a blanket mandatory examination, believing it damaging to individual freedom and also difficult to manage, due to the scarce availability of personnel qualified in genetic counseling. They proposed instead to limit the obligatoriness to only currently manifesting contagious illnesses, which could be easily identified by provincial laboratories of hygiene and prophylaxis, without any particular problem of organization or funding.[147] The commission—referring, on the suggestion of Barrai, to the statements of the World Health Organization—strongly stressed the necessity that the staff of counseling centers should include personnel specialized in problems of human and medical genetics.[148] In addition, the superior authority (health ministry or department) should consult "a special commission of experts that must include geneticists," in order to ensure that counseling centers had all the necessary useful structures for genetic diagnoses.[149]

To these specifications—the optional character of the premarital examination and the presence of geneticists in counseling centers—another was added. Luigi Luca Cavalli-Sforza,[150] and in general the secular component of the commission, strongly petitioned for the indication of a specific role of counseling in the field of family planning:

> As regards the juridical and moral advice that the counseling center can offer, it seems implicit that it should include, among other things, the responsibility of the newlyweds for the future offspring, also in the field of family planning.[151]

146 *Note dell'Associazione Genetica Italiana alla proposta di legge n. 1656 (Camera dei Deputati) su "Obbligatorietà della visita prematrimoniale e istituzione di consultori matrimoniali,"* attached to the letter from Giuseppe Sermonti, president of AGI, to the members of the commission and the directing committee, 28 May 1971, AM, b. 76, f. 6.

147 *Note dell'Associazione Genetica Italiana*, AM, b. 76, f. 6.

148 *Note dell'Associazione Genetica Italiana*, AM, b. 76, f. 6.

149 *Note dell'Associazione Genetica Italiana*, AM, b. 76, f. 6.

150 Draft attached to the letter from Luigi Luca Cavalli-Sforza to Benedetto Nicoletti, 22 December 1970, AM, b. 76, f. 6. The text continued until point 6: "The advantages to include, in the breadth of the premarital examination, are also those of effecting counselling on the relative problems of family planning in terms of number."

151 *Note dell'Associazione Genetica Italiana*, AM, b. 76, f. 6.

In May 1971, the AGI commission attempted to resolve the basic ambiguity of all the legislative proposals for premarital counseling that had been outlined until that time: the geneticists, in particular, stressed the difference between "infective" or "contagious" diseases (particularly venereal diseases), and genetic diseases. Regarding the latter, they demanded the recognition of their specific and irreplaceable professional skills. The main problem consisted evidently in the situation of serious backwardness of medical genetics in Italy. This concern also pervaded the final report of the AGI scientific meeting held in Pavia in September 1972. After reaffirming the fundamental role of "medical specialists in genetic diseases,"[152] the last part of this report explicitly denounced the retardation of Italian medical genetics:

> We can say that general genetics, as much as human genetics, medical genetics and molecular biology are, qualitatively, on an international level; specialized medical genetics however must still develop in Italy. In other terms, though we have solid bases on which to construct a series of schools of specialized medical genetics, these in practice, do not exist.[153]

The development of genetic counseling centers would therefore be a useful initiative, "both to prevent the birth of abnormal babies and to direct couples to make their decisions on a scientific basis rather than an emotional one." However, it had to be preceded by the constitution, with an "urgent nature" and "absolute priority," of a "school" to train personnel for the centers.[154] However, although believing the opening of genetic counseling centers to be appropriate, the AGI again confirmed its opposition to a mandatory premarital examination:

> While we believe that it is appropriate to open centers, once qualified staff has been trained, we think that the obligatoriness of the premarital examination is absolutely unadvisable. In fact, such a visit would constitute a notable limitation of individual freedom. We believe therefore that every citizen should have a genetic service available, and not a mandatory examination.[155]

[152] Associazione Genetica Italiana, *Consultorio di genetica medica* (Pisa: ETS, 1972), 7.
[153] Associazione Genetica Italiana, *Consultorio di genetica medica,* 8–9.
[154] Associazione Genetica Italiana, *Consultorio di genetica medica,* 10.
[155] Associazione Genetica Italiana, *Consultorio di genetica medica,* 11.

A few days after the publication of this document, on 18 October 1972, the Chamber of Deputies approved the bill on the reform of Italian Family Law: in the first chapter, where the physical acquaintance of both spouses was required for marriage, an optional premarital medical examination was inserted in article seven. In the Senate, two years later, on 30 May 1974, the "optional" character of the visit, together with the entire content of article seven, was newly rejected, leaving such work to sanitary regulations as its most natural place.[156]

Moreover, the premarital certificate seemed by now an obsolete sanitary instrument, compared to the new possibilities of prenatal diagnosis. The latter was welcomed by the popular science journal *Sapere*, for the first time in Italy, in March 1972, as a practice destined to revolutionize the cure of genetic diseases, both through the means of "selective" and "therapeutic" abortion, and through "euphenic" corrective therapy, adopted before birth.[157] Three years later, in 1975, in Sardinia, the research group led by Antonio Cao, professor of pediatrics at the University of Cagliari, devised the first method of prenatal diagnosis of the beta-thalassemic phenotype. In 1977, a program of voluntary screening and genetic counseling was set up, in which prenatal diagnosis was side by side with a large spectrum communicative work, coordinated with general practitioners, family planning associations and patient associations. Starting from a frequency of live homozygous births of about 1 in 250, in Sardinia the campaign managed to bring the frequency to 1 in 1000 in the first decade, getting to 1 in 4000 in 1997.[158]

3. Eugenics and Catholic Medical Genetics: Luigi Gedda and the "Gregorio Mendel" Institute

In 1951, Luisa Gianferrari, director of the Milan Study Center in Human Genetics, and Luigi Gedda, physician and important political exponent of the Italian Catholic right, created a new association: the Italian Society of Medical Genetics (*Società italiana di Genetica Medica*), presided over by

---

156 Giacomo Perico, "Aspetti medico-sociali della 'visita prematrimoniale,'" *La Civiltà Cattolica*, 2983 (October 1974): 58.

157 Fiorella Nuzzo, "La diagnosi prenatale," *Sapere* 746 (March 1972): 11.

158 De Sio and Capocci, "Southern genes," 812–15.

the physiologist Carlo Foà, and as opposed to the Montalenti–Barigozzi–Buzzati line as Gini's SIGE. On 6–7 September 1953, just a week after the Bellagio Congress, the First International Symposium of Medical Genetics (*Primum Symposium Internationale Geneticae Medicae*) was held in Rome, under the auspices of Pius XII. The convention was organized in collaboration with the Italian Society of Medical Genetics and coincided with the inauguration of the Institute of Medical Genetics and Twin research "Gregorio Mendel," founded in Rome, with headquarters in Piazza Galeno, and directed by Luigi Gedda.

The level of conflict existing between the Italian geneticists—Barigozzi, Buzzati and Montalenti—and Gedda and Gianferrari, is well represented by the few lines that Barigozzi wrote to Montalenti, in the midst of organizing the Bellagio Congress: "Gianferrari and Gedda are fighting against Gini and Jucci, because they want to not only form a sort of congress of medical genetics, but also an anti-SIGE association."[159] Still more incendiary, several months later, was the Buzzati-Traverso's quip to Montalenti:

> And what do you think of those S.O.Bs (if you don't know what it means, ask the nearest American) Gedda and Gianferrari, who are putting together a symposium of medical genetics, without saying even one word to the organizers of the congress? With this, they also make us look stupid, regarding those who would have been invited, who will conclude that usually in Italy, we gently lead each other to the gallows.[160]

The clash was, above all, of a scientific nature: the geneticists intended to impede the advance of those clinicians who were involved in the eugenic fascist past, and were currently attempting to present their constitutional, genealogical and twin analyses on the heredity of physiological, psychological and pathological traits under the label of "genetics." At the inaugural ceremony of the "Gregorio Mendel" Institute, Carlo Foà deliberately attacked the so-called "pure" geneticists, energetically restating the right of medicine to address human genetics:

> Let us be frank; our Society has not been created without any opposition, and now finds itself in a polemical phase. On one side, the major part of general

[159] Barigozzi to Montalenti, 26 May 1952, AM, b. 28, f. 9.
[160] Buzzati-Traverso to Montalenti, 2 February 1953, AM, b. 28, f. 9.

> genetics experts hesitates to admit that human genetics (and even less, medical genetics) has the right to an autonomous life. On the other side, physicians have taken the study of the hereditariness of physiological, psychological and pathological characteristics of the human species upon themselves.[161]

Only medicine could provide geneticists with that "verification of the most subtle clinical symptoms," necessary for the study of human heredity:

> "Genetics is one," I have heard it said. I agree: its laws hold true for all the living beings and represent the doctrinal basis of every specialized investigation, but this cannot be accomplished, except by the specialists of single branches of biological science, including clinical. Who, if not the specialized clinician, can discover how illnesses of the skeleton, blood, metabolism, organs of sense, the psyche, are propagated in descendents, if they do not have specific knowledge of each of these arguments?[162]

But the reply of the "pure" geneticists was not long in coming. It was Buzzati-Traverso who dedicated a vitriolic review to Gedda's symposium, in *Science*, denouncing the isolation of the initiative in terms of international scientific context, and inviting both Gedda and Foà to at least learn the "correct use of the terminology" before occupying themselves with genetics.[163]

Besides the scientific dimension, a political and academic opposition further aggravated the situation. In fact, Luigi Gedda's debut in the field of genetics was marked by a scandal that identified the harsh confrontation between the Catholic and the secular components of Italian medical genetics. In 1953, for the qualification exam sessions for the chair in human genetics, the Ministry of Public Education (*Ministero della Pubblica Istruzione*) consulted the First Section of the High Council (*Sezione I del Consiglio Superiore*) of Public Education regarding the composition of the deciding committee. The High Council proposed the following names: as permanent members, Claudio Barigozzi (professor of genetics in Milan),

161 Carlo Foà, "Discorso pronunciato nella cerimonia inaugurale dell'Istituto G. Mendel il 6 settembre 1953," in Luigi Gedda, ed., *Genetica Medica. Primum Symposium Internationale Geneticae Medicae Roma 6–7 settembre 1953* (Rome: Edizioni dell'Istituto Gregorio Mendel, 1953), 447.

162 Foà, "Discorso pronunciato nella cerimonia inaugurale dell'Istituto G. Mendel il 6 settembre 1953," 447–48.

163 Adriano Buzzati-Traverso, review of Luigi Gedda, ed., *Genetica Medica. Primum Symposium Internationale Geneticae Medicae Roma 6–7 settembre 1953* (Rome: Edizioni dell'Istituto Gregorio Mendel, 1953), in *Science* 122, 3161 (July 1955): 206.

Giuseppe Montalenti (professor of genetics in Naples) and Alfonso Giordano (professor of anatomy and pathological histology in Pavia); as substitute members, Adriano Buzzati-Traverso (professor of genetics in Pavia) and Umberto D'Ancona (professor of zoology in Padua). Without taking this recommendation into account, on 15 June 1953, the Christian Democrat Antonio Segni, Minister of Public Education, proposed an alternative: the three professors of genetics disappeared from the committee, and in their places Segni nominated, as permanent members, Luigi Gedda, Luisa Gianferrari and Giovanni Di Guglielmo (professor of general clinical medicine and medical therapy in Rome), and, as substitute members, Alfonso Giordano and Giovanni Dall'Acqua (professor of specialized medical pathology and clinical methodology in Bari).[164]

The first reaction to Segni's interference came from the Faculty of Medicine and Surgery of the University of Turin, which approved, on 4 July 1953, a motion of condemnation.[165] After having petitioned in vain for the Ministry to reconsider its choice, with a letter sent 15 July, Barigozzi, Buzzati-Traverso and Montalenti adopted the strategy of a frontal attack: the first two appealed, on 27 August, to the State Council (*Consiglio di Stato*),[166] while the third denounced Segni's decision directly to the President of the Italian Republic in December.[167] The accusation of the geneticists pointed the finger at the illegitimacy of the ministerial decision. The composition of a deciding committee should be, in fact, an act of acute technical discretion, possessed to the maximum by the High Council as the consultative organ expressly created to that end: the Ministry had not only ignored the recommendation of the High Council, but had not given any justification for its interference. The deciding committee had to be, in addition, composed of technicians, and therefore professors, of the relevant discipline (in this case, genetics) or of related disciplines. This last criterion was followed by the High Council but not by the Ministry, who had excluded the three pro-

[164] Ministerial decree by Antonio Segni, 15 June 1953, ACS, MPI, DGIS, Divisione I, Commissione libere docenze 1938–1953, b. 74, f. 1052.

[165] Faculty of Medicine and Surgery of the University of Turin, verbal extract of the Faculty Board, 4 July 1953, ACS, MPI, DGIS, Divisione I, Commissione libere docenze 1938–1953, b. 74, f. 1052.

[166] Claudio Barigozzi and Adriano Buzzati-Traverso appeal to the State Council, 27 August 1953, ACS, MPI, DGIS, Divisione I, Commissione libere docenze 1938–1953, b. 74, f. 1052.

[167] Appeal by Giuseppe Montalenti to the Head of state, 14 December 1953, in ACS, MPI, DGIS, Divisione I, Commissione libere docenze 1938–1953, b. 74, f. 1052.

fessors of the mother-discipline (genetics), completely neglecting the professors of "general biology and zoology, including genetics," and confirming, as a substitute member, the professor of anatomy and pathology first designated as a permanent member. On the contrary, three pathologists and a clinician had been included.

State Advocacy (*Avvocatura generale dello Stato*) was appointed to defend the Ministry and, on 23 March 1954, presented a written deposition to refute the charges. According to this report, it was up to the Ministry to nominate the deciding committee and the recommendation of the High Council was certainly not binding. Neither did the Ministry need to offer a motivation to explain a decision which contrasted with the advice of the High Council. As for the choice of specialties for the composition of the deciding committee, State Advocacy did not believe that it violated any law, substantially for two reasons. First, "there could not be full professors of human genetics, because it is a specialization which is not currently taught in universities." Second, "only three were tenured professors of similar disciplines—genetics—and their inclusion in the committee was not seen as appropriate because they were professors of genetics, a discipline of the Faculty of Science, according to the current university categorization, while centers of human genetics are above all in the Faculty of Medicine and Surgery." For these reasons, according to the Ministry of Public Education, "the inclusion in the committee of noted scholars was necessary—even if they were not professors—from centers of human genetics. It seems that we find ourselves in front of one of those typical cases in which the law allows us to have recourse to research fellows."[168]

Considering all sides of the controversy, the sixth session of the State Council, at its jurisdictional seat on 7 April 1954, found in favor of the geneticists and annulled Segni's decree of June 1953. The sentence was expressed in two parts. In the first, the State Council recognized the formal validity of the appeal. In the second, it reaffirmed, contrary to the position of State Advocacy, the function of the High Council and the limits of eventually different decisions.[169] The Ministry of Public Education, having received the

168 Memo from the State Advocacy in response to the appeal by Barigozzi and Buzzati-Traverso, 23 March 1954, ACS, MPI, DGIS, Divisione I, Commissione libere docenze 1938–1953, b. 74, f. 1052.

169 Sentence of the State Council, 7 April 1954, ACS, MPI, DGIS, Divisione I, Commissione libere docenze 1938–1953, b. 74, f. 1052.

recommendation of the High Council, could also have chosen not to follow it completely, or in part, but it had to justify its interference, demonstrating that reasons of public interest were incorrectly or not sufficiently valued by the High Council. As this had not happened, the State Council judged Segni's decree as illegitimate.[170] The public importance of this judiciary case is well exemplified by the title of the article that appeared in the journal of the Italian Communist Party, *L'Unità* on 14 April 1954[171]—"Gedda rejected."

Actually, the progressive advancement of Luigi Gedda in the field of medical genetics had started some years before, in 1952, with the publishing of the quarterly *Acta geneticae medicae et gemellologiae,* and in 1953, with the inauguration of the "G. Mendel" Institute, in the presence of Pius XII and the Prime Minister, Giuseppe Pella.

In the article opening the first number of *Acta geneticae medicae,* which he himself directed, Gedda included his approach to genetics in a general framework of methodological reformulation of medical constitutionalism. The title chosen—*Genetics, medicine and constitution*—was significant in itself. Gedda's discourse, in fact, started by criticizing the traditional *forma mentis* of the physician, caught between "Virchowian localism" and "Pasteurian esogenism":

> "Virchowian localism" and "Pasteurian esogenism" have dominated medical knowledge in the first half of the 20th century, determining "the mode" of scientific research and professional exercise, which has concentrated the fire of its attention on the anatomical-pathological framework and on the external pathogenic *noxa,* leaving causality and phenomena of a different order in half-light.[172]

According to Gedda, the three different schools of constitutional medicine—morphological, functional and neuro-endocrinal—had tried to resolve such dichotomies, but with little success. Only genetics could, in

---

170 Sentence of the State Council, 7 April 1954, ACS, MPI, DGIS, Divisione I, Commissione libere docenze 1938–1953, b. 74, f. 1052. The new commission, nominated 14 July 1954, with Barigozzi, Gedda and Gianferrari as permanent members, and Montalenti and Giordano as substitute members, assigned the position of professor of human genetics to Angelo Cresseri, Giuseppe Morganti (student of Gianferrari), Ruggero Ceppellini (student of Barigozzi), Amleto Maltarello (student of Gedda) and paediatrician Ignazio Gatto, although these last two exceeded the maximum number of nominations for the chair. See ACS, MPI, DGIS, Divisione I, Commissione libere docenze 1938–1953, b. 74, f. 1052.

171 E. Modica, "Gedda bocciato," *L'Unità* (14 April 1954).

172 Luigi Gedda, "Genetica, medicina e costituzione," *Acta geneticae medicae et gemellologiae* 1, no. 1 (January 1952): 2.

fact, allow a synthesis between "synchronic" (form and function in action) and "diachronic" (individual anamnesy) studies of the phenotype and analysis of the "family stock."[173]

In Gedda's opinion therefore, medicine had arrived at a "turning point," because, due to the decisive contribution of genetics, the focus of scientific and professional interest was shifting "from the recognition of the imprint of illness on the phenotype and from the knowledge of esogenic moments of illness," to the "endogenic moments, that is, to constitution."

In his discourse at the inauguration of the Mendel Institute, Gedda, after having listed the three methods on which medical genetics had to be based (familial anamnesy, twin research and the genetic study of the population), repeated the connection between genetics and constitutionalism:

> The problem of the constitution must be confronted using concepts, terms and laws of genetics to find a true, convincing and useful solution. In this framework we can completely understand the concept of "diathesis," which means the receptivity or reactivity that is specifically hereditarily conditioned, and the concept of "ground" [*terreno*], which qualifies the current or realizable constitutional resistance that an organism opposes in a certain moment to a certain morbid agent. The doctrine of the constitution is a corollary of medical genetics.[174]

The work of medical genetics was to "carry its help to the clinic to study, diagnose and cure the phenotype," but also to "make the phenotype as translucent as crystal, so that we can transparently see what is happening on the level of genotype and can provide for the individual and his offspring."[175] From here came "the prevention of the hereditary disease of the single individual and its cure without fatalism and purely symptomatic therapy, the treatment of diathesis, eugenics at the service of the individual rights and duties of the human person, and even premarital counseling." In Gedda's view, genetics must become the common heritage of family medicine, newly called to seize the "invisible fabric that links the illness of man to the history of his blood." In addition to family medicine, Gedda main-

173 Gedda, "Genetica, medicina e costituzione," 5.

174 Luigi Gedda, "Profilo scientifico della genetica medica," in *Genetica Medica. Primum Symposium Internationale Geneticae Medicae*, 13–14.

175 Gedda, "Profilo scientifico della genetica medica," 6.

tained the necessity of new specialized centers "where the problem can be posed and resolved through all the means that the insurance companies, military and sport medicine and other institutions that carry out collective medical assistance, can today arrange."[176]

Pope Pius XII confirmed Gedda's program, giving a long speech at the inauguration of the Mendel Institute, which, on one hand, approved the general problem of eugenics, judged "irreproachable" from a moral point of view. On the other, he strongly condemned certain defensive measures in genetics and eugenics.[177] Sterilization, the "prohibition of marriage," the segregation of defectives and therapeutic abortion were, therefore, all placed on the same plane and were equally rejected in the name of respect for the dignity of the human person, according to Catholic teachings.[178] Genetics, Pius XII concluded, could not regard the human being in the same way as other animal and vegetable species:

> The practical aims being pursued by genetics are noble and worthy of recognition and encouragement. Would that your science, in weighing up the means destined to achieve those ends, could only remain always conscious of the fundamental difference that exists between the animal and vegetable world on the one hand, and man on the other! In the first case, the means of bettering the species and race are entirely at the disposal of science. On the other hand, in the domain of man, genetics are always dealing with personal beings, possessing inviolable rights, with individuals who for their part are bound by inflexible moral laws in the exercise of their power to raise up a new life. Thus the Creator himself has established certain barriers in the moral domain, which no human power has authority to remove.[179]

With the strength of such papal investiture, the scientific activity of the Mendel Institute focused, in the following years, on "eugenic" counseling and twin research. Family and twin studies were published in *Acta geneticae medicae et gemellologiae*[180] and were presented at the International Sympo-

176 Gedda, "Profilo scientifico della genetica medica," 6.

177 See "Discorso di S.S. Pio XII ai partecipanti al "Primum symposium internationale geneticae medicae," 418.

178 "Discorso di S.S. Pio XII ai partecipanti al "Primum symposium internationale geneticae medicae," 419.

179 "Discorso di S.S. Pio XII ai partecipanti al "Primum symposium internationale geneticae medicae," 419–20.

180 See for example, Luigi Gedda, Giuseppe Del Porto and Adriana Del Porto-Mercuri, "Sindrome di Werdnig-Hoffmann familiare che include una coppia di gemelli MZ concordanti (un caso di Consulenza Eugenica)," *Acta geneticae medicae et gemellologiae* 11, no. 2 (April 1962): 113–21.

sium of Medical Genetics and at the 2nd International Congress of Human Genetics, organized by Luigi Gedda in 1953 and 1961.

Both events are particularly relevant because they reveal an international eugenics network, orbiting around the Mendel Institute and its president.

A first branch of Gedda's post-war *liaisons dangereuses* consisted in post-Nazi German eugenics: the most representative figures in this sense were undoubtedly the geneticist Othmar von Verschuer, head of the department of human heredity of the KWI for Anthropology, Human Genetics and Eugenics, and the physician Hans Grebe, Verschuer's assistant at Frankfurt and KWI in Berlin.[181] After the war, Verschuer, who came out unscathed from the purging trials, thanks to his academic connections and his close ties with the ecclesiastical environment, was appointed professor of human genetics at the University of Münster in 1951, became president of the *Deutsche Gesellschaft für Anthropologie* in 1952, and of the Faculty of Medicine of Münster, in 1954. His close links with Gedda were exemplarily reassumed, along with numerous scientific collaborations, by the pompous homage the Italian geneticist dedicated to him in 1956, on the occasion of his sixtieth birthday, in the pages of *Acta geneticae medicae et gemellologiae*. The title of Gedda's article, *A master and an example* completely summarized the apologetic nature of the contribution. After a detailed exposition of Verschuer's scientific production, the article concluded with a few eloquent lines:

> Master of well-known fame and forger of men, who dedicated himself to scientific research with a spirit of vocation, he is also an example of industry and method for all scientists, and especially for all geneticists, beyond the borders of his School and his Nation. It is our duty to recognize Verschuer's prominence, taking the opportunity of his birthday, profoundly convinced as we are that the best praise is this: Master and Example.[182]

[181] On Grebe and Verschuer see, in particular, among others, Schmuhl, *The Kaiser Wilhelm Institute for Anthropology, Human Heredity and Eugenics, 1927–1945*.

[182] Luigi Gedda, "Un maestro e un esempio," *Acta geneticae medicae et gemellologiae* 5, no. 3 (July 1956): 244. On the collaboration between Gedda and Verschuer, see Otmar von Verschuer, "Die Erbanlage als bestimmende Kraft auf dem Lebenswege," in Gedda, ed., *Genetica Medica*, 132–52; Otmar von Verschuer, "Die Häufigkeit von krankhaften Erbmerkmalen beim Menschen," in *Proceedings of the Second International Congress of Human Genetics (Rome, September 6–12, 1961)* (Rome: Istituto Gregorio Mendel, 1963), 1, 168–75; Otmar von Verschuer, "Ein altes und ein neues Problem der Zwillingsforschung," *Acta geneticae medicae et gemellologiae* 1, no. 2 (May 1952): 180–90.

Hans Grebe[183] was also frequently in contact with Gedda, as Grebe himself stated in an interview released by Benno Müller-Hill:

> I have always said that race is only the sum total of certain traits. But human genetics is not so simple. The Church is very interested in the subject. In 1953, I attended the First Congress of Human Genetics, which was held in Rome. The Director of the Institute of Human Genetics in Rome, Professor Gedda, explained to me why the Church is so interested in twin research. Do twins have two souls or one? The Holy Father received us in audience. He came up to me and said: "I have good news for you: Adenauer has been re-elected." Eugenics had its high and low points. The Holy Father spoke about this. But we should continue to aspire to the heights.[184]

In addition to connections with German post-Nazi eugenics, Gedda's eugenics network also involved Anglo-American racial anthropology. The successive chapter will focus more deeply on Gedda's collaboration with the International Association for the Advancement of Ethnology and Eugenics (IAAEE). Here it is perhaps worth mentioning the friendship between Gedda and the botanist and anthropologist Reginald Ruggles Gates, a significant figure for nearly four decades (from the 1920s to the 1960s) in Anglo-American scientific racism, inveterate advocate of biological differences between the human races and of the natural inferiority of the "blacks" in respect to the "whites."[185] Articles by Ruggles Gates abounded in *Acta geneticae medicae et gemellologiae,* dedicated to questions of "racial genetics."[186]

---

183 See Benno Müller-Hill, *Murderous Science. Elimination by Scientific Selection of Jews, Gypsies, and Others in Germany, 1933–1945* (Cold Spring Harbor: Cold Spring Harbor Laboratory Press, 1988), 163–68.

184 Müller-Hill, *Murderous Science,* 167–68. On the collaboration between Gedda and Grebe, see Hans Grebe, "Erbpathologie des Skelettsystems," in Gedda, ed., *Genetica Medica,* 188–222; H. Grebe, "Genetik und morphologische Variation," in *Proceedings of the Second International Congress of Human Genetics* 1, 355–68; Hans Grebe, "Diskordanzursachen bei erbgleichen Zwillingen," *Acta geneticae medicae et gemellologiae,* 1, no. 1 (January 1952): 103–107; Hans Grebe, "Über besondere Zwillingskonkordanzen," *Acta geneticae medicae et gemellologiae,* 5, no. 2 (May 1956): 138–54; Hans Grebe, "Familienbefunde bei letalen Herzmissbildungen," *Acta geneticae medicae et gemellologiae* (supplementum primum): 257–93; Hans Grebe, "Sportfamilien," *Acta geneticae medicae et gemellologiae* 5, no. 3 (September 1956): 418–26; Hans Grebe, "Zwergwuchs als genetisches Problem," *Acta geneticae medicae et gemellologiae* 6, no. 4 (October 1957): 429–36; Hans Grebe, "Biemond-Syndrom in einer Sippe mit Iriskolobomen, Hüftgelenksdysplasie und Epilepsie," *Acta geneticae medicae et gemellologiae* 9, no. 2 (April 1960): 197–210.

185 See Elazar Barkan, *The Retreat of Scientific Racism* (Cambridge: Cambridge University Press, 1992), 168–76.

186 See Reginald Ruggles Gates, "Records of Y-inherited Hairy Ears in India," *Acta geneticae medicae et gemellologiae* 6, no. 1 (January 1957): 103–108; Hans Grebe, "The African Pygmies," *Acta geneticae medicae et gemellologiae* 7, no. 2 (April 1958): 159–218; Hans Grebe, "The Genetics of the Australian Aborigines," *Acta geneticae*

Even more representative of his relationship with Gedda is perhaps the obituary, which appeared in the journal in January 1963:

> A year after he participated, accompanied by his wife, with great enthusiasm and notable scientific contributions, in the Second International Conference of Human Genetics in Rome, Prof. R. Ruggles Gates has died at the age of 80, in London. In that capital he was professor of Botany from 1921 to 1942. During his academic career he was oriented always more toward the study of genetics, with particular regard to racial and population genetics. Author of 380 publications, including books and articles, he took part in our treatise *De Genetica Medica* writing an original 128-page work, titled "Race Crossing." Also, he asked the Mendel Institute to collaborate on his research on hairy ears, of which trait he studied the hereditary transmission. [...] Brisk and youthful spirit, he experienced sacrifices and inevitable confrontations for science, conserving the impetus and enthusiasm of the first hour. Generous to the young, cordial with his friends, ingenious in his studies, pioneer of the genetic revision of anthropology, his exemplary spirit of researcher and master remains among us.[187]

In addition to these international contacts, several research interests highlight the involvement of Gedda's Mendel Institute in eugenics. A significant example is the investigation on the heredity of "sporting talent," conducted between the mid-1950s and the celebration of the Rome Olympics in 1960. At the Congress on Sports Medicine, organized in 1960 by the Olympic Executive Committee (presided over by the Christian Democratic politician Giulio Andreotti) and by the Surgical Clinic of the University of Rome, directed by Pietro Valdoni, Luigi Gedda presented a paper that synthesized the results of the genealogical and twin research conducted by the Mendel Institute since 1955. According to Gedda, the investigations on family pedigrees, as much as the study of twins, demonstrated the "true genetic roots of sporting athleticism": the "precious genotypes responsible for sporting talent" were transmitted "through domi-

medicae et gemellologiae 9, no. 1 (January 1960): 7–50; Hans Grebe, "Studies in Race Crossing. Crosses of Australians and Papuans with Caucasians, Chinese, and the Other Races," *Acta geneticae medicae et gemellologiae* 9, no. 2, (April 1960): 165–84; Hans Grebe, "The Melanesian Dwarf Tribe of Aiome," *Acta geneticae medicae et gemellologiae* 10, no. 3, (July 1961): 277–311. See also the participation of Ruggles Gates at the Congress of Human genetics of 1961, presided over by Gedda in Rome, see Reginald Ruggles Gates, "Inheritance of Racial and Sub-racial Traits," in *Proceedings of the Second International Congress of Human Genetics* 1, 369–70.

187 Luigi Gedda, "Prof. R. Ruggles Gates (in memoriam)," *Acta geneticae medicae et gemellologiae* 12, no. 1 (January 1963): 95.

nant Mendelian mechanisms."[188] Gedda went so far as to hypothesize the existence of a sporting "phenotype" and "genotype."[189] In an investigation on the athletes awarded with gold or silver medals until 1955 by the Italian National Olympic Committee (*Comitato Olimpico Nazionale Italiano*, CONI), Gedda and his collaborators deduced a so-called "index of familial sportingness," with the aim of identifying diverse "hereditary conditioning" of various sports.[190] Once the origin of the sporting talent was ascertained in "the hereditary constitutional variability of the individual," the role of medical genetics in the selection of athletes obviously assumed a fundamental centrality:

> The geneticist must advise that the selection of athletes should take maximum notice of the familial sporting anamnesy, [...] both to avoid the repeat of failures, and to orientate the subject toward those sporting goals for which presumably his hereditary constitution presents some *atouts* which it would be wise to consider.[191]

With this view, it is not surprising that Gedda considered the Olympics in Rome as an extraordinary laboratory of genetic analysis of sporting activity. In 1959, he presented CONI with a project relative to the adoption of an official scientific program for the Olympic Games, based on the following premise:

> 1) The Olympic athletes represent for the most part the fruit of a long and precise selection, which is realized in their country of origin, with the aim of presenting at the Olympic Games those sportingly endowed subjects who have the highest probability of victory. And so, not just the winners of the Olympic competitions, but all the Olympic athletes have, from a somatic-psychic point of view, a high level of representativeness;
> 2) the representativeness of the Olympic athletes is specific, that is, it enhances the morphological and functional characteristics of any sport to the highest level [...];

188 Luigi Gedda, "L'importanza della genetica nella selezione degli sportivi," in Tommaso Lucherini and Claudio Cervini, eds., *Medicina dello sport* (Rome: Società Editrice Universo, 1960), 78.
189 Gedda, "L'importanza della genetica nella selezione degli sportivi," 85.
190 Gedda, "L'importanza della genetica nella selezione degli sportivi," 89.
191 Gedda, "L'importanza della genetica nella selezione degli sportivi," 90–91.

3) data collection could not be completed, not even mostly, during the Olympic Games, because organizational and psychological reasons would make the athletes unapproachable and intractable on those days, and far from the ideal conditions of scientific research;
4) the progressive breaking of records in the results of the successive Olympics, fruit of increasingly vast selection and increasingly efficient training, postulates a scientific testing of the *homo olympicus*, every four years, as interesting scientific fact not only for sport, but for all sciences dealing with the human being and the development of human civilization;
5) the scientific investigation can not therefore be reduced to a team that operates in the place and time of the Olympic Games, but must result from the scientific collaboration of an international Olympic medical-scientific commission with national medical-scientific commissions, which must be conveniently planned, well in advance.[192]

CONI approved Gedda's program and decided the organization of a medical-scientific committee, presided over by Gedda, which was inaugurated on 27 November 1959. According to the program, CONI would adopt an "Olympic athlete card" as a "basic document for the scientific research," compiled by Gedda and sent "in useful time" to all National Olympic Committees in order to "solicit and orientate them in the gathering of necessary data for the scientific program during the pre-Olympic period." A health center was installed in the Olympic village, equipped "for the performance of requested medical cures and physiotherapeutic treatments, and also for the development and the control of official and voluntary scientific research." Finally, the "centralization of official scientific research" would take place at the Mendel Institute in Rome.

Gedda prepared two forms (for the male and female athletes), in five languages (Italian, English, French, Spanish and German) and sent them to all National Olympic Committees, nine months prior to the Olympic Games. The forms consisted of 73 questions, divided into four pages and several subgroups: genealogical tree, physiological and pathological anamnesy, clinical exams, anthropometric data, sporting anamnesy, and psychophysiological examination. Question number four (immediately after the indications of sur-

192 Luigi Gedda, Marco Milani-Comparetti and Gianni Brenci, *Rapporto scientifico sugli atleti della XVII Olimpiade. Roma 1960* (Rome: Istituto di Medicina dello Sport, 1968), 9–10.

name, name, place and date of birth) asked the athlete to specify their "race," choosing between "white," "negro," "mongloid," "American-Indian," "Indian," "mixed" or "other." Numbers 11 and 12 asked the athlete to specify if they were a "smoker" or a "drinker." Question number 26 investigated the "success of marriage" with a choice between "good," "medium" or "bad." In the "psychophysiological examination" questions, as well as studies completed, languages spoken, profession, and preferences in reading, spare time, color and type of design, the athlete was asked to evaluate their "temperament" in the "sexual sphere": here the options varied from "+++" to "-."[193] The analysis of the data obtained from the responses of 5192 athletes was undertaken at the Mendel Institute based on four analytical orientations, which proposed the definition of the relationship between sporting performance and the place of birth of the athlete; the characteristics of the family origin and the growth of the athlete; the normal phenotypic traits; age and pathological anamnesy; training and psychical and behavioral characteristics.[194]

In order to judge the nature of Gedda's research, it is interesting to read some of his conclusions:

> Manual ability (right-handed, left-handed, ambidextrous) does not appear to be associated with any differential value of performance (tab. 43);
> social conditions (tab. 44) seem to associate a certain better performance with less well-off social conditions;
> the level of instruction is highest for the athletes of fencing and field hockey, and lowest for those of football and boxing (tab. 45)
> the frequency of reading is highest in relation to water polo, fencing and range-shooting, and lowest for pentathlon, boxing, canoe and rowing (tab.46);
> the condition of smoker or non-smoker does not appear associated in the overall athlete body with any condition of advantage in performance (tab. 47);
> the use of alcoholic drinks appears associated with an improved performance, particularly in the case of wine and beer (tab. 48).[195]

In general, the program was a total failure, both because only 20% of the athletes agreed to compile the form (and not all the questions) and because the delegations of the various countries, particularly Britain, revolted

[193] Gedda, Milani-Comparetti and Brenci, *Rapporto scientifico sugli atleti della XVII Olimpiade*, 65–71.
[194] Gedda, Milani-Comparetti and Brenci, *Rapporto scientifico sugli atleti della XVII Olimpiade*, 24.
[195] Gedda, Milani-Comparetti and Brenci, *Rapporto scientifico sugli atleti della XVII Olimpiade*, 62–63.

against what they saw as a brazen and embarrassing violation of the athletes' intimacy. The following ironic account was published in the review *Il Ponte*, in June 1960:

> There are 300 English athletes at the Olympics, and according to today's news, they were "advised against" answering. Leslie Tuelove, the manager of the British Olympic delegation, today declared: "The initiative of Prof. Gedda was a fantastic example of brazenness. Our athletic association was never informed of anything and I will do everything to make sure my athletes refuse to respond." English Olympic runner Derek Ibbotson, 27, married, commented on the questionnaire with this dry phrase: "Prof. Gedda will receive only rude answers." Brian Hewson, European champion of the 1500 meters, who is also married, said "It is incredible that he is asking me if in my love life I am cold or passionate. I will certainly not tell him." The graceful Margaret Edwards, 21, swimming champion, declared: "I will not give him information on my intimate life. I don't like people who poke their noses into these things. It's ridiculous. How can I know if I am cold or passionate? Soon I'll be engaged; I'm sure that my fiancé wouldn't be happy if I answered these questions."[196]

Gedda himself could not deny the undignified results of the research, but tried to attribute the responsibility to scientific immaturity, the lack of adequate structures and the bad taste of certain newspapers:

> The forms, sent to the 84 participating countries in the period of Olympic athlete selection, were not received everywhere with the serene comprehension and sense of responsibility that scientific research requires.
> That could be expected on the part of promoters because it lacked a tradition, as it was the first time in which they were asked to overcome the commitments and emotion of sporting competition with the calm and objective vision of the scientific eye.
> Additionally, many nations were not equipped to respond to the questions of the inquiry due to the lack of health structures, or of personnel adept at data collection, or due to lack of time, absorbed by late training and the trip.
> Several newspapers also showed the bad taste to joke about this work, making it more difficult. As often happens with new initiatives, it is easier to ridicule then it is to evaluate it.[197]

[196] See A. P., "Gedda vuole la firma," *Il Ponte* 16, no. 6 (June 1960): 990–91.

[197] Gedda, Milani-Comparetti and Brenci, *Rapporto scientifico sugli atleti della XVII Olimpiade*, 16.

As if all this were not enough, Italian newspapers, in May 1960, on top of the Olympic scandal, published revelations of the maneuvers which, in the same year, had helped Luigi Gedda, ex-president of the Catholic Action, to attain a professorship of medical genetics at the University of Rome. With a convention signed on 19 November 1959, the University of Rome had instituted the chair of medical genetics, completely financed by ONMI, for 3,200,000 lire annually. The position, it goes without saying, was offered by the company *San Tommaso Apostolo*, proprietor and managing entity of the Mendel Institute.[198]

The competition for the position that was held a few months later, in 1960, already had an assured winner, but the way in which it was carried out—reconstructable thanks to the correspondence found in Montalenti papers—demonstrates the political and ideological context which marked Gedda's academic rise in the field of medical genetics.

On 12 November 1960, the famous Italian histologist Giuseppe Levi[199] wrote to Montalenti, indignant that the professorship had gone to Gedda, and determined to denounce the fact:

> Dear Montalenti,
> It has been reported to me that in the competition for the professorship of human genetics [sic] at the University of Rome, the number one proposal was Gedda. You know that no tenured professor of genetics was part of the deciding committee, and instead Lambertini took part! Wouldn't it be appropriate if this news were communicated to a "moderate" newspaper, such as "Il Mondo"? Or perhaps better, to "Il Ponte," that deals more specifically with problems pertaining to culture? In "Il Ponte," the news could appear in the *Ritrovo* column. Would you like to place the news yourself, without comments? If you do not wish to do it, I ask you to tell me all the information: names of the committee, names of the applicants (I know that the number two proposal was Ceppellini, but number three I don't know).
> Naturally it won't do any good, but that doesn't matter; it is good for the public to know.[200]

---

198 "Una cattedra universitaria per il prof. Luigi Gedda," *Paese Sera* (3–4 May 1960).

199 Professor of human anatomy at the University of Turin, Giuseppe Levi introduced the method of in-vitro tissue culture to Italy. His students were future Nobel prize winners, forced to leave Italy after the promulgation of the racial laws in 1938: Rita Levi Montalcini, Renato Dulbecco and Salvatore Luria (naturalised American with the name of Salvador Edward Luria). See Claudio Pogliano, "Le scienze biomediche," in Antonio Casella, ed., *Una difficile modernità. Tradizioni di ricerca e comunità scientifiche in Italia, 1890–1940* (Pavia: Università degli Studi di Pavia, 2000), 257–86.

200 The letter is preserved in AM, b. 33, f. 18.

A few days later, Montalenti answered Levi's question, likewise scandalized and also inclined to bring attention to what had happened, but without personal exposure:

> What has occurred, which has been long in preparation with the creation of the Mendel Institute (largely supported by the Vatican) and with the convention between that Institute and the University of Rome, is truly scandalous. I give you all the details in the attached paper.
> For various reasons that you will understand (among others, we are colleagues at the University of Rome this year, and it could seem as if I were jealous of him) I would prefer that my name does not appear. But I agree with you that the scandal must be denounced, even if, as you say, it will not do any good.[201]

The anonymous document, attached by Montalenti to his letter, is worth citing entirely, because of the precision and bitter irony with which it describes the organization and results of the competition:

> The voting of the Faculty of Medicine for the deciding committee of the competition for the professorship of medical genetics in Rome had the following results (in order of number of votes): Luigi Condorelli, professor of clinical medicine at University of Rome; Gastone Lambertini, professor of normal human anatomy, Naples; Luigi Turano, professor of medical radiology, Rome; Antonio Lanedei, professor of medical pathology, Florence; Giov. Federico De Gaetani, professor of general pathology, Turin.
> None of these has the least competency in human genetics, nor in general genetics: they would all be rejected if they presented themselves for a professorship or even a university exam in human genetics.
> Many votes demonstrated however that the body of professors of medicine were far from unanimous. Many voting cards had to be cancelled because they expressed votes such as the following: disgusting; Gedda (the candidate); Cardinal Siri or Cardinal Ottaviani; Gregorio Mendel; Pius XII or John XII.
> Additionally, there were about forty blank voting cards. 13 votes went to Giuseppe Montalenti, professor of genetics in the Faculty of Science in Naples; eight to Claudio Barigozzi, professor of genetics in Milan; and five to Adriano Buzzati-Traverso, professor of genetics in Pavia.

201 Montalenti to Levi, 21 November 1960, AM, b. 33, f. 18.

> The results of the competition were the following: winner Luigi Gedda, with unanimity, with a relation acclaiming his capacity as a geneticist. The three clinicians wished to leave the other two posts vacant. However the recommendation of the other members of the committee prevailed, and so the second post was covered by Ruggero Ceppellini, lecturer in Turin, human geneticist of great value, who however received only three votes; one vote for second place went to Marcello Siniscalco, of the Institute of Genetics of Naples, another competent person of value. In third place, with three votes, was L. L. Cavalli-Sforza, who was short-listed third of three in genetics the previous year, but did not receive the professorship [...].
> Therefore: the only competent people, able to give a judgment of the value of the applicants (that is, the professors of genetics Montalenti, Barigozzi and Buzzati, some of whom also have a direct and specific competence in human genetics) were excluded from the deciding committee. A domesticated committee was created, made up of physicians incompetent in genetics, all dutiful to the commands of the Vatican. In this way they achieved the aim of offering the professorship to the ex-president of the Catholic Action, a name whose scientific value is nil, and whose only foreseeable future activity is politics.[202]

In the same day in which he sent this scorching document to Giuseppe Levi, about to transfer from Naples to Rome, Montalenti formally congratulated Gedda, who responded cordially:

> Dear Prof. Montalenti,
> I am very grateful for your congratulations and best wishes so kindly expressed and very dear to me. I am looking forward to your arrival in Rome, when I can consult with you more easily.[203]

On the letterhead of the Mendel Institute, under the title "Director," a new title now appeared: "Professor of Medical Genetics, University of Rome."

202 Anonymous document attached to the letter from Montalenti to Levi, 21 November 1960, AM, b. 33, f. 18.
203 Gedda to Montalenti, 8 November 1960, AM, b. 33, f. 18.

CHAPTER VII

AGAINST UNESCO

ITALIAN EUGENICS AND AMERICAN SCIENTIFIC RACISM

The fight against racism has been a constituent aspect of UNESCO's actions since its inception. In 1946, while defining the philosophical guidelines of the young UN affiliated organization, UNESCO's first director general, British naturalist Julian Huxley, set the conciliation of the ethical and political principles of equality with the biological fact of diversity as an objective. In the following years, staff at UNESCO's Paris headquarters found themselves involved in an attempt to defeat racial prejudice by demonstrating the lack of scientific base for the very concept of race. This proved an arduous task that would ultimately bring forth a struggle within the international scientific community and that would culminate in the publication of two "Statements on Race" within a short period of time, in 1950 and in 1951.[1]

Scholars have highlighted a substantial lack of academic consideration within Italy of UNESCO's two "Statements on Race," which went practically unnoticed by a scientific body still permeated by the legacy of fascism.[2] Deeper research, however, suggests this to be deliberate silence from the Italian scientific community as a result of outright adversity toward UNESCO's policy. If, for instance, the "Statements" never raised the attention of either the *Archivio per l'Antropologia e l'Etnologia* [Journal of anthropology and ethnology] or the *Rivista di Antropologia* [Review of Anthropology]—organs of the Florentine and Roman schools, respectively[3]—it must be considered notable that relevant Italian circles of medical genetics and social

1 For an in-depth reconstruction of the whole matter, see Pogliano, *L'ossessione della razza*, 145–210.
2 Pogliano, *L'ossessione della razza*, 191.
3 Pogliano, *L'ossessione della razza*, 191.

sciences nevertheless objectively converged on the positions of Anglo-American scientific racism.

By selecting scientific arguments as the core of its anti-racist campaign, UNESCO had, for all intents and purposes, suggested to American and European racist movements the possibility of a new camouflage strategy: racism and the pursuit of "white supremacy," just like anti-racist ideologies, had to be based on scientific evidence, threatened as they were by civil rights campaigns in the USA and steady decolonization in Africa and Asia. The main expression of such scientific racism was represented by the establishment of the International Association for the Advancement of Ethnology and Eugenics (IAAEE)[4] and its publication *The Mankind Quarterly*.

1. The IAAEE and *The Mankind Quarterly* (1959–1965)

The IAAEE was founded on 24 April 1959 in Baltimore. Its chairman was Robert E. Kuttner, the secretary was Anthony James Gregor, and the treasurer was Donald A. Swan. The executive committee comprised Robert Gayre, Reginald Ruggles Gates, Henry E. Garrett, Charles C. Tansill, Heinrich Quiring and the Italian demographer and statistician Corrado Gini. The first issue of *The Mankind Quarterly*, organ of the IAAEE based in Edinburgh, was published in June 1960, with Robert Gayre as editor, and Garrett and Gates as associate editors.

The segregationist scientists in the IAAEE shared some common traits. First, in many cases they held important academic positions. For example, Henry E. Garrett had been chairman of the American Psychological Association in 1946, was a member of the US National Research Council and from 1941 to 1955 was head of the Psychology Faculty at Columbia University.[5]

---

4 On the IAAEE, see Barry Mehler, "Foundations for Fascism: The New Eugenics Movement in the United States," *Patterns of Prejudice* 23 (1989): 17–25; William H. Tucker, *The Science and Politics of Racial Research* (Urbana: University of Illinois Press, 1994); Michael Billig, *Psychology, Racism and Fascism* (Birmingham: Searchlight, 1979); John P. Jackson, Jr., *Science for Segregation. Race, Law and the case against Brown v. Board of Education* (New York: New York University Press, 2005).

5 William H. Tucker, *The Funding of Scientific Racism. Wickliffe Draper and the Pioneer Fund* (Urbana–Champaign: University of Illinois Press, 2002), 79.

Similarly, Reginald Ruggles Gates, botanist, geneticist and anthropologist, professor at King's College London and Harvard University, had been an outspoken advocate of morphological, biological and psychological differences between human races since the 1930s.[6] Second, they all had relationships with the neo-Nazi and neo-fascist extreme rightwing in the US and Europe. Kuttner and Garrett, for example, contributed to publications of the Liberty Lobby, a far-right organization founded by Willis Carto in 1955.[7] Robert Gayre of Gayre and Nigg was a Scottish anthropologist, an expert in heraldry and a supporter of Madison Grant's Nordicism, close to the racist and anti-Semitic organizations of Arthur K. Chesterton. Anthony James Gregor, an Italian-American by origin (his original name was Anthony Gimigliano), gained a PhD at Columbia University with a thesis on the scientific and philosophical ideas of Giovanni Gentile. Between 1952 and 1956 he wrote for Oswald Mosley's "The European,"[8] then intensified his relationship with Italian neo-fascism during the 1960s and popularized the works of historians such as Ernst Nolte and Renzo De Felice in the USA.[9] Donald Swan was contributing by the late 1950s to the *Truth Seeker* and was the most outspoken admirer of Hans F. K. Günther. Finally, Charles Tansill, an historian at Georgetown University, was a member of the Nazi "Viereck Circle," which during World War II had suggested an alliance between the USA and Hitler's Germany.[10] Moreover, dating from the famous 1954 Supreme Court sentence *Brown vs. Board of Education*, the IAAEE fought constantly against the integrationist process in the USA. In fact all these scientists benefited from the donations of textile tycoon Wickliffe Draper's Pioneer Fund, an organization that from 1937 made ample contributions to economically sustain the main adversaries of the American integrationist system, and continues even today to support anti-egalitarian race scientists.[11]

---

6 For a biographical sketch of Gates, see Barkan, *The Retreat of Scientific Racism*, 168–76.

7 Tucker, *The Funding of Scientific Racism*, 79–86.

8 Oswald Mosley (1896–1980) was a British politician, known principally as the founder, in 1932, of the British Union of Fascists. The monthly journal *The European* (1953–59) was edited by Mosley's wife.

9 B. Mehler's biographies of Gayre and Gregor, included in *Institute for the Study of Academic Racism-Bibliographies*, can be consulted for free online at *www.ferris.edu/isar/bibliography/homepage.html.*

10 Tucker, *The Funding of Scientific Racism*, 87–88.

11 Tucker, *The Funding of Scientific Racism*.

From the first issue of *The Mankind Quarterly*, four Italians were members of the advisory board: Luigi Gedda, Corrado Gini, Gaetano Martino and Sergio Sergi. Of these, Gedda and Gini were most closely involved in Italian eugenics and in the *liaisons dangereuses* with the IAAEE.

2. *Meticciato di Guerra*: Luigi Gedda and Reginald Ruggles Gates

The link between Luigi Gedda, physician and director of Rome's "Gregorio Mendel" Institute, and *The Mankind Quarterly* occurred through the mediation of Reginald Ruggles Gates and essentially developed around a work titled *Il meticciato di guerra e altri casi* [The hybrids of war and other cases] and published in 1960 by the "Gregorio Mendel" Institute, in which an explicit stance in favor of the scientific legitimacy of "racial genetics" was presented.

It was Ruggles Gates himself, a personal friend of Luigi Gedda,[12] who wrote the preface to *Meticciato di Guerra*, which he welcomed as an important contribution to the development of a "genetics of races": "The studies on interracial breeding are presently assuming a new meaning. From the occasional or systematic studies conducted in many parts of the world, a science of Racial Genetics is slowly but steadily stemming, the fundamental principles of which are already visible."[13] In the second issue of *The Mankind Quarterly*, again Ruggles Gates signed the volume's review, the contents of which he enthusiastically indicated as "a model": "This work will serve as a model for future studies on the hybrids of war. It is of crucial interest for anyone involved with the study of races."[14]

Gedda was not entirely new to the study of interracial breeding. In 1938, for instance, he welcomed the fascist laws against race crossing in the pages of the catholic journal *Vita e Pensiero*, declaring the crosses between "very different races" as unfavorable:

---

12 On this issue see also Gates' obituary, written by Gedda himself: see ch. 6.

13 Luigi Gedda, Angelo Serio and Adriana Mercuri, *Il Meticciato di Guerra e Altri Casi* (Rome: Edizioni dell'Istituto Gregorio Mendel, 1960), VI.

14 Reginald Ruggles Gates, "Il Meticciato di Guerra," *The Mankind Quarterly*, 2 (October 1960): 218.

> As a general rule, and in this case, nature loves orderly, gradual processes, "Natura non facit saltus," and for this reason crossbreeding among highly different races is usually unfortunate. On the other hand, the mix of kindred races, thus similar, far from hurting, can produce new, valuable matches and, in the end, improve the stock [...] It is the mix of very different, distant—or, as we also say—divergent races which will end up being very damaging for the human stock; an example can be seen in the hybrids which result from the crossing between white and negro races; a type of mix that, using appropriate measures, must be strongly recommended against.[15]

Perhaps remembering these sentences, in his preface to *Meticciato di Guerra*, Gedda quickly drew a distinction between racism—which he condemned without hesitation—and the scientific study of human races, made more urgent and relevant by the increase in racial mixing that resulted from the rapid development of transportation and means of communication. Gedda's claim of the scientific value of "race genetics" revealed an implicit polemic purpose, which combined under the same negative title every political intervention on race issues, regardless of whether it came from Hitler's Germany or from UNESCO's Statements:

> The study of races is a consequence of our times and as such, is destined to develop, even if an arrest of the regular process of scientific development has been caused by the ill-advised use of racial phenomena in political and social activities as a criteria for discrimination, barring or war. Racism is not good science, and equally, is not good politics. Such an arbitrary transfer of scientific hypothesis and analysis into the incubator of politics has not furthered our knowledge of the argument of race, and instead has damaged it by making it appear as an arbitrary hype, alien to science and detrimental to ethical, individual and social values, and also as a source of controversies and rigidities, in contrast with the custom of scientific research, which avoids any passions and requires a spirit of cooperation to ensure the necessary control.[16]

In this specific case, fitting into the plentiful eugenic literature of "racial hybridism" analysis—largely quoted in his pages—Gedda's work (assisted

15 Luigi Gedda, "A Proposito di Razza," *Vita e Pensiero* 29, no. 9 (September 1938): 416.

16 Gedda, Serio and Mercuri, *Il Meticciato di Guerra e Altri Casi*, 6.

by two of the Institute's physicians, Adriana Mercuri and Angelo Serio) concerned 44 "hybrids of war," aged between eight and twelve: 34 males, in-patients at Anzio's *Instituto SS. Cuori*; and 10 females, in-patients at Rome's *Instituto S. Cuore in Borgata del Trullo*; children of "European Italian mothers" whose fathers were "colored" soldiers from occupying forces in Italy in the years 1943–1948.

Gedda's reference to genetics here was nothing but an attempt to linguistically modernize a research methodology of racial anthropology, based on anthropometric measurements, IQ tests, genealogical researches and clinical examinations. His definitions of three hybrid groups, for example, were reconstructed from the unknown "paternal race," starting with the "exotic genotype"; in other words, from identifying "non-European racial traits present in the hybrid."[17]

From the research on "hybrids of war," Gedda drew three conclusions. First, through anthropometric surveys, a positive evaluation of racial crossbreeding emerged, which in some cases presented forms of "heterosis" or "hybrid vigor," demonstrating the creative energy of racial mixing.[18] Secondly, the use of mental tests seemed to indicate psychological inferiority of hybrids, due to hereditary factors as well as to environmental influences.[19] Third, drawing on an argument used—within the IAAEE—by Anthony J. Gregor and psychologist Clairette Armstrong,[20] Gedda justified the segregation of hybrids as a form of "protection" in a hostile social context. Only isolation in the boarding schools of the Childhood Protection Agency (*Ente per la Protezione del Fanciullo*) could defend the hybrid from surrounding racial prejudice and guarantee normal psychological development: "There's no doubt that this not only postpones contact between the hybrid and the leucodermic world; it is also true that contact will occur at an age less delicate and thus more apt to overcome and sublimate inferiority complexes."[21]

The research conducted by Gedda, Serio and Mercuri soon sparked heated debate that directly involved *The Mankind Quarterly* and the IAAEE group. Not at all coincidentally, in Italy it was Corrado Gini who extensively

---

17 Gedda, Serio and Mercuri, *Il Meticciato di Guerra e Altri Casi*, 275–76.
18 Gedda, Serio and Mercuri, *Il Meticciato di Guerra e Altri Casi*, 278.
19 Gedda, Serio and Mercuri, *Il Meticciato di Guerra e Altri Casi*, 279.
20 Tucker, *The Funding of Scientific Racism*, 85.
21 Gedda, Serio and Mercuri, *Il Meticciato di Guerra e Altri Casi*, 214.

reviewed *Meticciato di Guerra* on the pages of *Genus*, concentrating his criticisms on the problem of "heterosis," an issue very dear to the statistician since the 1930s. In Gini's opinion, there were essentially two unresolved problems undermining Gedda's claims. First of all, colored soldiers in Italy did not represent the populations they belonged to, because they had been through numerous selection processes, making "the characteristics of the offspring not comparable to those of their peers from the parent races."[22] Moreover, the literature on "racial hybrids"—and Gini quoted, apart from his own works, also the data of Davenport and Steggerda on Jamaican race-crossings[23]—demonstrated the impossibility of conceiving "heterosis" as a common or generalizable phenomena: on the contrary, "as far as the cross-breeding between whites and negroes is concerned, various and reliable testimonies bear witness against it."[24] These same arguments are found in a letter sent in January 1961 from Gini to Gayre, the editor of *The Mankind Quarterly*, to propose an essay specifically dedicated to the problem of interracial mixing. Both the Italian statistician and the Scottish editor shared a negative opinion on hybridization between whites and blacks, and Gini did not hesitate to take a clear stance against the process of integration that was taking place in the USA, thus revealing the political core of the issue: "Apart from the scientific matters," he wrote, "I believe that this isn't the most appropriate moment to promote hybridization between negroes and whites."[25] On scientific grounds, the reference to Gedda's research and to the problem of heterosis was explicit:

> I don't know if you've seen the recent book of our colleague Prof. Gedda on war hybrids in Italy. He comes to the conclusion that there is an [sic] heterosis in the mulattos, what is contrary to all the previous results. This conclusion can well be attributed to the selection of the fathers and probably also of the mothers, which makes their children not comparable to those of the general populations.[26]

22 Corrado Gini, "Eterosi nei Meticci di Guerra?," review of *Il Meticciato di Guerra e Altri Casi*, by Gedda, Serio and Mercuri, *Genus* 16, no. 1–4 (1960): 168.

23 See Charles B. Davenport and Morris Steggerda, *Race Crossing in Jamaica* (Washington: Carnegie Institution, 1929). For a critical analysis of this research, crucial in the history of American eugenics, see Barkan, *The Retreat of Scientific Racism*, 162–68.

24 Gini, "Eterosi nei Meticci di Guerra?," 168.

25 Corrado Gini to Robert Gayre, 30 January 1961, ACS, Gini Papers (from now on, AG), b. b.6.

26 Gini to Gayre, 30 January 1961, ACS, AG, b. b.6.

As for the biological negativity of race crossings between "whites" and "negroes," there was substantial agreement from Gayre:

> I think that Professor Ruggles Gates will be of your opinion as he tends on the whole, I think I am right in saying, to deprecate the tendency to look for heterosis in human beings. In my own case, I have thought that some of the energy generated by the Americans is due to heterosis, not of course heterosis due to crossings of specific types, but within the various races of the one stock.
> Concerning Professor Gedda's theory, I think that you are probably quite right, and that there may well be a selection taking place when this kind of hybridisation occurs. The American negro soldiers that were sent to Italy, if I remember rightly, were specially selected. I was there at the time. On the whole also, they were definitely themselves to be classified more as mulattos than Negroes in a vast number of cases. In fact, the pure negro among the American negro troops, seems to be a rarity. Therefore, I am entirely in agreement with you that the results that Professor Gedda is getting are not necessarily due to heterosis at all.[27]

However, despite this theoretical convergence, the inappropriateness of opening a critical debate within the IAAEE, which would have opposed Gini and Gedda on the issue of racial breeding, drew a curtain over the idea of publishing the essay. This seemed even wiser as *Meticciato di Guerra* at that time was also at the center of heated polemics within the Anglo-American scientific community. Indeed, in 1962 the geneticist Leslie C. Dunn—editor of the first UNESCO "Statement" and among the authors of the second[28]—strongly attacked Gates and Gedda in the *Eugenics Review*, openly accusing them of racism:

> There are still reminders of the uncritical use of what look like genetic methods applied to racial anthropology. What shall one say, for example, when three authors, after anthropometric examination of forty-four Italian war orphans of whom the fathers were unknown but assumed to be "colored," draw sweeping conclusions concerning heterosis ("established with certainty"), inheritance of erythrocyte diameter ("very convincing") and other statements not supported by evidence? Yet these are statements made in 1960 by Luigi

27 Gayre to Gini, 3 February 1961, ACS, AG, b. b.6.

28 On Dunn's anti-racist commitment, see Barkan, *The Retreat of Scientific Racism*, 266–68. See also Melinda Gormley, "Scientific Discrimination and the Activist Scientist: L.C. Dunn and the Professionalization of Genetics and Human Genetics in the United States," *Journal of the History of Biology* 42 (2009): 33–72.

> Gedda and his co-workers Serio and Mercuri in their recent book *Meticciato di Guerra*. R.R. Gates, who writes an introduction in English to this elaborate book, refers to it as an important contribution to what he calls "racial genetics." Others will have greater difficulty in detecting any contribution to genetics, but may see in it, as I do, a reflection in 1960 of the uncritical naïveté of that early period of human genetics which delayed its progress. (...) Truly the past is not yet buried, and human genetics, in spite of its recent evidences of new life, is still exposed to old dangers.[29]

Gedda did not respond to the criticism directly, instead it was Gayre himself, the editor of *The Mankind Quarterly* who intervened in his defense, thus reasserting once more the deep ties between the catholic geneticist and the IAAEE's eugenicists. According to Gayre, Dunn's opinion was factious, outrageous, lacked scientific objectivity and was purely ideological:

> The hallmark of the witch-hunter is the use of such terms as *racist* and *racialism*—used here in connection with Professor Gedda and Doctors Serio and Mercuri, as well as Professor Ruggles Gates; *The Mankind Quarterly* and its editors and contributors are, therefore, in good company. But the people who use these terms abusively are motivated by an almost hysterical hatred of anyone who recognizes, or anything which establishes, the existence of different and great racial groups, with all their differences in heredity (whether biological or sociological).[30]

Unlike Dunn's statement, Gayre argued, there was no contradiction whatsoever between genetics and racial anthropology. On the contrary, the former had come to justify the latter:

> But frequency genetics has not in any way altered basic biological facts. Frequency studies can add very little when we consider those fundamental characters which anthroposcopically distinguish the major human stocks. [...] We might well go over a lengthy list of human characteristics which in the past have been used for racial classification, and find that they are equally valid.[31]

29 Leslie C. Dunn, "Cross Currents in the History of Human Genetics," *The Eugenics Review* 2 (July 1962): 74.

30 Robert Gayre of Gayre, "L. C. Dunn on Luigi Gedda, Angelo Serio, Adriana Mercuri, R. Ruggles Gates and 'The Mankind Quarterly,'" *The Mankind Quarterly* 1 (July–September 1962): 49–50.

31 Gayre, "L. C. Dunn on Luigi Gedda, Angelo Serio, Adriana Mercuri, R. Ruggles Gates and 'The Mankind Quarterly,'" 49.

Here Gayre supported an *evolutionist* interpretation of the history of genetics, which blended the acquisitions of modern science with all previous ideas on inheritance, from Aristotle onwards, against the *revolutionary* hypothesis of Dunn, who believed true genetics only started with Mendel. Therefore, according to Gayre, neither Gedda nor *The Mankind Quarterly* had a past they should be ashamed of:

> Because of Gedda, Serio, Mercuri, Gates and *The Mankind Quarterly*, we are told that the past is not yet buried and human genetics is still exposed to old dangers! We might ask *what* past is not yet buried? What are the old dangers? And to what or to whom? To the old school of cytological geneticists? Or to civilization?[32]

The debate between Gayre and Dunn, an emblematic moment of the clash between UNESCO's anti-racism and the racist eugenics of the IAAEE, marked the point of Gedda's highest visibility in *The Mankind Quarterly*. From then on, no other essay was published regarding the Italian physician, although his name always remained highly visible on the magazine's front page among the members of the honorary advisory board.

3. Corrado Gini and the "Guerrilla War" against UNESCO

Corrado Gini's contributions to *The Mankind Quarterly* span from the magazine's first issue until 1965, and were characterized mainly by two aspects: first, the development of a scientific and organizational exchange with the members of the IAAEE; second, the embracing of a personal strategy in conducting the battle against the anti-racism of UNESCO.

First of all, Gini co-opted the IAAEE's most prominent members for the International Institute of Sociology (IIS), which he chaired from 1950, and made the pages of its journal, *Genus*, available to them. In particular, his relationship with A. J. Gregor grew most intensely. It was Gregor who opened the IAAEE's doors to Gini[33], and again Gregor who translated his essays into English. In the United States, Gregor was a fervent advocate of Gini's organicism, to which he devoted a number of essays (in collaboration

---

32 Gayre, "L. C. Dunn on Luigi Gedda, Angelo Serio, Adriana Mercuri, R. Ruggles Gates and 'The Mankind Quarterly,'" 50.

33 Anthony J. Gregor to Corrado Gini, 3 July 1960, ACS, AG, b. b.5; Gini to Gregor, 11 July 1960, ACS, AG, b. b.5.

with the sociologist Michele Marotta)[34] and a seminar at the Johns Hopkins University.[35] The scientific collaboration with Gini allowed Gregor to become a member of the International Institute of Sociology and to attend its 19th (Mexico City, 1960)[36] and 20th (Córdoba, 1963)[37] Congresses. For his part, Gini asked Gregor if the leaders of *The Mankind Quarterly* would be willing to become members of the IIS: "Do you think—he wrote in a letter—that any of *Mankind*'s managers would like to be elected members of the Institute?"[38] In 1963 Gregor became chairman of the Research Committee on Intergroup Relations created within the IIS.[39] The following year, due to Gregor's mediation, the IAAEE became a sponsor of Gini's new edition of the *Revue Internationale de Sociologie,* for which printing costs would be shared between the University of Rome and the American organization.[40] Like Gregor, Gayre was nominated as a member of the International Institute of Sociology: Gini was particularly interested in Gayre's studies on "Nordic racial origins" and therefore proposed that he become a member of the IIS Committee, instituted in order to verify the validity of De Tourville's theories on the influence of the Nordic family on modern society.[41]

34 Anthony J. Gregor, "Corrado Gini and the Theory of Race Formation," *Sociology and Social Research* 45 (January 1961): 175–81; Anthony J. Gregor and Michele Marotta, "Sociology in Italy," *Sociological Quarterly* 2 (July 1961): 215–21; Anthony J. Gregor, review of Corrado Gini, "Corso di Sociologia," *Mankind Quarterly* 2, no. 1 (April–June 1961): 298–300; Anthony J. Gregor, review of Vittorio Castellano, "Studi in Onore di Corrado Gini," *Sociology and Social Research* 46 (July 1962): 501; Anthony J. Gregor, "Corrado Gini, the Organismic Analogy and Sociological Explanation," *Sociological Quarterly* 8 (spring 1967): 165–72.

35 Gregor to Gini, 3 May 1961, ACS, AG, b. b.5.

36 Anthony J. Gregor, "Sociology and the Anthropobiological Sciences," *Mémoire du XIXe Congrès International de Sociologie – Communications,* (Mexico: Comité Organisateur, 1960), 2, 83–107.

37 Anthony J. Gregor and Angus D. McPherson, "Sociology and Mental Testing of Non-Industrial Peoples," in *La Sociología y las Sociedades en Desarrollo Industrial: Communications before the XXth International Congress of Sociology* (Córdoba: Universidad de Córdoba, 1963), 2, 337–50; Anthony J. Gregor and Angus D. McPherson, "Sociology and the Assimilation of Non-Industrial Peoples," in *La Sociología y las Sociedades en Desarrollo Industrial,* 2.

38 Gini to Gregor, 3 October 1960, ACS, AG, b. b.5, followed by an affirmative answer on 6 October 1960. Gregor directly suggested the names of Charles Galton Darwin (Gregor to Gini, 18 February 1961) and George A. Lundberg (Gregor to Gini, 19 November 1962).

39 Gregor to Gini, 21 September 1963, ACS, AG, b. b.5.

40 Gini to Gregor, 25 October 1964, ACS, AG, b. b.5; Gregor to Gini, 5 November 1964, ACS, AG, b. b.5.

41 Gayre also joined the "International Committee for the Study of Hairy Humanoids" (*Comitato internazionale per lo studio degli umanoidi pelosi*), promoted by Gini within the International Institute of Sociology. On this, see "Comitato Internazionale per lo Studio degli Umanoidi Pelosi," *Genus* 18, no. 1–4 (1962): 1–4. On Gini's interests on the Abominable Snowman, see John P. Jackson Jr., "'In Ways Unacademical': The Reception of Carleton Coon's *The Origin of Races*," *Journal of the History of Biology* 34 (2001): 247–85. On this topic, see also: Brian Regal, "Amateur versus Professional: the Search for Bigfoot," *Endeavour* 32, no. 2 (June 2008): 53–57.

The idea of the Celtic-Irish origin of pre-Colombian America represented a point of particular agreement between Gini and Gayre.[42]

Finally, Gregor, as well as other contributors of *The Mankind Quarterly* such as Kuttner and Swan, published their essays, which shared strong racist arguments, in the pages of *Genus*.[43] Therefore, if the main contributors to *The Mankind Quarterly* often appeared in *Genus*, and were frequently members of the International Institute of Sociology, equally Corrado Gini—a member of the honorary advisory board since 1960 and an assistant editor since 1962—published two essays in the *The Mankind Quarterly*. One was in 1960 (*The Testing of Negro Intelligence*)[44] and one in 1961 (*Psychic and Cultural Traits and the Classification of Human Races*):[45] both were English translations of essays that first appeared in *Genus*, in 1960[46] and in 1955, respectively.[47]

The first essay was a review of Audrey M. Shuey's book, also titled *The Testing of Negro Intelligence*. Shuey was a teacher of psychology at the Randolph-Macon Women's College (in Lynchburg, Virginia) and a member of the honorary advisory board of *The Mankind Quarterly*. The book had been financed by the Pioneer Fund, prefaced by Garrett, and it aimed to demonstrate—through the use of IQ tests—the mental inferiority of "Negroes."[48] According to Gini, Shuey's work was the ultimate demonstration of the existence of those innate racial differences in mental attitudes so strongly denied in UNESCO's "Statements on Race":

> In my opinion it is probable that the volume will arouse objections and discussions because the techniques and the employment of mental tests involve,

42 Gayre to Gini, 8 December 1960; Gini to Gayre, 26 December 1960; Gayre to Gini, 2 January 1961; Gini to Gayre, 9 January 1961, all in ACS, AG, b. b.6.

43 Anthony J. Gregor, "The Logic of Race Classification," *Genus* 14, no. 1–4 (1958): 150–61; Anthony J. Gregor, "The Biosocial Nature of Prejudice," *Genus* 18, no. 1–4 (1962): 116–28; Robert E. Kuttner, "Cultural Selection of Human Psychological Types," *Genus* 16, no. 1–4 (1960): 1–4; Robert E. Kuttner, "Eugenic Aspects of Preventive Therapy for Mental Retardation," *Genus* 19, no. 1–4 (1963): 1–9; Donald Swan, "Genetics and Psychology," *Genus* 20, no. 1–4 (1964): 23–35.

44 Corrado Gini, "The Testing of Negro Intelligence," *The Mankind Quarterly* 1, no. 2 (October–December 1960): 120–25.

45 Corrado Gini, "Psychic and Cultural Traits and the Classification of Human Races," *The Mankind Quarterly* 1, no. 4, (April–June 1961): 236–41.

46 Corrado Gini, "Sulle differenze innate tra i caratteri mentali delle varie popolazioni," review of *The Testing of Negro Intelligence*, by Audrey M. Shuey, *Genus* 16, no. 1–4 (1960):161–66.

47 Corrado Gini, "Possono e devono i caratteri psichici e culturali essere tenuti presenti nella classificazione delle razze umane?," *Genus* 11, no. 1–4 (1955): 71–77.

48 Tucker, *The Funding of Scientific Racism*, 74.

> for the time being, very subjective elements—but in any event it is possible to say that, because of the abundance of the material collected and objectively reported, the volume constitutes a milestone in this area. After its publication, the burden of proof rests upon those who maintain the non-existence of the stated differences.[49]

In the wake of Shuey's book, Gini suggested a theory that summed up racist differentialism:

> If, in a stable environment, two groups of individuals differentiate themselves by virtue of a character which is at least partly hereditary, and which, at least in one of the two groups, is subject to natural selection, the differences observed between the two groups are, at least in part, innate.[50]

In other words, if two human groups live in different environments and, in at least one of them, the characteristic taken into consideration allows for natural selection, this will differentially eliminate certain modalities of that characteristic while favoring others in the two groups; and if such modalities are partly hereditary, the two groups will display innate differences. As a consequence—Gini concluded—it is possible to reckon that "under the influence of natural selection, innate mental attitudes differ among various population groups."[51] Behind the differentialist paradigm of Gini's racist discourse it is easy to recognize traditional hierarchical and inferiority logic. In particular the argument that when "negro races [are] compared to the white ones" natural selection favors physical characteristics over mental ones: hence the physical superiority of "Negroes," but also their innate intellectual inferiority.[52]

On the other side of the Atlantic, Gini's review attracted the barbs of *Man*, the authoritative journal of the British Royal Anthropological Institute. If the "theorem" presented by Gini meant anything—wrote G. Ainsworth Harrison—it signified that "there is a necessary relation between the way one difference is determined in one population and the way it is deter-

49 Gini, "The Testing of Negro Intelligence," 122.
50 Gini, "The Testing of Negro Intelligence," 122.
51 Gini, "The Testing of Negro Intelligence," 164.
52 Gini, "The Testing of Negro Intelligence," 164.

mined among two populations." But, he went on to say, this "is not a theorem": while a relationship often existed in reference to characteristics that presented a certain environmental weakness, such a relation "is certainly not necessary, as clearly indicated by experimental evidences."[53] In private, Gini's essay also provoked the disapproval of the illustrious geneticist Walter Landauer (University of Connecticut, Department of Animal Genetics), who reprimanded Gini for the "innatism" (and implicitly, the racism) of his "theorem" on the mental differences among populations:

> It seems to me further that your "theorem" constitutes a rather astonishing tautology. I should think that in this statement the words "hereditary" and "innate" are to all intents synonymous.
>
> I have the impression that the *Mankind Quarterly* is an attempt to forget Mendelian genetics and to return to the nineteenth century and Galton. I hope, of course, to be wrong and may judge hastily after seeing only one issue.[54]

Gini's reply substantially confirmed his anti-UNESCO racist differentialism:

> My point is that, if a characteristic is not only hereditary but also subject to natural selection (as it is usually the case) then two groups, living in different conditions, become innately differentiated relatively to such a characteristic.
>
> Then we may conclude that the differences between human groups may be, and practically are, in part innate and not only cultural as the Unesco Statement declared. Let me think that it is a conclusion of some bearing especially in the present epoch.[55]

The second essay—published in *Genus* in 1955 and in the *The Mankind Quarterly* in 1961—epitomizes Gini's main objections to UNESCO's "Statements on Race."[56] Setting off from a neo-Lamarckian theoretical base, Gini supported, in a dispute with UNESCO's anti-racism, the existence of a parallel between environmental and racial differences. Each environment, in substance, would have its matching race:

---

53 Geoffrey Ainsworth Harrison, "Reviews—The Mankind Quarterly," *Man* 61 (September 1961): 164.

54 Walter Landauer to Corrado Gini, 31 January 1961, ACS, AG, b. b.6.

55 Gini to Landauer, 19 February 1961, ACS, AG, b. b.6.

56 It must be remembered that Gini alone in Italy, had written a review of UNESCO's First *Statement on Race*: see Corrado Gini, review of *Statement on Race*, by Ashley Montagu, *Genus* 10, no. 1–4 (1953–1954): 192–94.

> It is to be observed, however, with respect to this proposition, that, even assuming that the diverse populations were originally identical with respect to innate mental characteristics, differences of environment (at first natural, then also social) in which their life developed would have inevitably impelled selection (natural, nuptial, reproductive) in a different direction for each race, in each favoring individuals possessed of traits better adapted to environmental conditions. And, since the individual differences with respect to the characteristics in question might be at times acquired but at other times innate, selection led, consequently, to the differentiation, in the adaptation to the environment, of the hereditary patrimony of the individual races.[57]

Beyond permanent physical differences, also psychic and cultural differences had to be considered. Contrary to UNESCO's "Statements," every race—purported Gini—is characterized by an innate disposition to work and saving, which marks the demarcation line between "primitive" and "civilized":

> Therefore, while there do not seem to be reasons as a consequence of which psychic and cultural characteristics should be excluded from the classification of races, a strong reason can be adduced which would counsel the adoption of the first even in preference to the second; it is the decisive importance that psychic traits exercise in determining the differences of human societies. This is to be said particularly with respect to the propensity [...] to labor and accumulation. For in this trait is found the fundamental difference between primitive populations, which, refusing to work beyond that strictly necessary to satisfy the most basic needs of existence, live on the margin of subsistence, and civilized populations in which individuals are disposed, even if in different measures, to make an effort which carries them beyond the subsistence level, and to conserve part of their produce with a view to future needs.[58]

The translation of this second essay as it appeared in *The Mankind Quarterly* presents an interesting hidden background, which outlines the heterodox nature of Gini's contribution with great clarity.

57 Gini, "Psychic and Cultural Traits and the Classification of Human Races," 236–37.

58 Gini, "Psychic and Cultural Traits and the Classification of Human Races," 239.

Archival evidence, in fact, reveals that editor Gayre, as was his habit, intervened brutally and without prior notice on Gini's text, erasing the following paragraph:

> To decide, in any case, whether cultural traits of a population have, at least in part, a hereditary base or whether they constitute simply acquired characteristics is in practice very difficult. But this difficulty does not arise only with respect to such characteristics. In point of fact, after the research of Boas on the European immigrants to America, those of Dorning on the Jewish immigrants to Berlin and above all after our own research with respect to the Albanian colonies in Calabria and the Ligurian colonies in Sardinia, it is very difficult to maintain that physical characteristics such as cephalic index, stature and also pigmentation, which constitute the basis for the classification of human races, are in fact hereditary and not, rather, acquired under the influence of the environment. Their permanent character, over a number of generations, would be, in the generality of populations, the effect of the constant conditions of the environment in which the population lives.[59]

Facing Gini's rather annoyed reaction, Gayre answered, specifying the reasons for the cut:

> The paragraph which I suggested should come out is one which is largely irrelevant to the whole of your main argument, and I thought would have the effect of marring your very excellent article by causing a certain amount of controversy to develop around your statement concerning Boas. As you perhaps know, Boas was very severely treated by Karl Pearson, Keith and others when he enunciated his doctrine. It is certainly one which most of us do not share, and I have written at some length, in a work I am now publishing, against it. Therefore I felt that it was better to avoid at this stage bringing in a controversial side-issue. If you wish to expound some new version of Boas in a complete article, that would be quite another matter, and it could be dealt with objectively as the principle matter under discussion.[60]

It appears evident that the controversy revolved around the interpretation of the research conducted by Franz Boas, Columbia University's pioneer of

[59] Gini, "Psychic and Cultural Traits and the Classification of Human Races," 237.

[60] Gayre to Gini, 25 January 1961, ACS, AG, b. b.6.

American cultural anthropology,[61] thus revealing how, apart from the editorial dispute, Gini and Gayre were engaged in a more general confrontation between the American hereditarian eugenics and the Italian environmentalist approach.

In 1911, following a suggestion of the U.S. Immigration Commission, Boas, with the help of thirteen assistants, had measured the height and the cephalic index of more or less eighteen thousand immigrants or children of immigrants in New York, coming to the conclusion that the various European types were not at all stable, as maintained by hereditarian racism, but—rather the opposite—had a tendency to uniformity, due to environmental influences, toward an average "American" type.[62] Boas' studies soon became a reference point in Italy for eugenicists, who used his results as a means to counter the fears of their American colleagues about the biological threat of miscegenation with Italian immigrants arriving on Ellis Island. Gini himself had followed Boas' line, directing, as of 1938, the research of the CISP on the Albanian community in Calabria and on the Ligurian-Piedmontese community in Sardinia. In summarizing the results in the early 1950s, Gini believed he had demonstrated the eventual "physical assimilation" of immigrants to the local environment:

> From all the above-mentioned research, one concludes that emigrated populations, even without crossbreeding, gradually lose their physical characteristics and acquire those of the autochthonous population. The peoples appear as the children of their land and it is indeed to be noted that, contrary to what is currently believed, assimilation, at least in some cases, happens more rapidly in relation to physical traits than to cultural ones [...]. Hence if we accept Boas' theory that there is, in the differential characters of a race, a hereditary nucleus and a fringe which varies with the environment, we must admit that the latter is such that, at least in Caucasian races, the hereditary nucleus will come down to not much at all.[63]

61 On Boas, see Pogliano, *L'ossessione della razza*, 290–96.

62 See Franz Boas, *Changes in the Bodily Form of Descendants of Immigrants* (Washington: Senate Document 208, 1911).

63 Corrado Gini, "L'assimilazione fisica degli immigrati," *Genus* 9, no. 1–4 (1950–52): 19 (lecture read on the Italian radio on 31 December 1951). The research of CISP on the physical assimilation of immigrants was the subject of Gini's contributions at various international conferences on eugenics and genetics between the end of the 1930s and the mid-1950s: specifically, at the 2nd International Congress of the Latin Eugenics Societies (Bucharest, 1939, never held because of the outbreak of World War II), at the 7th, the 8th and the 9th International Congresses of Genetics (held in Edinburgh, 1939, Stockholm, 1948 and Bellagio, 1953 , respectively), and at the 1st International Congress of Human Genetics (Copenhagen, 1956).

Boas represented—for Gini and, more generally, for Italian eugenics—confirmation of the environment's role in varying racial characteristics. Conversely, for the IAAEE's segregationist scientists, who strongly advocated hereditarian eugenics, the "school of Boas"—including, among others, the father of the first "Statement on Race," M. F. Ashley Montagu[64]—instead embodied the ghost of that "Jewish–Communist" conspiracy which had led the United States to abandon Jim Crow's laws. As a consequence, two opposite and confronting theoretical stances surrounded the *Boas case*, despite sharing a common enemy in UNESCO's "Statements on Race": on one side, there was Gayre's "Mendelian" racism, biological and hereditarian; on the other, Gini's "neo-Lamarckian" racism, psychological and environmentalist.

To demonstrate how these two positions, as different as they were on epistemological grounds, were in fact objectively converging, it is worth quoting the words with which Gini, while rejecting Gayre's objections, gave his ultimatum regarding the editorial line of *The Mankind Quarterly*:

> You insist upon the elimination of one paragraph of my article because it is controversial, with the view of getting the unanimous support of everyone of your way of thinking.
> Now I think that the facts mentioned in the paragraph in question cannot be denied, while their interpretation is controversial. But this is, for me, not a reason for eliminating it but on the contrary a reason for insisting—*as I insist*—on its publication.[65]

In the name of his long and "non-conformist" scientific career, Gini insisted on the need to separate the responsibilities of the editor from those of the author, and to guarantee a minimal pluralism of points of view. Finally, he threatened to withdraw:

> I am very jealous of the integrity of my thought, and, as a strict principle, I cannot accept any modification of my writings except for material mistakes.
> I understand very well that the others—and you especially—may have different views, but my writings are signed by me and imply only my scientific responsibility.

64 On the multi-faceted and long-lived activities of Montagu, see Andrew P. Lyons, "The Neotenic Career of M. F. Ashley Montagu," in Larry T. Reynolds and Leonard Lieberman, eds., *Race and Other Misadventures. Essays in Honor of Ashley Montagu in His Ninetieth Year* (Dix Hills NY: General Hall, 1996).

65 Gini to Gayre, 7 February 1961, ACS, AG, b. b.6.

> I suppose indeed that you—as it is usual for the editors of the scientific journals—do not feel yourself scientifically responsible for all what is published in the *Mankind Quarterly*. Otherwise I should to my regret renounce to collaborate to it, because with all the respect that I have for your scientific views—that, on the other hand, I know only in a small part—I cannot bound myself to follow them.[66]

Gini went on to suggest that if *The Mankind Quarterly* were to adopt the pluralist line exemplified by *Metron* or *Genus*, it would in reality be possible to collect "a more numerous and varied and higher standing group of collaborators." In conclusion, Gini further clarified the specific character of his adhesion to the IAAEE in the name of a common strategy against UNESCO:

> I quite agree with you that—according to what you wrote to me in your letter of January 14—the time has come when people who are more soundly grounded in science than some of the people who signed the Unesco document should make their views known (for my part I have already done that), but this does not imply that their views must be uniform. In scientific field the fights—in my opinion—must be combated with the system of *guerrillas*, which does not exclude coordination but allows personal initiative. Scientific thought is difficult to concile [sic] with regimentation.[67]

Not "regimentation," then, but scientific "guerrillas" against UNESCO: this was Gini's justification for his own role within the IAAEE and for his contribution to *The Mankind Quarterly*. In the end, Gayre was forced to give in, although he did not miss his chance for one last, ironic jab: "Of course, I am quite willing to publish the article as it stands, although I still am of the opinion that a slight modification of unnecessary material is always an advantage."[68]

Carried out between January and March 1961, the diatribe between Gayre and Gini finally appeared to reach a clarification and a relative differentiation of stances. Indeed, from this moment on, other situations allowed Gini to affirm his heterodox line within the common and agreed scientific "guerrilla" approach against UNESCO.

---

66 Gini to Gayre, 7 February 1961, ACS, AG, b. b.6.
67 Gini to Gayre, 7 February 1961, ACS, AG, b. b.6.
68 Gayre to Gini, 2 March 1961, ACS, AG, b. b.6.

A crucial test occurred on the occasion of Garrett and Gayre's suggestion to write a collective preface to Carleton Putnam's book, *Race and Reason: a Yankee View*.[69] Sustained by a massive advertising campaign, and financed by the Pioneer Fund, Putnam's volume was none other than a racist pamphlet which revolved around two arguments repeated obsessively: the mental inferiority of the "Negroes," as demonstrated by the scientific results of IQ tests, and an interpretation of the anti-racist battle as the umpteenth expression of the "Jewish–Communist" conspiracy.[70] The anti-UNESCO intent of the preface promoted by the IAAEE had already been openly declared by Gayre to Gini himself:

> I have read it through, and while it is of course on a political-social problem, it is basically relevant to anthropology. I am sending you herewith a copy of the foreword which Professor Henry E. Garrett has proposed, and where I have marked "A," I propose that the piece I have written should go in. If you agree with these two drafts, would you please be good enough to indicate that you are, and then we will add your name to the signatories. Professor Garrett is most anxious that as many scientists as possible, in the short time available, should sign this foreword. *It is felt that the time has come when people who are more soundly grounded in science than some of the people who signed the UNESCO document should make their views known.*[71]

However, in the same letter in which Gini harshly rejected Gayre's interventions on his essay, he also rejected the suggestion of joining the initiative. A similar, collective declaration against UNESCO, he objected, would eventually mirror the vagueness and the approximation of the "Statements":

> I am also reluctant to sign joint declarations. In order to reach a text which satisfies all the signatories, every one must renounce a part of his own thought, and the Minimum Common Denominator that is attained cannot be but vague and colorless. (By the way I think that if the signatories of the Unesco "Statement"—among whom there were also very distinguished scholars—had been

69 Carleton Putnam, *Race and Reason: A Yankee View* (Washington: Public Affairs Press, 1961). On Carleton Putnam and the publication of *Race and Reason*, see Tucker, *The Funding of Scientific Racism*, 101–11.

70 On this matter, see the slating by Barton J. Bernstein, "Race and Reason: Review," *The Journal of Negro History* 1 (January 1963): 58–60.

71 Gayre to Gini, 14 January 1961, ACS, AG, b. b.6; italics added.

> invited to give their individual advice, we would have had a much more valuable document).[72]

As a consequence, the first edition was published with a preface signed by Gates, Garrett, Gayre, and, in Gini's place, Wesley Critz George, a professor of anatomy at the University of North Carolina School of Medicine and an advocate of racial segregation even before the *Brown* decision.[73] Shortly thereafter, in light of the 200,000 copies sold and of the twelve reprints in eighteen months, it was Putnam himself who once again asked Gini for a contribution for the pocket edition:

> As you may know, a panel of four scientists headed by the late R. Ruggles Gates signed the introduction to the first edition. I would be greatly honored if I might add your name to this panel in preparing the pocketbook edition. The tide seems to be turning in the United States, and I believe we may soon have the integrationists and "scientific" propagandists on the defensive. I solicit your aid in rallying here the forces with which I believe you are in sympathy.[74]

Although declaring that he shared Putnam's line of thought, Gini again refused to endorse any collective declaration. In the scientific field, he argued, it is not possible to reach an effective interpretative "common denominator" on the issue of race. On the contrary, scientific manifestos always end up obscuring the value of those who signed them. Authorities such as Haldane, Dahlberg or Dunn—all of whom Gini personally knew and appreciated—had sacrificed the complexity of their research on the altar of UNESCO's "Statement on Race," and Gini—from an opposite standpoint—did not want to make the same mistake:

> Naturally there are not two scholars who have exactly the same opinion on a scientific field of a certain extent while a common declaration must constitute a minimum common denominator of the thought of all the signers neglecting the particular aspects which characterize the scientific personality of the various signers. I think that Haldane, Dahlberg, Dunn and the other signers of the manifesto of Unesco that you and I deplore (all people, in my opin-

72 Gini to Gayre, 7 February 1961, ACS, AG, b. b.6.

73 On the figure of Wesley C. George, see Tucker, *The Funding of Scientific Racism*, 69–78 and 105–09.

74 Carleton Putnam to Corrado Gini, 12 December 1962, ACS, AG, b. b.6.

ion, of a remarkable scientific value whom I know personally) would have written much more reasonable things should they have written their declarations freely and independently from the others.[75]

Gini's objective to articulate and broaden the spectrum of anti-UNESCO "guerilla" action promoted by *The Mankind Quarterly* can also be clearly seen in his attempt to involve the geneticist Cyril D. Darlington in the IAAEE. Upon the death of Ruggles Gates in August 1962, Gini accepted the role of substitute for Gates as honorary associate editor of *The Mankind Quarterly*, but asked Darlington to join him. His reasons for the choice were laid out in a letter dated 18 October 1962:

> The reasons for which I think desirable that you be an honorary associate editor of *The Mankind Quarterly* are several:
> 1°) because, so far as I know, this offer had already been made to you in the time and I think that it should be maintained;
> 2°) because you are a scholar of very high reputation and your name as associate editor will certainly be useful to the journal;
> 3°) because you have a wide field of scientific interests and I, although approving the main lines of *Mankind Quarterly*, think that it will be advisable to enlarge the field of the subjects treated in its papers.[76]

Darlington responded with a brief but dense note, in which, after having expressed his doubt on the scientific value of Gayre and reminded Gini of his inability to put up with Gates ("I always thought him an irresponsible investigator and writer"), he clearly asserted his perplexity on the scientific neutrality of the IAAEE and *The Mankind Quarterly*. He asked Gini directly to clarify the nature of the financing and political links of the association:

> I am, right or wrongly, apprehensive of the methods of organizations connected with racial or political propaganda and controlling large funds of unknown origin. How much genuine scientific and academic support or driving force have they? Or is their support and driving force largely political? You can perhaps tell me.[77]

[75] Gini to Putnam, 24 December 1962, ACS, AG, b. b.6.

[76] Corrado Gini to Cyril Darlington, 18 October 1962, ACS, AG, b. b.6.

[77] Darlington to Gini, 24 October 1962, ACS, AG, b. b.6.

In his reply, Gini first all defended the editor, Gayre:

> I know a little more from the personal point of view Dr. Gayre. He was an officer with important functions in the occupation army of Italy, a fact which excludes, I think, that his racial views are of the Nazi tendency. In that capacity he made several friends here also between important persons and he comes pretty often in Italy. I had him twice at my home and from a personal point of view he is quite agreeable and gives a good impression.[78]

Therefore, Gini explicitly justified his attempt to involve Darlington as a measure to give greater authority and depth to the scientific position of *Mankind Quarterly*: "I would be very glad that we will be associate editors, because I think that you and I, we may exercise an useful influence in order that the scope of the review may become larger and more scholarly." Finally, Gini confronted the burning question of political and financial backing of the journal, obviously claiming its absolute independence and scientific correctness:

> I do not think that Gayre and his circle has a political basis. They represent, in my opinion, the reaction of the Unesco policy (which has certainly a political character) to put at the same level all the races. In my opinion, a reaction is quite justified also from a scientific point of view, but it is necessary that every participant in the movement preserve his full independency of thought because it is difficult that two persons have exactly the same opinion in all the details of the question.
>
> For the origins of funds, I have the impression that Gayre is a pretty rich man. Other funds are collected by the International Association for the Advancement of Ethnology and Eugenics, in which Garrett, Gregor and Swan have prominent influence, but I think that its publications are independent from *Mankind Quarterly* and its sources are in any case in my opinion not political in character.[79]

After having denied any political interference in the editorial position of *Mankind Quarterly,* Gini hoped that Darlington would join them as an associate editor, helping to prevent any "political degeneration" of the journal and its related initiatives.

---

78 Gini to Darlington, 27 October 1962, ACS, AG, b. b.6.

79 Gini to Darlington, 27 October 1962, ACS, AG, b. b.6.

Darlington however was not convinced by Gini's "candid letter," as he ironically called it. A fresh occurrence contributed to the darkening of the atmosphere that surrounded *The Mankind Quarterly*. In November 1960, Bozo Skerlj, a Slovenian anthropologist, had resigned from *Mankind Quarterly*'s honorary advisory board, explaining, on the pages of *Man*, that the abuse of anthropology in the interest of racial prejudice was offensive to him not just as a scientist but also as a former prisoner at Dachau. A year later, Gayre and Garrett decided to denounce *Man*, voice of the Royal Anthropological Institute (RAI), for having published Skerlj's protest, who had meanwhile disappeared in November 1961.[80] Gayre communicated the news to Darlington, who decided at this point to refuse Gini's offer, and not to align himself with the IAAEE or with RAI. In the name of scientific neutrality, Darlington chose, therefore, to not choose:

> In these circumstances I should much prefer not to associate myself with either Gayre or the RAI. I think that they are both ill-considered in their views. Both have pre-conceived ideas with a strong emotional element. I think what we all need now is a disentanglement, a withdrawal, from these strong emotional positions. We need time for reflection and opportunities for cool discussion.[81]

It was the "Skerlj episode," together with his personal conflict with Gayre in 1961 and the refusal of Darlington, that in November 1962 led Gini to play a new card in his attempt to differentiate the framework of collaborators of *The Mankind Quarterly*. In November 1962—following a suggestion from Sergio Sergi, himself a member of the honorary advisory board of the journal—Gini proposed the front page inclusion of a declaration that would sanction the different viewpoints represented within the common conviction of physical and psychical difference between human races.[82] The suggested text, which was accepted by Gayre and published on the first issue of 1963, read as follows:

---

80 Bozo Skerlj, "Correspondence. 'The Mankind Quarterly,'" *Man* 60 (November 1960): 172. Skerlj was Gini's assistant at the University of Rome from August to December 1941. See also Gini to Darlington, 21 November 1962, in ACS, AG, b. b6.

81 Darlington to Gini, 9 November 1962, ACS, AG, b. b.6.

82 Gini to Gayre, 23 November 1962, ACS, AG, b. b.6.

> *The Mankind Quarterly* exists to discuss the subjects which are included in its title and sub-titles. It is the view of the Editors (as would seem to them to be manifestly true and generally accepted to be true by the vast majority of observers past and present) that human races are physically and/or psychically different. The question of whether any particular race or racial group is superior to another *in the totality of all its characteristics* is not accepted by the Editor, and, as far as is known by the other associate and assistant editors.
> The views expressed in articles which appear in *The Mankind Quarterly* and the associated series of *Mankind Monographs* are those of the authors, and the editors and the Honorary Advisory Board of *The Mankind Quarterly* do not necessarily accept responsibility for the views so expressed.
> We believe, however, that it would be a disservice to science to refuse to publish an article or monograph just because the views expressed by the author were not accepted by the Editor, or one of the other editors, or of some members of the Honorary Advisory Board, and we are certain that none of these persons would wish to take the responsibility of stifling the expression of such views.[83]

In the following issues, again upon Gini's insistent request, a more synthetic sentence was included: "The articles bind the authors and not the editors."

The papers which, from then on, Gini sent to Gayre, should probably be interpreted in the same line of differentiation within the IAAEE's offensive against UNESCO. For instance, Gayre favorably accepted the idea of translating and publishing Gini's contribution to the First International Congress on Human Genetics in 1956:[84] a paper based on the theory of "sub-Lamarckism"—very far, as we have seen, from the views of the editor—that culminated, nonetheless, in a racist differentialism that was perfectly compatible with the general orientation of *The Mankind Quarterly*.

Actually, neither this last essay, nor two of Gini's other proposals presented between 1962 and 1965—the publication of the essay *Alla soglia dell'umanità* [At the threshold of humanity][85] and the translation, to be published in the *Mankind Monographs*, of his 1940 book *Le rilevazioni*

83 Gini to Gayre, 23 November 1962, ACS, AG, b. b.6. (enclosed).

84 Corrado Gini, "The Physical Assimilation of the Descendants of Immigrants," in Tage Kemp, Mogens Hauge and Bent Harvald, eds., *Proceedings of the First International Congress of Human Genetics*, vol. 2 (Nasel, NY: S. Karger, 1958), 400–403.

85 Corrado Gini, "Alla soglia dell'umanità," *Rivista di Politica Economica* 64, no. 11 (November 1964): 1475–505.

statistiche fra le popolazioni primitive [Statistical surveys in primitive populations]—were realized, due to Gini's sudden death in 1965. Their findings, made possible by the retrieval of the original correspondence, contribute, however, to highlighting the complexity and the importance of the ideological and scientific relationship between Gini and the IAAEE. It was certainly a relationship marred by tensions and clashes between different stances, but reaffirmed until the end, in the name of the struggle against the common enemy: the egalitarianism and the anti-racism upheld by UNESCO and by its "Statements on Race."

4. Epilogue: *Race and Modern Science*

In 1967, Reginald Ruggles Gates' project to organize a "manifesto" against UNESCO took shape in a collection of essays titled *Race and Modern Science.* The polemic intent of the volume was comprised in its title, which echoed UNESCO's previous publication, *The Race Question in Modern Science.* The editorial enterprise was managed by Robert Kuttner and dedicated to the memory of Ruggles Gates, "who suggested, and helped put together, this book." Anthropologists, sociologists and psychologists who belonged to the ideological reservoir of *The Mankind Quarterly* crowded its pages in the attempt to demonstrate the scientific value of the concept of race and the legitimacy of racism: Bertil Lundman, Jan Czekanowski, J.D.J. Hofmeyr, Ilse Schwidetzky, David C. Rife, Clarence P. Oliver, Robert Kuttner, Cyril D. Darlington, Anthony James Gregor, George A. Lundberg, Friedrich Keiter, Frank McGurk, R. Travis Osborne, and Stanley D. Porteus. Two Italian contributions, whose authors may be easily guessed, must also be added to this catalogue. The first is a translation of a part of Luigi Gedda's *Meticciato di Guerra*;[86] the second, by Corrado Gini, is a collection of passages from his sociology lessons at the University of Rome, published in 1957.[87]

86 Luigi Gedda, "A Study of Racial and Subracial Crossing," in Robert E. Kuttner, ed., *Race and Modern Science* (New York: Social Sciences Press, 1967), 123–40.

87 Gini, "Race and Sociology," in *Race and Modern Science*, 261–76.

In the same year, *Race and Modern Science, Challenge to the Court: Social Scientists and the Defense of Segregation, 1954–1966,*[88] an essay by historian Idus A. Newby, was published in the US. For the first time, historiography pointed its finger against the IAAEE and *The Mankind Quarterly*: it would not be the last.[89]

88 Idus A. Newby, *Challenge to the Court: Social Scientists and the Defense of Segregation, 1954–1966* (Baton Rouge: Louisiana State University Press, 1967).

89 The controversy on scientific racism in the United States has erupted again after the publication of the best-seller by Charles Murray and Richard J. Herrnstein, *The Bell Curve. Intelligence and Class Structure in American Life* (New York: Free Press, 1994). The "evidence" shown by the authors to prove racial differences in intelligence on genetic bases is taken, not surprisingly, from *The Mankind Quarterly*. For authoritative critiques of the scientific case for racial differences in IQ, see the articles collected in Ned J. Block and Gerald Dworkin, eds., *The IQ Controversy* (New York: Pantheon, 1976) and in Jefferson M. Fish, ed., *Race and Intelligence: Separating Science from Myth* (Mahwah, NJ: Lawrence Erlbaum, 2002).

CONCLUSIONS

The eugenic gospel spread in Italy in the late nineteenth and early twentieth centuries, providing a scientific solution for the profound political, economic and social problems characterizing a country that had achieved political unification only in 1871. The construction of a national identity, social cohesion, and the problem of emigration were as central to eugenics as they were to other social and political movements.

In 1912, the Italian delegation at the First International Eugenics Congress in London represented the whole spectrum of positivist science: from legal medicine (Raffaele Garofalo) to physical anthropology (Giuseppe Sergi and Vincenzo Giuffrida-Ruggeri); from psychiatry (Enrico Morselli and Antonio Marro) to political economy (Archille Loria); from political sociology (Roberto Michels) to statistics (Alfredo Niceforo and Corrado Gini). Despite the heterogeneity of these disciplinary perspectives, several elements indicate the specificity of Italian eugenics within the international context right from the start: an anti-Mendelian, and often neo-Lamarckian hereditary theory; a natalist approach to population policies; a prevalently favorable evaluation of social exchange and racial breeding; a widespread emphasis on positive eugenic measures. In 1913 the First Italian Committee of Eugenic Studies was established, headed by Giuseppe Sergi: the members numbered just eighty-three, including anthropologists, legal physicians, military physicians, psychiatrists, demographers, statisticians, gynecologists and hygienists.

The First World War—simultaneously a biological threat and an immense human laboratory—deeply influenced the Committee's activities, favoring the development of a new phase, which was protracted until the First Congress of Social Eugenics in 1924. The postwar period saw a process of intense institutionalization of Italian eugenics, producing debates of an extreme variety and on a number of issues. In 1919, the Italian Society for Genetics and Eugenics (SIGE) was founded, directed by gynecologist Ernesto Pestalozza. In 1921, the Italian Society for the Study of Sexual Questions (SISQS) was established, and its members were deeply involved in the issue of premarital certification. In 1922, the Institute for Public Welfare and Assistance (IPAS) was constituted which, due to the commitment of its director and founder, Ettore Levi, led the campaign in favor of birth control as main instrument of eugenic selection. During this period, even the introduction of some form of eugenic sterilization found support, from Angelo Zuccarelli's criminal anthropology to Umberto Saffiotti's experimental psychology; from Gaetano Pieraccini's neo-Lamarckian social medicine to Paolo Enriques' Mendelian biology.

In December 1923, the results of the meeting between Pietro Capasso, leader of the Neapolitan Eugenics Group and Benito Mussolini, announced the beginning of a new phase in the history of Italian eugenics. Mussolini's refusal to introduce a form of premarital certification in Italy was confirmed a year later, in 1924, in the rather moderate nature of the conclusions of the First Italian Congress of Social Eugenics, both from the theoretical and from the practical point of view.

An evident political turning point came with Mussolini's "Ascension Day speech" in 1927, which inaugurated the fascist natalist demographic campaign, and with the publication of Pope Pio XI's encyclical "On Christian Marriage" in 1930. The impact these developments had on Italian eugenics was immediate. The principal institutions involved in "qualitative" eugenics—Aldo Mieli's SISQS and Ettore Levi's IPAS—were quickly "fascistized." SIGE passed into the control of the statistician and demographer Corrado Gini—president of the National Central Institute of Statistics (ISTAT) between 1926 and 1931, and of the Italian Committee for Population Problem Studies (CISP) after 1928—and of the physician and psychologist Agostino Gemelli, founder and chancellor of the Catholic University of Milan. At the international level, between 1927 and 1932, CISP controversially separated from the International Union for the Scientific Investigation of Population Problems (IUSIPP), and SIGE abandoned the International Federation of Eugenic Organizations (IFEO). On 14 July

1933, when Nazi Germany approved the "Law on the Prevention of Genetically Deficient Progeny," fascist Italy expressed its severe ideological and scientific criticism.

The strengthening of the ideological and political alliance between fascist Italy and Nazi Germany during the second half of 1930s produced new tensions and contradictions in the field of Italian eugenics, which exploded in 1938 with the publication of the so-called "Manifesto of racist scientists," the official document enshrining fascist state racism in Italy. Between 1938 and 1943, new conflicts emerged, characterized by the opposition between hereditarian, "Nordic," Mendelian, philo-German eugenics, which influenced the biological and esoteric-traditionalistic currents of fascist racism, and an environmentalist, neo-Lamarckian and "Latin" eugenics, which instead distinguished the nationalist current of fascist racism.

After the Second World War, in a period spanning from 1948 to the first half of 1970s, the last complex phase in the history of Italian eugenics unfolded. During these decades, the internal dynamics of eugenics in Italy was articulated on at least three different levels.

First of all, the development of Italian genetics after the Second World War led to the internal schism of SIGE, between 1948 and 1950, followed by the birth of the Italian Genetics Association in 1953: the International Congress of Genetics held in Bellagio, in 1953, represented a catalyzing event in this respect.

Second, between the 1950s and 70s, the recontextualization of eugenics within the new cultural, ideological and political framework was intertwined with the radical conflict concerning the methodology and clinical practices of medical genetics. The Study Center for Human Genetics, headed in Milan from 1940 by the zoologist Luisa Gianferrari, and the Institute of Human Genetics and Twin Research "Gregorio Mendel" in Rome, directed, after 1953, by the physician Luigi Gedda, proposed a methodological approach still focused on social hygiene, and a craft knowledge of heredity, interpreting "eugenic" counseling in terms of preventive medicine and reproductive hygiene. On the other side, Italian human genetics, adopting the lines of research introduced in Great Britain by Lionel Penrose during the 1930s, developed a medicalized form of eugenics based on Mendelian statistical analysis of modes of transmission, construction and examination of genealogical trees indicating genetic transmission, and diagnosis of transmission or recurrence probabilities. The collaboration between geneticist Giuseppe Montalenti and clinicians Ezio Silvestroni and Ida Bianco on the genetics of thalassemia and on the implementation of a widespread prevention campaign between 1949 to 1955 represents a significant example of this latter approach. This contraposition in the field of medical genetics was not only theoretical and methodological, but also affected the question of professional boundaries, feeding the struggle between clinical physicians and experimental biologists. Furthermore, other sources of tension arose from the broader ideological and political context in which the Catholic, familist and natalist positions encountered secular, birth control and family planning-oriented ones.

Third, racist *mainline* eugenics, represented above all by Gini's theories, did not disappear with the fall of the fascist regime, but remained active after the Second World War, offering an important—and heterodox—contribution to the racist campaign organized by the International Association for the Advancement of Ethnology and Eugenics (IAAEE) and *The Mankind Quarterly*. From the historiographical point of view, the history of Italian eugenics constitutes an important case study on several relevant issues that would merit further research and study.

First of all, it allows us to consider the introduction of the label "Latin eugenics" as a form of scientific, political and ideological demarcation. On this subject, the centrality of Corrado Gini's role provides a fundamental connection between the constitution of the Latin Federation of Eugenic Societies in 1935; the natalist and neo-Lamarckian paradigm of Italian eugenics since the late 1920s; and the ideological, political and cultural strategies of Italian Fascism on the international stage.

Secondly, the Italian case demonstrates an original interconnection between eugenics and genetics after the Second World War. In fact, starting from the end of the 1940s, the process of institutionalization and autonomization of genetics developed in a context of radical refusal of fascist eugenics, embodied by Gini's SIGE. Such a break coincided with the internal conflict concerning the definition of another label—"genetics"—between different social actors (geneticists, clinicians, statisticians, agronomists), as well as the reformulation of the concept of "eugenics" in relation to the development of medical genetics in the 1950s and 1960s.

Last but not least, Italian eugenics was undoubtedly distinguished by the institutional, political and ideological influence of Catholicism. Catholic eugenics not only had a function of opposition to negative eugenic measures, but also elaborated a precise alternative model. The latter was theoretically based on medical constitutionalism and biotypological holism, and institutionally articulated in the international network of Catholic physicians and in the activity of research institutions, such as Nicola Pende's Biotypological Institute during the fascist period, and the "Gregorio Mendel" Institute in Rome, directed by Luigi Gedda, after the Second World War.

The analysis of the international impact of Catholic, "Latin" and biotypological eugenics, and the Italian role in this history, constitutes a fascinating topic, which certainly deserves further historiographic attention.

BIBLIOGRAPHY

Archives

State Central Archive, Rome
State Central Archive, Rome
 Papers of Corrado Gini
Institute of History of Medicine, Rome
 Papers of Adriano Buzzati-Traverso
 Papers of Giuseppe Montalenti
Catholic University Archive, Milan
 Papers of Agostino Gemelli
American Philosophical Society, Philadelphia
 Papers of Franz Boas
 Papers of Charles B. Davenport
 Papers of Raymond Pearl

Primary sources

PROCEEDINGS:

Problems in Eugenics: Papers Communicated to the First International Eugenics Congress Held at the University of London July 24th to 30th (London: Eugenics Education Society, 1912).

Atti della Società Italiana di Genetica ed Eugenica (Rome: Tipografia del Senato di G. Bardi, 1920).

Davenport, Charles B., et al., eds., *Scientific Papers of the Second International Congress of Eugenics, (Held at the American Museum of Natural History, New York, Settembre 22–28, 1921)*, vol. 1, *Eugenics, Genetics, and the Family* (Baltimore: Williams and Wilkins, 1921); vol. 2., *Eugenics in Race and State* (Baltimore: Williams & Wilkins, 1923).

Atti del Primo Congresso italiano di Eugenetica sociale (Milano, 20–23 September 1924) (Rome: Stabilimento Poligrafico dello Stato, 1927).

Atti del Secondo Congresso italiano di Genetica ed Eugenica (Roma, 30 settembre – 2 ottobre 1929) (Rome: Failli,1932).

A decade of progress in eugenics: Scientific papers of the Third International Congress of Eugenics (Baltimore: The Williams & Wilkins Company, 1934).

Gini, Corrado, ed., *Atti del Congresso internazionale per gli Studi sulla Popolazione (Roma, 7–10 settembre 1931)* (Rome: Istituto Poligrafico dello Stato, 1934).

Fédération Internationale Latine des Sociétés d'Eugénique, *Ier Congrès Latin d'Eugénique. Rapport* (Paris: Masson et C., 1937).

"Atti della Terza riunione della Società Italiana di Genetica ed Eugenica (Bologna, 5–7 settembre 1938)," *Genus* 3, no. 3–4 (June 1939): 1–371.
Atti del Convegno per studi di assistenza sociale (Milan: Marzorati, 1947).
Atti del convegno dedicato a "I recenti contributi della genetica umana alla medicina" (Milan: Istituto Sieroterapico Milanese S. Belfanti, 1949).
Profilassi pre-matrimoniale (problemi medico-sociali). Atti ufficiali del Convegno internazionale per la trattazione dei problemi medico-sociali di profilassi pre-matrimoniale (Bologna: Cappelli, 1949) 52–53.
Atti del IV Congresso internazionale dei medici cattolici (Roma, 24 settembre – 2 ottobre 1949) (Rome: Orizzonte Medico, 1950).
Gedda, Luigi, ed., *Genetica Medica. Primum Symposium Internationale Geneticae Medicae Roma 6–7 settembre 1953* (Rome: Edizioni dell'Istituto Gregorio Mendel, 1953).
Montalenti, Giuseppe and Alberto Chiarugi, eds., *Atti del IX Congresso internazionale di genetica. Bellagio (Como, 24–31 agosto 1953)* (Florence: Florentiae, 1954).

Articles, books:

Abel, Wolfgang, "Meticci di Renania," *La Difesa della razza* 3, no. 10 (20 March 1940): 26–30.
A. L., "Il razzismo nord-americano," *La Difesa della razza* 2, no. 1 (5 November 1938): 22–23.
A. P., "Gedda vuole la firma," *Il Ponte* 16, no. 6 (June 1960): 990–91.
Agostini, Cesare, "Sulla utilizzazione degli epilettici in zona di guerra," *Giornale di medicina militare* 1 (1918): 24–33.
-----------, "Per un trattamento più razionale ed economico degli alienati di mente," *Quaderni di psichiatria* 10, no. 9–10 (September–October 1923): 193–95.
Ainsworth Harrison, G., "Reviews – The Mankind Quarterly," *Man* 61 (September 1961): 163–64.
Alberti, Salvatore, *La mortalità antenatale* (Milan: Vita e Pensiero, 1934).
Alimena, Bernardino, "Concludendo sulla violenza carnale e il 'diritto all'aborto'," *La Scuola Positiva* 6, no. 8 (August 1915): 673–75.
Allevi, Giovanni, "Lavoro ed Eugenetica," in *Atti del Primo Congresso italiano di Eugenetica sociale*, 395–400.
Almirante, Giorgio, "Una razza alla conquista di un continente," *La Difesa della razza* 2, no. 1 (5 November 1938): 20–21.
Ambrosoli, Gian Angelo, "Le malattie della pelle in rapporto all'Eugenetica," in *Atti del Primo Congresso italiano di Eugenetica sociale*, 341–44.
Angeli, D., "I non desiderati," *Giornale d'Italia*, (23 February 1915).
Antonini, Giuseppe, "Alcoolismo ed Eugenetica," in *Atti del Primo Congresso italiano di Eugenetica sociale*, 117–20.
Arcari, Paola Maria, *Le lingue nazionali della Confederazione Elvetica ed i loro spostamenti attraverso il tempo* (Rome: Tip. C. Colombo, 1930).
Arthos [J. Evola], "Sui limiti del razzismo: il problema dell'eredità," *La Vita Italiana*, 28, no. 323 (February 1940): 178–83.
Artom, Cesare, "Principi di genetica," *Rivista di antropologia* 19, no. 1–2 (1914): 281–410.
-----------, "Per gli studi di genetica ed eugenica," in *Atti della Società Italiana di Genetica ed Eugenica*, 11–14.
-----------, "Indicazioni sommarie sugli studi di genetica," in *Atti della Società Italiana di Genetica ed Eugenica*, 15–20.
Ascoli, Vittorio, "Visita prematrimoniale obbligatoria?," *Difesa sociale* 6, no. 4 (April 1927): 2–3.
Associazione Genetica Italiana, *Consultorio di genetica medica* (Pisa: ETS, 1972).
"Atti del Comitato Italiano per gli Studi di Eugenica," *Rivista di antropologia* 18 (1913): 543–44.
Bachi, Roberto, "Gli indici della attrazione matrimoniale," *Il Giornale degli Economisti* 69 (November 1929): 894–938.

Baglioni, Silvestro, *Principii di eugenica* (Naples: Edizioni del Pensiero sanitario, 1926).
------------, "Funzioni somatiche e genetiche," in *Atti del Secondo Congresso italiano di Genetica ed Eugenica*, 153–60.
------------, "Problemi eugenici e demografici nei riguardi del rafforzamento della razza," in Lucio Silla, ed., *Atti della SIPS: XXVI riunione (Venezia, 12–18 settembre 1937)* (Rome: SIPS, 1938), vol. 1, 363–96.
------------, "Continuità della razza," *La Difesa della razza* 3, no. 4 (20 December 1939): 6–12.
Baldassari, Vittore, "Alcuni dati statistici della Clinica ostetrica della R. Università di Genova," *Atti del Primo Congresso italiano di Eugenetica sociale*, 253–56.
Balestra, Duilio, "La preparazione dell'indagine antropometrica sugli iscritti in una classe di leva in Italia," in Gini, ed., *Atti del Congresso internazionale per gli Studi sulla Popolazione*, vol. 3, 7–34.
Banu, Gheorghe, "Les facteurs dysgéniques en Roumanie: principes d'un programme pratique d'eugénique," in Fédération Internationale Latine des Sociétés d'Eugénique, *Ier Congrès Latin d'Eugénique*, 296–319.
Barbara, Mario, "Leon Battista Alberti precursore di Galton," *Le Opere e i Giorni* 7, no. 11 (November 1928): 86–92.
Barbera, Mario, *Ortogenesi e Biotipologia* (Rome: La Civiltà Cattolica, 1943).
Barduzzi, Domenico, "Sul certificato sanitario prematrimoniale," *Rassegna di studi sessuali* 3, no. 1 (January–February 1923): 44–46.
Barigozzi, Claudio, "I nuovi orizzonti della citogenetica," *Genus* 3, no. 3–4 (June 1939): 35–72.
------------, *La stanza di genetica* (Luino: Francesco Nastro, 1981).
Bellezza, Luigi, "Educazione sessuale ed Eugenetica," in *Atti del Primo Congresso italiano di Eugenetica sociale*, 281–84.
Bellini, Angelo, "Effetti vicini e lontani della blenorragia nell'uomo e nella donna," in *Atti del Primo Congresso italiano di Eugenetica sociale*, 345–54.
Benedetti, Piero, "Contributo alla ricerca dei rapporti tra fecondità e costituzione," in Gini, ed., *Atti del Congresso internazionale per gli Studi sulla Popolazione*, vol. 3, 116.
Benini, Rodolfo, "Le combinazioni simpatiche in demografia," *Rivista italiana di sociologia* 2, no. 2 (March 1898): 152–171.
------------, *Principi di demografia* (Florence: G. Barbera, 1901).
------------, "Sulla rappresentazione in diagramma cartesiano di fenomeni classificati secondo caratteri qualitativi," *Rendiconti della R. Accademia dei Lincei—Classe di scienze morali, storiche e filologiche. Serie V*, 24, no. 12, (December 1915).
------------, "Gruppi chiusi e gruppi aperti in alcuni fatti collettivi di combinazioni," *Bulletin de l'Institut International de Statistique* (Le Caire: Imprimerie Nationale, 1928): 362–83.
Bergamaschi, Carlo, "L'Opera Nazionale Maternità e Infanzia per la difesa sociale," *Razza e civiltà* 1, no. 1 (23 March 1940): 91–98.
Bernstein, Barton J., "Race and Reason: Review," *The Journal of Negro History* 1 (January 1963): 58–60.
Bertolini, Pietro, "Assicurazioni operaie e provvidenze sociali," *Nuova Antologia* 1107–08 (1918): 3–30; 149–76.
Bianchi, Leonardo, "Medicina preventiva e malattie nervose e mentali," *Difesa sociale* 1, no. 6 (June 1922): 3–7.
------------, *Eugenica, igiene mentale e profilassi delle malattie nervose e mentali* (Naples: Idelson, 1925).
------------, "Che vuol dire 'sano di mente'?," *L'igiene mentale* 6, no. 2 (1926): 3.
------------, "Iperpopolazione ed eugenica," *Il pensiero sanitario* 3 (1928): 12–16.
Bilancioni, Guglielmo, "Questione di civiltà," *Il Resto del Carlino* (2 February 1927).
Binding, Karl and Alfred Hoche, *Die Freigabe der Vernichtung lebensunwerten Lebens, ihr Mass und ihre Form* (Leipzig: F. Meiner, 1920).
Biondi, Emerico, "Il parto podalico e sua influenza sulla vita dei bambini," in *Atti del Primo Congresso italiano di Eugenetica sociale*, 202–10.
Boas, Franz, *Changes in the Bodily Form of Descendants of Immigrants* (Washington: Senate Document 208, 1911).

Boas, F. and Helene M. Boas, "The Head Forms of the Italians as Influenced by Heredity and Environment," *American Anthropologist* 15, no. 2 (April–June 1913): 163–88.

Boldrini, Marcello, "Sulle famiglie con pazzi e sulla variabilità del primo nato – ricerche statistiche," *Rivista di Antropologia* 19, no. 1–2 (1914): 411–31.

------------, "I cadaveri degli sconosciuti. Ricerche demografiche e antropologiche sul materiale della Morgue di Roma," *La Scuola Positiva* 1, no. 7–8 (July–August 1920): 323–47.

------------, "Gli studi statistici sul sesso. Le traviate," *Rassegna di studi sessuali* 1, no. 2 (March–April 1921): 69–81.

------------, *Sviluppo corporeo e predisposizioni morbose* (Milan: Vita e Pensiero, 1925).

------------, "Qualità e quantità," *Rassegna di studi sessuali, demografia ed eugenica* 10, no. 4 (October–December 1930): 256–84.

------------, *La fertilità dei biotipi* (Milan: Vita e Pensiero, 1931).

------------, "Biotipi e classi sociali," in Lucio Silla, ed., *Atti della SIPS. XX riunione (Milan, 12–18 September 1931)* (Rome: SIPS, 1932), vol. 1, 63–73.

------------, "Tipi e attitudini costituzionali e sostituzione delle aristocrazie (XII Congresso dell'Istituto Internazionale di Sociologia, Bruxelles 25–29 augusto 1935)," in *Contributi del Laboratorio di Statistica. Serie IV* (Milan: Vita e Pensiero, 1936), offprint.

------------, "Constitution et Eugénique," in Fédération Internationale Latine des Sociétés d'Eugénique, *Ier Congrès Latin d'Eugénique*, 228–31.

------------, "La fertilità degli individui deficienti e difettosi," *Genus* 3, no. 3–4 (June 1939): 279–321.

------------, "Costituzione ed eugenica," in *Contributi del Laboratorio di Statistica. Serie V* (Milan: Vita e Pensiero, 1939), 183–211.

Boldrini M. and Aldo Alberti, "Il patriziato italiano nelle categorie dirigenti," in *Contributi del Laboratorio di Statistica. Serie IV* (Milan: Vita e Pensiero, 1946), 183–230.

Boldrini, M. and Aldo Crosara, "Sull'azione selettiva della guerra tra gli studenti universitari italiani," *Metron* 2, no. 3 (1923), 554–67;

Bonola, Francesco, "Soluzione negativa," *Il Resto del Carlino* (9 February 1927).

Boschi, Alfredo, "Visita e certificato medico prematrimoniale," *Palestra del Clero* 3 (1 March 1952): 193–204 and *Palestra del Clero* 11 (1 June 1952): 489–500.

Bossi, Luigi Maria, "In difesa delle donne belghe e francesi violentate dai soldati tedeschi. Una grave questione d'eugenetica e di giustizia," *La ginecologia moderna* 8, no. 1–4 (January–April 1915): 90–97.

------------, "I pericoli e le vittime della cultura tedesca nel campo ginecologico," *La ginecologia moderna* 8, no. 5–8 (May–August 1915): 145–66.

------------, "Per la difesa della donna e della razza," *La ginecologia moderna* 10 (1917): 128.

Brousseau, Albert, "De la viabilité et de la fécondité des insuffisants intellectuels," in Fédération Internationale Latine des Sociétés d'Eugénique, *Ier Congrès Latin d'Eugénique*, 187–97.

Businco, Lino, "Salute della famiglia, forza della razza," *La Difesa della razza* 2, no. 4 (20 December 1938): 37–39.

------------, "Individuazione e difesa dei caratteri razziali," *La Difesa della razza* 2, no. 10 (20 March 1939): 15–17.

Buzzati-Traverso, Adriano, "I nuovi orizzonti della radiogenetica," *Genus* 3, no. 3–4 (June 1939): 73–130.

------------, "Il pedigree umano," *L'Europeo* 5, no. 41 (9 October 1949).

------------, ed., *Symposium on Genetics of Population Structure. Pavia, Italy, August 20–23, 1953* (Pavia: Tip. succ. Fusi, 1954).

------------, review of Luigi Gedda, ed., *Genetica Medica. Primum Symposium Internationale Geneticae Medicae Roma 6–7 settembre 1953* (Rome: Edizioni dell'Istituto Gregorio Mendel, 1953," Science 122, 3161 (July 1955): 206.

Calderini, Giulio, "Sulla sorte dei feti nati da gravide albuminuriche," in *Atti del Primo Congresso italiano di Eugenetica sociale*, 273–80.

Camavitto, Dino, *La decadenza delle popolazioni messicane al tempo della Conquista* (Rome: Failli, 1935).

------------, "Premiers résultats d'une recherche anthropologique sur les Zambos de la Costa Chica (Guerrero, Mexique)," in Fédération Internationale Latine des Sociétés d'Eugénique, *Ier Congrès Latin d'Eugénique*, 40–60.

Campione, Francesco, *Per i germi della specie* (Bari: Laterza, 1920).

Cantoni, Giuseppe, "Su la consanguineità nelle valli alpestri della Venezia Tridentina," in Gini, ed., *Atti del Congresso internazionale per gli Studi sulla Popolazione*, vol. 2, 309–14.

Capasso, Pietro, *L'assistenza di oggi e l'assistenza di domani* (Naples: Stab. Tipografico Morano, 1920).

------------, "Intorno al certificato prematrimoniale," *Rassegna di studi sessuali* 3, no. 3 (May–June 1923): 187–89.

------------, "Il certificato prematrimoniale," *Rassegna di studi sessuali* 3, no. 6 (November–December 1923): 376–80.

------------, *Pressione demografica, emigrazione ed eugenica* (Naples: Edizioni del Pensiero sanitario, 1926).

------------, "Ettore Levi," *Il pensiero sanitario* 14 (1932): 11.

Carelli, Augusto, "Visita prematrimoniale obbligatoria?," *Difesa sociale* 6, no. 4 (April 1927): 6.

------------, "Valore della sterilizzazione eugenica nel miglioramento della razza umana," *Difesa sociale* 7, no. 10 (October 1928): 341–45.

------------, "A proposito di sterilizzazione eugenica," *Difesa sociale* 7, no. 11 (November 1928): 398.

------------, "Quanti e quali individui dovrebbero essere sottoposti alla sterilizzazione eugenica?," *Difesa sociale* 7, no. 12 (December 1928): 436–40.

------------, review of Charles Wicksteed Armstrong, *The Survival of the Unfittest* (Daniel, London 1927), *Difesa sociale* 8, no. 3 (March 1929): 124–25.

------------, "Il presunto aumento dei deficienti e malati mentali fra le popolazioni," in *Atti del Secondo Congresso italiano di Genetica ed Eugenica*, 97–106.

Carrara, Mario, "Il VII Congresso Internazionale d'Antropologia Criminale," *Archivio di antropologia criminale, psichiatria e medicina legale* 32 (1911): 664.

------------, *Lezioni di medicina legale* (Turin: Litografia A. Viretto, 1913).

------------, "La difesa sociale nel Diritto privato," *Archivio di antropologia criminale, psichiatria e medicina legale* 44 (1924): 1–49.

------------, "Influenze della biologia sulle leggi," *La Parola* (September 1925), offprint.

------------, "Primo congresso di Eugenetica sociale," *Archivio di antropologia criminale, psichiatria e medicina legale* 45 (1925): 72.

------------, review of Ludwig Altmann, *Die Fruchtabtreibung* (Hölder–Pichler–Tempsky, Wien 1926), *Archivio di antropologia criminale, psichiatria e medicina legale* 46 (1926): 731.

------------, "Le leggi dell'eredità in una storica famiglia italiana," *Difesa sociale* 5, no. 4 (April 1926): 6–9.

------------, review of Guarany Sampaio, *A estarilizaçäo eugenica e a deontologia medica* (Irmãos-Ferrez, Sao Paulo 1928), *Archivio di antropologia criminale, psichiatria e medicina legale* 49 (1929): 732.

------------, review of Otto Kankeleit, *Die Unfruchtbarmachung aus rassenhygienischen und sozialen Gründen* (Lehmanns, München 1929), *Archivio di antropologia criminale, psichiatria e medicina legale* 50 (1930): 787.

Carrel, Alexis, *Man the Unknown* (New York: Harper & Brothers, 1935).

Casagrandi, Oddo, "Tentatives microscopiques et biologiques en vue de l'identification de certaines tares organiques séminales, héréditaires et acquises," in Fédération Internationale Latine des Sociétés d'Eugénique, *Ier Congrès Latin d'Eugénique*, 147–49.

Castaldi, Luigi, "Nonni, figli e nipoti. Eredità dell'indice cefalico," *La Difesa della razza* 2, no. 2 (20 November 1938): 11–12.

------------, "Eredità delle attitudini psichiche," *La Difesa della razza* 3, no. 3 (5 December 1939): 26–31.

Castrilli, Vincenzo, "La nuptialité et la fécondité des diplômés de l'enseignement secondaire en Norvège," in Fédération Internationale Latine des Sociétés d'Eugénique, *Ier Congrès Latin d'Eugénique*, 110–19.

Cattaneo, Cesare, "Influenza della vitaminosi ed avitaminosi sul divenire della razza," in *Atti del Primo Congresso italiano di Eugenetica sociale*, 347–50.

Cazzamalli, Ferdinando, "Problemi eugenetici del domani. Guerra e degenerazione etnica," *Quaderni di psichiatria* 3, no. 7–8 (July–August 1916): 165–77.
------------, "Una riforma della Spedalità psichiatrica," *Quaderni di psichiatria* 6, no. 5–6 (May–June 1919).
------------, "La guerra e le malattie nervose e mentali," in Giulio Casalini, ed., *Almanacco igienico popolare* (Rome: L'igiene e la vita, 1920), 197–209.
Centanni, Eugenio, "La eredità dei tumori," in *Atti del Primo Congresso italiano di Eugenetica sociale*, 211–24.
Cerletti, Ugo, "Necessità biologica delle malattie," in *Atti del Primo Congresso italiano di Eugenetica sociale*, 387–90.
Cesari, Luigi, "Una questione di bonifica della razza: per i figli dei tarati neuropsichici," *Razza e civiltà* 1, no. 1 (23 March 1940): 75–82.
------------, "Considerazioni sulle cause del nomadismo," *Razza e civiltà* 2, no. 1 (23 March 1941): 109–22.
Cesarini Sforza, Widar, "Perché approviamo la 'visita prematrimoniale,'" *Il Resto del Carlino* (17 February 1927).
Champy, Christian, "L'importance des variations raciales de sensibilité aux hormones dans l'appréciation de la valeur sexuelle de l'individu," in Fédération Internationale Latine des Sociétés d'Eugénique, *Ier Congrès Latin d'Eugénique*, 150–53.
Chessa, Federico, *La trasmissione ereditaria delle professioni* (Turin: Fratelli Bocca, 1912).
Ciarla, Ernesto, "Per l'istituzione di un servizio provinciale di profilassi delle malattie mentali," *Quaderni di psichiatria* 11, no. 9–10 (September–October 1924): 189–93.
Ciarrusso, Gesualdo, "Risposta affermativa," *Il Resto del Carlino*, (30 January 1927).
Cipriani, Lidio, *Considerazioni sopra il passato e l'avvenire delle popolazioni africane* (Florence: R. Bemporad & F., 1932).
------------, "Unità spirituale degli italiani," *Corriere della Sera* (5 August 1938).
------------, "La razza e la vita delle Nazioni," *Corriere della Sera* (3 December 1938).
------------, "Le scienze antropologiche nella Germania hitleriana," *Gerarchia* (December 1939): 787–791.
CISP-Commissione di demografia storica, *Fonti archivistiche per lo studio dei problemi della popolazione fino al 1848*, 11 vols. (Rome: Tip. Luigi Proja, 1933–1941).
CNPDS, *Prevenzione matrimoniale e introduzione di un certificato prematrimoniale obbligatorio nella legislazione italiana. Relazione della Commissione di studio – art. 7 del progetto di legge del sen. Monaldi* (Milan: CNPDS, 1951).
Colloridi, Franco, "La donna media lombarda come campione antropometrico per le indagini ostetrico-ginecologiche in Lombardia," *Annali di Ostetricia e Ginecologia* (1934).
------------, "Il tipo costituzionale nelle donne portatrici di fibromiomi uterini," *Annali di Ostetricia e Ginecologia* (1934).
Consiglio, Placido, "Studii di Psichiatria Militare; parte I," *Rivista sperimentale di freniatria* 38 (August 1912): 370–407.
------------, "Studii di Psichiatria Militare; parte II," *Rivista sperimentale di freniatria* 39 (December 1913): 792–840.
------------, "La pretesa rieducabilità dei pregiudicati militari in guerra," *Rivista di psicologia* 9, no. 4 (July–August 1913): 333–52.
------------, "Problemi di eugenica," *Rivista italiana di sociologia* 18, no. 3–4 (May–August 1914): 444–66.
------------, "Studii di Psichiatria Militare; parte III," *Rivista sperimentale di freniatria* 40 (December 1914): 881–97.
------------, "Studii di Psichiatria Militare; parte IV," *Rivista sperimentale di freniatria* 41 (March 1915): 35–73.
------------, "Le anomalie del carattere dei militari in guerra," *Rivista sperimentale di freniatria* 42 (October 1916): 131–72.
------------, "Nuovi studi sulle anomalie del carattere dei militari in guerra," *Rivista sperimentale di freniatria* 42 (December 1917): 529–44.

------------, "Come difenderci dagli anormali e dai degenerati nell'ambiente militare," *Difesa sociale* 2, no. 10 (October 1923): 5–10.

Corberi, Giuseppe, "L'ereditarietà nella epilessia," in *Atti del Primo Congresso italiano di Eugenetica sociale,* 431–34.

Costa, Romolo, "Opportunità della reazione novocaino-formalinica prima del matrimonio," in *Atti del Primo Congresso italiano di Eugenetica sociale,* 295–96.

Costanzo, Alessandro, "Costituzione e mortalità," in *Contributi del Laboratorio di Statistica. Serie III* (Milan: Vita e Pensiero, 1934), 403–30.

------------, *Costituzione e mortalità* (Milan: Vita e Pensiero, 1935).

Croci, P., "Angosciosi problemi della guerra. L'innocente," *Corriere della Sera,* (10 March 1915).

D'Addario, Raffaele, "L'azione selettiva della guerra in un gruppo di studenti universitari italiani," *Archivio scientifico del R. Istituto Superiore di Scienze economiche e commerciali di Bari* (1926–27 and 1927–28).

Dal Prato, Antonio, "Basta la pratica igienica," *Il Resto del Carlino,* (6 February 1927).

D'Anna Botta, Raffaele, "Meticciato," *La Difesa della razza* 5, no. 12 (20 April 1942): 21–22.

------------, "La pseudo-scienza ebraico-massonica contro il genio italiano," *Il Tevere* (30–31 March 1939): 3.

D'Arenzano, Bonaventura, "La visita prematrimoniale," *Orientamenti pastorali* 3 (March 1960): 44–46.

Darwin, Leonard, "Eugenics and the Criminal," in *Atti del Primo Congresso italiano di Eugenetica sociale,* 151–58.

Davenport, Charles B., "Sono utili gli incroci di razza?," in *Atti del Secondo Congresso italiano di Genetica ed Eugenica,* 59–60.

Davenport, C. B. and Morris Steggerda, *Race Crossing in Jamaica* (Washington: Carnegie Institution, 1929).

------------, "Mulatti di Giamaica," *La Difesa della razza* 3, no. 10 (20 March 1940): 18–24.

Davicini, Giovanni, *Lex-Legislazione italiana* 42, July–December (Turin: UTET, 1956): 1254–59.

De Agazio, Vincenzo, "Attrazione matrimoniale," *La Difesa della razza* 2, no. 20 (20 August 1939): 9–10.

De Berardinis, Luigi, "La profilassi anticeltica nell'esercito," in *Atti del Primo Congresso italiano di Eugenetica sociale,* 333–40.

De Blasio, Ada, "Frenastenie e meticciato," *La Difesa della razza* 6, no. 1 (5 November 1942): 17.

De Giovanni, Giuseppe and Mario Mazzeo, *L'eugenica* (Naples: Pelosi, 1924).

De Lisi, Lionello, "Profilassi delle malattie ereditarie in Psichiatria," in Lucio Silla, ed., *Atti della SIPS. XXVII riunione (Bologna, 4–11 Settembre 1938)* (Rome: SIPS, 1939), vol. 6, 124–38

Del Monte, Genesio Eugenio [with the pseudonim "Eudemon"], "Il fenomeno degli incroci nel pensiero di Corrado Gini," *L'Azione Coloniale* (25 February 1937).

------------, [with the pseudonim "Eudemon"], "Il fenomeno degli incroci," in *L'Azione Coloniale,* (4 March 1937).

De Molinari, Gustave, *La viriculture* (Paris: Guillaumin et Cie, 1897).

De Napoli, Ferdinando, "Lue, maternità, eugenica e guerra in rapporto alla Politica Sanitaria," *Il Policlinico-Sezione pratica* 45 (1919): 1322–27.

------------, "Visita prematrimoniale," *Rassegna di studi sessuali* 3, no. 1 (January–February 1923): 46–54.

------------, "Difendiamo la stirpe," *Il Resto del Carlino* (26 January 1927).

------------, *Da Malthus a Mussolini. La guerra che noi preferiamo* (Bologna: Cappelli, 1934).

De Sanctis, Sante, "L'organizzazione scientifica del lavoro mentale," *Rivista italiana di sociologia* 20, no. 5–6 (September–October 1916): 496–530.

------------, "Visita prematrimoniale obbligatoria?," *Difesa sociale* 6, no. 3 (March 1927): 2

------------, *Igiene Mentale* (Turin: Paravia, n. d.), 6–7.

------------, "I problemi di rieducazione," *Gerarchia* 12 (December 1928): 960–71.

Dettori, Giovanni, "Di alcuni caratteri dei neonati secondo l'ordine di generazione e l'età della madre," *Rivista di Antropologia* 19, no. 1–2 (1914): 443–572.

Devoto, Luigi, "La famiglia del lavoratore del piombo," in *Atti del Primo Congresso italiano di Eugenetica sociale,* 409–10.

De Zuani, Ettore, "Problemi razziali nell'America Latina," *La Difesa della razza* 2, no. 18 (20 July 1939): 11–13.
La Direzione, "Il lavoro degli anormali psichici e la Guerra," *Quaderni di psichiatria* 4, no. 3–4 (March–April 1917): 79–80.
-----------, "Per la psichiatria nel dopo-guerra," *Quaderni di psichiatria* 6, no. 3–4 (March–April 1919): 98.
-----------, "Per la psichiatria del dopo-guerra. Proposta di un Congresso Alienistico pel Dopo-guerra," *Quaderni di psichiatria* 6, no. 5–6 (May–June 1919): 144.
-----------, "I nuovi indirizzi della assistenza neuro-psichiatrica," *Quaderni di psichiatria* 14, no. 5–6 (May–June 1927): 108.
Donaggio, Arturo, "La visita medica prematrimoniale obbligatoria," *Difesa sociale* 6, no. 4 (April 1927): 4.
Dossena, Gaetano, "Il peso dei feti nati da madri tubercolose," in *Atti del Primo Congresso italiano di Eugenetica sociale*, 365–66.
Dunn, Leslie C., "Cross Currents in the History of Human Genetics," *The Eugenics Review* 2 (July 1962): 74.
Enriques, Paolo, "Eugenica e diritto," *Studi sassaresi* 1 (1921), offprint.
-----------, *L'eredità nell'uomo* (Milan: Vallardi, 1924).
Ermolli, Luciano, "Un problema di Eugenetica operaia," in *Atti del Primo Congresso italiano di Eugenetica sociale*, 411–14
Evola, Julius, "Razza e nascita, ovvero: gli isterici trascendentali," *Il Regime Fascista*, 14 (15 March 1939): 3.
-----------, "Problemi della razza. Lo spirito e gli epilettici," *Il Corriere Padano*, 18 (27 December 1940).
-----------, *Indirizzi per una educazione razziale* (Naples: Conte, 1941).
-----------, *Sintesi di dottrina della razza* (Milan: Hoepli, 1941).
-----------, *Il mito del sangue* (1937; repr., Milan: Hoepli, 1942).
-----------, "La razza, l'ideale classico e gli 'spostati spirituali,'" *Roma Fascista* 19, no. 9 (22 January 1942): 3.
Fanfani, Amintore, "I mutamenti economici nell'Europa moderna e l'evoluzione costituzionalistica delle classi dirigenti," in *Contributi del Laboratorio di Statistica. Serie IV* (Milan: Vita e Pensiero, 1936): 137–56.
Federici, Nora, "Mortalité infantile et mortalité prénatale chez les familles nombreuses italiennes," in Fédération Internationale Latine des Sociétés d'Eugénique, *Ier Congrès Latin d'Eugénique*, 278–82.
-----------, "La curva di sviluppo individuale presso alcune popolazioni isolate," *Genus* 3, no. 3–4 (June 1939): 323–44.
Ferrari, Giulio Cesare, "Il prossimo avvenire dell'Assistenza psichiatrica in Italia," *Quaderni di psichiatria* 10, no. 5–6 (May–June 1923): 112–15.
-----------, "La lega italiana per l'igiene mentale," *Difesa sociale* 3, no. 6 (June 1924): 4–6.
Ferri, Enrico, "Visita prematrimoniale obbligatoria?," *Difesa sociale* 6, no. 4 (April 1927): 1–2.
Ferrio, Carlo, "Nota conclusiva sull'Assistenza Psichiatrica non coattiva in Italia," *L'igiene mentale* 1 (1936): 101.
Ficai, Giuseppe, "S.O.S. degli antirazzisti," *La Difesa della razza* 2, no. 11 (5 April 1939): 38–39.
Fioretti, Arnaldo, "Lieo bifronte," *Razza e civiltà* 1, no. 8 (October 1940): 565–92.
Fischer, Eugen, "Die gegenseitige Stellung der Menschenrassen auf Grund der mendelden Merkmale," in Gini, ed., *Atti del Congresso internazionale per gli Studi sulla Popolazione*, vol. 3, 179–88
-----------, "La realtà della razza," *La Difesa della razza* 3, no. 1 (5 November 1939): 11–17.
-----------, "I bastardi di Reoboth," *La Difesa della razza* 3, no. 10 (20 March 1940): 12–17.
Foà, Carlo, "L'eredità dei caratteri normali e patologici. 1," *Gerarchia* 9 (1925): 609–13.
-----------, "L'eredità dei caratteri normali e patologici. 2," *Gerarchia* 10 (1925): 677–82.
-----------, "L'eredità dei caratteri normali e patologici. 3," *Gerarchia* 11 (1925): 745–50.
-----------, "Conseguenze sociali dell'eredità biologica," *Gerarchia* 12 (1925): 815–19.
-----------, "Eugenetica e diritto," *Gerarchia* 1 (1926): 58–61.
-----------, "Opere e leggi di medicina sociale," *Gerarchia* 2 (1927): 148–54.

------------, "I fattori biologici della diminuzione delle nascite," in *Atti del Secondo Congresso italiano di Genetica ed Eugenica*, 173–94.

------------, I fattori biologici della diminuzione delle nascite, in Gini, ed., *Atti del Congresso internazionale per gli Studi sulla Popolazione*, vol. 2, 9–56.

------------, "Eugenica e matrimonio italiano," *Politica sociale* 4 (1932): 191–200.

------------, "Discorso pronunciato nella cerimonia inaugurale dell'Istituto G. Mendel il 6 settembre 1953," in Gedda, ed., *Genetica Medica. Primum Symposium Internationale Geneticae Medicae Roma 6–7 settembre 1953* (Rome: Edizioni dell'Istituto Gregorio Mendel, 1953), 446–49.

Foà, C. and Nicola Pende, *La fisiologia e la clinica degli increti* (Milan: Istituto Biochimico Italiano, 1927).

Fortunati, Paolo, "Le métabolisme social d'après des recherches sur les étudiants de l'Université de Padoue," in Fédération Internationale Latine des Sociétés d'Eugénique, *Ier Congrès Latin d'Eugénique*, 79–90.

Francioni, Carlo, "Le anomalie costituzionali e diatesiche dell'età infantile in rapporto coll'Eugenetica," in *Atti del Primo Congresso italiano di Eugenetica sociale*, 87–110.

Franzì, Leone, "Il meticciato. Insidia contro la salute morale e fisica dei popoli," *La Difesa della razza* 1, no. 4 (20 September 1938): 29–31.

Gabbi, Umberto, "Sentimento e necessità," *Il Resto del Carlino* (28 January1927).

------------, "La battaglia per la natalità," *Archivio fascista di medicina politica* 2 (1928): 267–68.

Galli, Giovanni, "L'Eugenetica di fronte all'ereditarietà delle malattie cardio-vascolari," in *Atti del Primo Congresso italiano di Eugenetica sociale*, 307–10.

Galton, Francis, *Inquiries into Human Faculty and its Development* (London: Macmillan, 1883).

------------, *Natural Inheritance* (London: Macmillan, 1889).

Garofalo, Raffaele, *Criminologia. Studio sul delitto, sulle sue cause e sui mezzi di repressione* (Turin: Bocca, 1885).

Gasteiner, Elio, "Un pericolo per la razza. La decadenza dei ceti superiori," *La Difesa della razza* 1, no. 2 (20 August 1938): 26–27.

Gayre, Robert, "L. C. Dunn on Luigi Gedda, Angelo Serio, Adriana Mercuri, R. Ruggles Gates and 'The Mankind Quarterly,'" *The Mankind Quarterly* 3, no. 1 (July–September 1962): 48–50.

Gedda, Luigi, "A Proposito di Razza," *Vita e Pensiero* 29, no. 9 (September 1938): 408–16.

------------, "Genetica, medicina e costituzione," *Acta geneticae medicae et gemellologiae* 1, no. 1 (January 1952): 1–6.

------------, "Profilo scientifico della genetica medica," in *Genetica Medica. Primum Symposium Internationale Geneticae Medicae*, 3–18.

------------, *I problemi della popolazione* (Rome: Staderini, 1955).

------------, "Un maestro e un esempio," *Acta geneticae medicae et gemellologiae* 5, no. 3 (July 1956): 244.

------------, "L'importanza della genetica nella selezione degli sportivi," in Tommaso Lucherini and Claudio Cervini, eds., *Medicina dello sport* (Rome: Società Editrice Universo, 1960), 73–91.

------------, "Prof. R. Ruggles Gates (in memoriam)," *Acta geneticae medicae et gemellologiae* 12, no. 1 (January 1963), 95.

------------, *Problemi di frontiera della medicina* (Turin: Borla, 1963).

------------, "Eugenetica e profilassi mentale," in *Sanità mentale ed assistenza psichiatrica. Atti del II Congresso italiano di Medicina forense* (Rome: Homo, 1962), 77–84.

------------, "A Study of Racial and Subracial Crossing," in Robert Kuttner, ed., *Race and Modern Science* (New York: Social Sciences Press, 1967), 123–40.

Gedda, L., Giuseppe Del Porto and Adriana Del Porto-Mercuri, "Sindrome di Werdnig-Hoffmann familiare che include una coppia di gemelli MZ concordanti (un caso di Consulenza Eugenica)," *Acta geneticae medicae et gemellologiae* 11, no. 2 (April 1962): 113–21.

Gedda, L., Marco Milani-Comparetti and Gianni Brenci, *Rapporto scientifico sugli atleti della XVII Olimpiade. Roma 1960* (Rome: Istituto di Medicina dello Sport, 1968).

Gedda, L., Angelo Serio and Adriana Mercuri, *Il Meticciato di Guerra e Altri Casi* (Rome: Edizioni dell'Istituto Gregorio Mendel, 1960).

Gemelli, Agostino, "Si ereditano le qualità psichiche?," *Vita e Pensiero* 1, no. 3 (1915): 273–83.

------------, "Eugenica e guerra," *Vita e Pensiero* 4, no. 3 (September 1916): 133–45.
------------, "Religione ed eugenetica," in *Atti del Primo Congresso italiano di Eugenetica sociale*, 53–66.
------------, "Le dottrine eugenetiche sul matrimonio e la morale cattolica," *Vita e Pensiero* 22, no. 3–4 (March–April 1931): 195–99.
------------, "Ancora della condanna della eugenetica. Echi e critiche alla enciclica 'Casti Connubii' sul matrimonio cristiano," *Vita e Pensiero* 22 (October 1931): 603–14.
------------, "La 'sterilizzazione coattiva e preventiva' nell'insegnamento degli studiosi italiani," *L'Economia Italiana* 11–12 (December 1933): 117–28.
------------, "Le vedute della psicologia e della psichiatria nel problema della natalità," in Gini, ed., *Atti del Congresso internazionale per gli Studi sulla Popolazione*, vol. 2, 343–46.
Genna, Giuseppe, "Correlazione fra i caratteri morfologici degli sposi," in Gini, ed., *Atti del Congresso internazionale per gli Studi sulla Popolazione*, vol. 4, 795–804.
------------, *I Samaritani – 1. Antropologia* (Rome: CISP, 1938).
------------, "Razza e sangue," *Razza e Civiltà* 1, no. 5–7 (July–September 1940): 455–78.
Ghigi, Alessandro, "Fecondità e sterilità nell'ibridismo e nella consanguineità," in *Atti del Secondo Congresso italiano di Genetica ed Eugenica*, 161–72.
------------, "Costituzione e fertilità," in Gini, ed., *Atti del Congresso internazionale per gli Studi sulla Popolazione*, vol. 3, 63–76.
Gianferrari, Luisa, "Effetti demografici e genetici della consanguineità," in Gini, ed., *Atti del Congresso internazionale per gli Studi sulla Popolazione*, vol. 2, 295–308.
------------, "Importanza, urgenza di ricerche genetiche in popolazioni endogame," *Atti e Memorie della Società Lombarda di Medicina* 5, no. 8 (1937): 581–84.
------------, "Il contributo dell'Università al Centro di studi di genetica umana," *Gli Annali della Università d'Italia* 3, no. 1 (29 October 1941): 24–28.
------------, "Sull'organizzazione e sull'attività svolta dal Centro di studi di genetica umana nel primo quadriennio dalla sua fondazione," *Natura* 35 (1944): 112–16.
------------, "Proposte per l'inquadramento della prevenzione eugenica prematrimoniale nell'organizzazione sanitaria italiana," *La settimana medica* 37, no. 21 (1949): 4–5.
------------, "Il Centro di Studi di Genetica umana dell'Università di Milano ed i Consultori di genetica umana dell'Università e del Comune di Milano," *Natura* 41 (1950): 75–81.
------------, "Piano per un'organizzazione eugenetica in Italia," *L'economia umana* 2 (1952): 5–7.
------------, "Introduzione alla profilassi delle malattie ereditarie," *Acta geneticae medicae et gemellologiae* 1, no. 2 (May 1952): 113–17.
------------, "Genetica e matrimonio," *Riflessi* 11, no. 1 (March 1959): 1–11.
Gianferrari, L. and Giuseppe Cantoni, *Manuale di Genetica con particolare riguardo all'Eredità nell'Uomo* (Milan: Vallardi, 1942).
Gianferrari, L. and Giuseppe Morganti, "Appunti per una organizzazione eugenica in Italia," *Acta geneticae medicae et gemellologiae* 1, no. 2 (May 1952): 212–14.
Gini, Corrado, *Il sesso dal punto di vista statistico. Le leggi della produzione dei sessi* (Milan–Palermo–Naples: Remo Sandron, 1908).
------------, "Il diverso accrescimento delle classi sociali e la concentrazione della ricchezza," *Giornale degli Economisti* 2, no. 37 (January 1909): 3–59.
------------, *Indici di concentrazione e di dipendenza* (Turin: UTET, 1911).
------------, *Variabilità e mutabilità. Contributo alla studio delle distribuzioni e delle relazioni statistiche* (Bologna: Tipografia Cuppini, 1912).
------------, *I fattori demografici dell'evoluzione delle nazioni* (Turin: Bocca, 1912).
------------, "Contributi statistici ai problemi dell'Eugenica," *Rivista italiana di Sociologia* 16, no. 3 (May–August 1912): 317–426.
------------, "L'uomo medio," *Giornale degli Economisti e Rivista di Statistica* 48, no. 1 (January 1914): 1–24.
------------, "Nuove osservazioni sui problemi dell'eugenica. La distribuzione dei professori delle Università secondo l'ordine di nascita," *Rivista italiana di Sociologia* 18, no. 2 (March–April 1914): 212–15.

------------, "Genetica e statistica rispetto all'eugenica," *Rivista italiana di Sociologia* 19, no. 2 (March–April 1915): 218–22.
------------, "La coscrizione militare dal punto di vista eugenico," *Metron* 1, no. 1 (1920): 83–112.
------------, *Problemi sociologici della guerra* (Bologna: Zanichelli, 1921).
------------, *The War from the Eugenic Point of View*, in *Scientific Papers of the Second International Congress of Eugenics*, vol. 2., *Eugenics in Race and State*, 1923: 430–31.
------------, "La guerra dal punto di vista dell'eugenica," in Roberto Almagià, ed., *Atti della SIPS. XI riunione (Trieste 9–13 settembre 1921)* (Rome: SIPS, 1922), 44–75.
------------, "On Birth Control," *Difesa sociale* 4, no. 3–4 (March–April 1925): 83–87.
------------, "Il neo-malthusianismo," *Difesa sociale* 1, no. 8 (August 1922): 170–73.
------------, "Le relazioni dell'Eugenica con le altre scienze biologiche e sociali," in *Atti del Primo Congresso italiano di Eugenetica sociale*, 3–26.
------------, "Prime ricerche sulla fecondabilità della donna," *Atti del Regio Istituto Veneto di Scienze, Lettere ed Arti* 83, part 2 (1924): 315–44.
------------, "Nuove ricerche sulla fecondabilità della donna," *Atti del Regio Istituto Veneto di Scienze, Lettere ed Arti* 84, part 2 (1925): 269–308.
------------, "Decline in the Birth-Rate and the Fecundability of Woman," *Eugenics Review* 17 (January 1926): 258–74.
------------, "Il numero come forza," *Critica Fascista* 6, no. 19 (1928): 363.
------------, "Le Comité Italien pour l'étude des problèmes de la population," *Bulletin de l'Institut International de Statistique* 23, no. 1 (1928): 204–06.
------------, "Problemi della popolazione," *Annali Istituto di Statistica dell'Università di Bari* (Bari: Tip. Cressati, 1928): 6–23.
------------, "The Cyclical Rise and Fall of Population," in Corrado Gini, Shiroshi Nasu, Robert R. Kuczynski, and Oliver E. Baker, *Population* (Chicago: Harris Foundation Lectures, The University of Chicago Press, 1929), 1–140.
------------, "Discorso di inaugurazione dell'Istituto Centrale di Statistica (14 luglio 1926)," *Annali di Statistica. Serie VI*, 2 (1929), 18–19.
------------, "The Italian Demographic Problem and the Fascist Policy on Population," *The Journal of Political Economy* 38, no. 6 (1930): 682–97.
------------, *Le basi scientifiche della popolazione* (Catania: Studio editoriale moderno, 1931).
------------, "Discorso d'apertura," in *Atti del Secondo Congresso italiano di Genetica ed Eugenica*, 17–26.
------------, "Prime indagini sulle famiglie numerose," in *Atti del Secondo Congresso italiano di Genetica ed Eugenica*, 289–338.
------------, "Nuovi risultati delle indagini sulle famiglie numerose," *Atti Istituto Nazionale Assicurazioni* 4 (1932): 7–46.
------------, "Alcuni risultati preliminari dell'indagine antropometrica sui soldati italiani," in Gini, ed., *Atti del Congresso internazionale per gli Studi sulla Popolazione*, vol. 3, 98.
------------, *Enquête démographique sur les familles nombreuses italiennes. Résultats des recherches* (Paris: Gembloux Imprimerie – J. Duculot Éditeur, 1933).
------------, "Response to the Presidential Address," in *A decade of progress in eugenics*, 25–26.
------------, "Remarks on the Explanation of Heterosis," in *A decade of progress in eugenics*, 421–24.
------------, "Gli effetti eugenici o disgenici della guerra," *Genus* 1–2 (1934): 29–42.
------------, "Researches on Population," *Scientia* 55, no. 265 (May 1934): 357–73.
------------, "Parole inaugurali del Prof. C. Gini, lette alla riunione delle Società di Eugenica dell'America Latina tenutasi a Città del Messico il 12 ottobre 1935," *Genus* 2, no. 1–2 (June 1936): 78.
------------, "Le rilevazioni statistiche fra le popolazioni primitive," *Supplemento statistico ai Nuovi problemi di politica, storia ed economia* 3, no. 1–2 (1937).
------------, "Une question importante pour la science des constitutions et pour la médecine militaire: comment juger si les proportions d'un individu sont normales?," *Revue de l'Institut International de Statistique* 5, no. 2 (July 1937): 107–14; no. 3 (October 1937): 203–11.

------------, "De quelques recherches sur les variations que présenteraient certains caractères suivant le nombre d'enfants de la famille," in Fédération Internationale Latine des Sociétés d'Eugénique, *Ier Congrès Latin d'Eugénique*, 262–69.
------------, "Prolificità e frequenza dei parti plurimi," *Genus* 3, no. 3–4 (June 1939): 279–96.
------------, *Le rilevazioni statistiche fra le popolazioni primitive* (Rome: Manuali Universitari – Facoltà di Scienze statistiche, demografiche ed attuariali, 1940).
------------, "I 'tradimenti' dei primitivi," *Genus* 5, no. 1–2 (December 1941): 1–28.
------------, "Caratteristiche e cause della primitività," *Genus* 5, no. 3–4 (1942): 1–44.
Gini, C. and Federici, N., *Appunti sulle spedizioni scientifiche del Comitato Italiano per lo studio dei problemi della popolazione (febbraio 1933 – aprile 1940)* (Rome: Tip. Operaia Roma, 1943).
------------, "L'assimilazione fisica degli immigrati," *Genus* 9, no. 1–4 (1950–52): 14–19.
------------, "The Physical Assimilation of the Descendants of Immigrants," in Tage Kemp, Mogens Hauge and Bent Harvald, eds., *Proceedings of the First International Congress of Human Genetics*, 2, (Nasel, NY: S. Karger, 1958), 400–03.
------------, "Alla soglia dell'umanità," *Rivista di Politica Economica* 64, no. 11 (November 1964): 1475–1505.
------------, review of *Statement on Race*, by Ashley Montagu, *Genus* 10, no. 1–4 (1953–1954): 192–94.
------------, "Possono e devono i caratteri psichici e culturali essere tenuti presenti nella classificazione delle razze umane?," *Genus* 11, no. 1–4 (1955): 71–77.
------------, "Sulle differenze innate tra i caratteri mentali delle varie popolazioni," review of *The Testing of Negro Intelligence*, by Audrey M. Shuey, *Genus* 16, no. 1–4 (1960):161–66.
------------, "Eterosi nei Meticci di Guerra?," review of *Il Meticciato di Guerra e Altri Casi*, by Gedda, Serio and Mercuri, *Genus* 16, no. 1–4 (1960): 167–68.
------------, "The Testing of Negro Intelligence," *The Mankind Quarterly* 1, no. 2 (October–December 1960): 120–25.
------------, "Psychic and Cultural Traits and the Classification of Human Races," *The Mankind Quarterly* 1, no. 4, (April–June 1961): 236–41.
Giorni, Secondo, *Come si prepara la classe del 1916. Il Neo-Malthusianismo e la guerra tra le nazioni* (1916; repr., Florence: Soc. Ed. Neomalthusiana, 1920).
Giuffrida-Ruggeri, Vincenzo, "Il problema fondamentale dell'eugenica," in *Atti della Società Italiana di Genetica ed Eugenica*, 31–36.
Graziani, Felice, "I meticci nella storia," *La Difesa della razza* 5, no. 15 (5 June 1942): 16–17.
Grebe, Hans, "The African Pygmies," *Acta geneticae medicae et gemellologiae* 7, no. 2 (April 1958): 159–218.
------------, "The Genetics of the Australian Aborigines," *Acta geneticae medicae et gemellologiae* 9, no. 1 (January 1960): 7–50.
------------, "Studies in Race Crossing. Crosses of Australians and Papuans with Caucasians, Chinese, and the Other Races," *Acta geneticae medicae et gemellologiae* 9, no. 2 (April 1960): 165–84.
------------, "The Melanesian Dwarf Tribe of Aiome," *Acta geneticae medicae et gemellologiae* 10, no. 3 (July 1961): 277–311.
------------, "Erbpathologie des Skelettsystems," in *Genetica Medica*, ed. L. Gedda, 188–222.
------------, "Genetik und morphologische Variation," in *Proceedings of the Second International Congress of Human Genetics*, 355–68.
------------, "Diskordanzursachen bei erbgleichen Zwillingen," *Acta geneticae medicae et gemellologiae* 1, no. 1 (January 1952): 103–07.
------------, "Über besondere Zwillingskonkordanzen," *Acta geneticae medicae et gemellologiae* 5, no. 2 (May 1956): 138–54.
------------, "Familienbefunde bei letalen Herzmissbildungen," *Acta geneticae medicae et gemellologiae* (supplementum primum): 257–93.
------------, "Sportfamilien," *Acta geneticae medicae et gemellologiae* 5, no. 3 (September 1956): 418–26.
------------, "Zwergwuchs als genetisches Problem," *Acta geneticae medicae et gemellologiae* 6, no. 4 (October 1957): 429–36.

------------, "Biemond-Syndrom in einer Sippe mit Iriskolobomen, Hüftgelenksdysplasie und Epilepsie," *Acta geneticae medicae et gemellologiae* 9, no. 2 (April 1960): 197–210.
Gregor, Anthony James, "The Logic of Race Classification," *Genus* 14, no. 1–4 (1958): 150–61.
------------, "Sociology and the Anthropobiological Sciences," *Mémoire du XIXe Congrès International de Sociologie – Communications* (Mexico: Comité Organisateur, 1960), vol. 2, 83–107.
------------, "Corrado Gini and the Theory of Race Formation," *Sociology and Social Research* 45 (January 1961): 175–81.
------------, review of Corrado Gini, "Corso di Sociologia," *The Mankind Quarterly* 2, no. 1 (April–June 1961): 298–300.
------------, "The Biosocial Nature of Prejudice," *Genus* 18, no. 1–4 (1962): 116–28
------------, review of Vittorio Castellano, "Studi in Onore di Corrado Gini," *Sociology and Social Research* 46 (July 1962): 501.
------------, "Corrado Gini, the Organismic Analogy and Sociological Explanation," *Sociological Quarterly* 8 (Spring 1967): 165–72.
Gregor, A. J. and Michele Marotta, "Sociology in Italy," *Sociological Quarterly* 2 (July 1961): 215–21.
Gregor, A. J. and Angus D. McPherson, "Sociology and Mental Testing of Non-Industrial Peoples," in *La Sociología y las Sociedades en Desarrollo Industrial: Communications before the XXth International Congress of Sociology* (Córdoba: Universidad de Córdoba, 1963), 185–205.
------------, "Sociology and the Assimilation of Non-Industrial Peoples," in *La Sociologia y la Sociedades en Desarrollo Industrial*, 2.
Gregoraci, Pier Nicola, "La mia nuova Dottrina," *La Vita Italiana* 23, no. 263 (February 1935): 149–55.
Gualco, Sellina and Antonio Nardi, *L'Istituto Biotipologico Ortogenetico di Roma* (Rome: Stab. Tip. Luigi Proja, 1941).
Haldane, John B. S., "La selezione naturale nell'uomo: Discussione," in *Atti del convegno dedicato a "I recenti contributi della genetica umana alla medicina,"* 62–70.
Harris, James A. and Albert Govaerts, "Note on Assortative Mating in Man with Respect to Head Size and Head Form," *American Naturalist* 56, no. 645 (July 1922): 381–83.
Hauschild, Rita, "Gli incroci negro-cinesi," *La Difesa della razza* 3, no. 10 (20 March 1940): 51–53.
Heuyer, Georges, "Constitution et Eugénique," in Fédération Internationale Latine des Sociétés d'Eugénique, *Ier Congrès Latin d'Eugénique*, 232–38.
Hirsch, Max, *Chi debbo sposare? Consigli di un medico* (Rome: Leonardo da Vinci, 1923).
Ilvento, Arcangelo, "Visita medica prematrimoniale?," *Difesa sociale* 6, no. 3 (March 1927): 3–4.
Interlandi, Telesio, "Cattolici sugli specchi," *Il Tevere* (23–24 July 1938).
------------, "Zone di dissidentismo," *Il Tevere* (23–24 April 1938).
Istituto Italiano di Medicina Sociale, *La consultazione prematrimoniale (Roma, 24 gennaio 1969)* (Rome: Tip. Loffari, 1969).
Jona, Raffaele, "Considerazioni cliniche e profilattiche sui rapporti fra tubercolosi ed Eugenetica," in *Atti del Primo Congresso italiano di Eugenetica sociale*, 311–18.
Jucci, Carlo and T. Amendola, "L'indice facciale come elemento di attrazione matrimoniale," in Lucio Silla, ed., *Atti della SIPS. XXI riunione (Roma, 9–15 Ottobre 1932)* (Rome: SIPS, 1933), vol. 3, 316–19.
Kaznelson, Reuben, *L'immigrazione degli Ebrei in Palestina nei tempi moderni* (Rome: Failli, 1931).
Kemp, Tage, "Malattie e difetti ereditari," in *Atti del convegno dedicato a "I recenti contributi della genetica umana alla medicina,"* 7–17.
Krogman, Wilton Marion, *L'antropologia fisica degli Indiani Seminole dell'Oklahoma* (Rome: Failli, 1936).
Kuttner, Robert E., "Cultural Selection of Human Psychological Types," *Genus* 16, no. 1–4 (1960): 1–4.
------------, "Eugenic Aspects of Preventive Therapy for Mental Retardation," *Genus* 19, no. 1–4 (1963): 1–9.
------------, ed., *Race and Modern Science* (New York: Social Sciences Press, 1967).
L. S. [Landra Silvio], "Ambiente razza e attitudini professionali," *La Difesa della razza* 4, no. 5 (5 January 1941): 13–17.

Lami, Guido, "Significati e moniti di un Congresso," *Studium* 31, no. 6 (June 1935): 362–65.
------------, "Il Congresso Internazionale dei Medici Cattolici a Vienna e il prossimo Congresso-Pellegrinaggio a Roma," *Studium* 32, no. 11 (November 1936): 628–31.
Landra, Guido, "I bastardi," *La Difesa della razza* 1, no. 1 (5 August 1938): 16–17.
------------, "L'ambiente non snatura la razza," *La Difesa della razza* 2, no. 3 (5 December 1938): 16–18.
------------, "L'ologenesi del Rosa," *La Difesa della razza* 2, no. 10 (20 March 1939): 11–14.
------------, "La razza italiana nella teoria dell'ologenesi," *La Difesa della razza* 2, no. 11 (5 April 1939): 9–11.
------------, "Gli studi razziali nell'Europa balcanica," *La Difesa della razza* 2, no. 16 (20 June 1939): 32–34.
------------, "Gli studi razziali in Polonia e in Russia," *La Difesa della razza* 2, no. 18 (20 July 1939): 14–17.
------------, "Studiosi americani di problemi razziali," *La Difesa della razza* 2, no. 20 (20 August 1939): 13–16.
------------, "Razza e nazionalità in Romania," *La Difesa della razza* 2, no. 21 (5 September 1939): 10–13.
------------, "Studi razziali in continenti extraeuropei," *La Difesa della razza* 2, no. 23 (5 October 1939): 34–37.
------------, "Gli studi razziali in Ungheria e in Bulgaria," *La Difesa della razza* 3, no. 3 (5 December 1939): 32–33.
------------, "L'eredità dei caratteri razziali," *La Vita Italiana* 28, no. 322 (January 1940): 29–31.
------------, "Studi italiani sul meticciato," *La Difesa della razza* 3, no. 10 (20 March 1940): 8–10.
------------, "L'eredità delle qualità psicologiche," *La Vita Italiana* 28, no. 324 (March 1940): 286–90.
------------, "Studio razziale delle impronte digitali," *La Difesa della razza* 3, no. 12 (20 April 1940): 40–41.
------------, "Studi razziali in Transilvania," *La Difesa della razza* 3, no. 14 (20 May 1940): 16–19.
------------, "Studi sulle mescolanze etniche della popolazione," *La Difesa della razza* 3, no. 15 (5 June 1940): 12–13.
------------, "Ereditarietà e ambiente," *La Vita Italiana* 28, no. 327 (June 1940): 648–51.
------------, "Gli studi di patologia ereditaria in Germania," *La Difesa della razza* 3, no. 17 (5 July 1940): 18–22.
------------, "Il metodo dei gemelli," *La Difesa della razza* 3, no. 18 (20 July 1940): 28–31.
------------, "I fattori ereditari dell'accrescimento," *La Difesa della razza* 3, no. 19 (5 August 1940): 36–39.
------------, "Sistematica antica e moderna delle razze umane," *La Difesa della razza* 3, no. 19 (5 August 1940): 23–28.
------------, "I metodi per lo studio delle razze umane," *La Difesa della razza* 3, no. 20 (20 August 1940): 29–35.
------------, "La forma del viso nelle razze umane," *La Difesa della razza* 3, no. 21–22 (5–20 September 1940): 51–54.
------------, "Le variazioni del sistema tegumentario nelle razze umane," *La Difesa della razza* 3, no. 23 (5 October 1940): 11–16.
------------, "Il problema dei meticci in Europa," *La Difesa della razza* 4, no. 1 (5 November 1940): 11–15.
------------, "Ricerche moderne sui gruppi sanguigni," *La Difesa della razza* 4, no. 2 (20 November 1940): 34–37.
------------, "Studi sull'aumento della statura in Scandinavia," *La Difesa della razza* 4, no. 5 (5 January 1941): 10–12.
------------, "La classificazione delle razze umane secondo von Eickstedt," *La Difesa della razza* 4, no. 6 (20 January 1941): 12–15.
------------, "Caratteri fisionomici identità razziale," *La Difesa della razza* 4, no. 9 (5 March 1941): 18–20.

------------, "Antropologia – Problemi di metodo per la definizione dei tipi razziali," *La Difesa della razza* 4, no. 10 (20 March 1941): 22–25.
------------, "Le razze dell'Asia meridionale e orientale," *La Difesa della razza* 4, no. 11 (5 April 1941): 18–20.
------------, "Antropologia – Forme esterne del corpo umano, variazioni nel sesso e nell'età," *La Difesa della razza* 4, no. 12 (20 April 1941): 18–20.
------------, "Antropologia – Ricerche e dottrine craniologiche," *La Difesa della razza* 4, no. 14 (20 May 1941): 26–29.
------------, "Le razze europee e il problema delle aristocrazie," *La Difesa della razza*, 4, no. 13 (5 May 1941): 12–15.
------------, "Lo scheletro facciale nelle razze umane," *La Difesa della razza* 4, no. 15 (5 June 1941): 24–26.
------------, "Il certificato prematrimoniale," *La Difesa della razza* 4, no. 16 (20 June 1941): 24–26.
------------, "Poligenismo e monogenismo," *La Difesa della razza* 3, no. 21 (5 September 1941): 27–29.
------------, "Antropologia – Ricerche craniologiche," *La Difesa della razza* 5, no. 3 (5 December 1941): 24–26.
------------, "Antropologia – Studi razziali sulle differenze razziali della faccia," *La Difesa della razza* 5, no. 5 (5 January 1942): 22–23.
------------, "Antropologia – Morfologia facciale," *La Difesa della razza* 5, no. 6 (20 January 1942): 28–29.
------------, "Fondamenti biologici del razzismo," *La Difesa della razza* 5, no. 10 (20 March 1942): 4–7.
------------, "Il problema degli incroci a Trinidad e nel Venezuela," *La Difesa della razza* 6, no. 2 (20 November 1942): 14–16.
Landucci, Francesco, "Sul nuovo regolamento riguardante l'assistenza degli esposti," in *Atti del Primo Congresso italiano di Eugenetica sociale*, 415–18.
La Torre, Felice, "I fondamenti dell'eugenica," *Rivista italiana di Sociologia* 19, no. 2 (March–April 1915): 196–218.
Lattes, Leone, "Dalla teoria alla pratica," *Il Resto del Carlino* (6 February 1927).
L'Eltore, Giovanni, "Contributo allo studio degli effetti selettivi della guerra dal punto di vista dell'eugenica," *Genesis* 11, no. 1–2 (January–June 1932): 49–62.
Leoncini, Francesco, "Relazione su la procurata sterilità di fronte alla morale e alla legge," *Studium-Quaderno dei Medici. Il II Convegno dei medici cattolici (Firenze, 16–18 ottobre 1932)*, suppl. no. 3 (March 1933): 38–64.
Lesné, Edmond-Alexandre, "Influence des régimes carencés et déséquilibrés, suralimentation et sous-alimentation, sur la natalité et la mortalité des petits rats," in Fédération Internationale Latine des Sociétés d'Eugénique, *Ier Congrès Latin d'Eugénique*, 144–46.
Letard, Étienne, "Les leçons de l'expérimentation animale dans le problème du métissage," in Fédération Internationale Latine des Sociétés d'Eugénique, *Ier Congrès Latin d'Eugénique*, 61–71.
Levi, Ettore, *La medicina sociale in difesa della vita e del lavoro* (Rome: La Voce, 1921).
------------, *I partiti e la salute della stirpe* (Rome: IPAS, 1921).
------------, *Central International Committee for the Coordination of the International Federations of Preventive Medicine and Social Relief* (Rome: IPAS, 1924).
------------, "Per l'avvenire della razza," *Difesa sociale* 1, no. 1 (January 1922): 7–10.
------------, "Alle radici dei mali sociali: il fascismo alla prova," *Difesa sociale* 2, no. 1 (January 1923): 1–4.
[------------], "Contenuto etico e sociale dell'Eugenica," *Difesa sociale* 4, no. 11 (November 1925): 13–15.
[------------], "La fecondità dei deficienti come problema di Eugenica," *Difesa sociale* 5, no. 1 (January 1926): 15.
------------, "Risposta al professore A. Zuccarelli, in tema di eugenica," *Il pensiero sanitario* 19 (1922): 3–4.
------------, "Il controllo delle nascite (neomalthusianismo)," *Rassegna di studi sessuali* 4, no. 1 (January–February 1924): 24–30.

------------, "Demografia ed eugenica in rapporto al movimento contemporaneo per il razionale controllo delle nascite," in Roberto Almagià, ed., *Atti della SIPS. XIV riunione (Pavia, 24–29 maggio 1925)* (Rome: SIPS, 1926), 99–122.

------------, "Le finalità eugeniche del controllo delle nascite," in *Atti del Primo Congresso italiano di Eugenetica sociale*, 257–72.

Levi della Vida, Giuseppina, "Le métabolisme social comme facteur de dégénération dans la société," in Fédération Internationale Latine des Sociétés d'Eugénique, *Ier Congrès Latin d'Eugénique*, 120–31.

Liggeri, Don Paolo, "A proposito di consultori prematrimoniali," *Riflessi* 2, no. 2 (1950): 6.

Livi, Livio, "Emigrazione ed Eugenetica," in *Atti del Primo Congresso italiano di Eugenetica sociale*, 37–52.

Lombroso, Cesare, *Genio e follia in rapporto alla medicina legale, alla critica e alla storia* (Turin: Bocca, 1882).

------------, *Troppo presto. Appunti al nuovo progetto di codice penale con appendici*, (1888; repr., Turin: Bocca, 1889).

Longhi, Silvio, "Le donne violentate in guerra e lo 'stato di necessità,'" *La Scuola Positiva* 6, no. 6 (June 1915): 481–87.

Loria, Achille, "L'antropologia sociale," in Achille Loria, ed., *Verso la giustizia sociale—(Idee, battaglie, apostoli)* (Milan: Società Editrice Libraria, 1908), 562–73.

------------, "The Psycho-physical Elite and the Economic Elite," in *Problems in Eugenics*, 179–83.

------------, "I confluenti economici dell'eugenismo," in *Atti della Società Italiana di Genetica ed Eugenica*, 37–40.

Lucchetti, Giuseppe, "Le difficoltà del certificato," *Il Resto del Carlino* (6 February 1927).

Lucidi, Giuseppe, "Il sangue. Individualità biologica di razza," *La Difesa della razza* 1, no. 5 (5 October 1938): 37–38.

------------, "Gruppi sanguigni e nuclei razziali. Necessità di un censimento del sangue," *La Difesa della razza* 2, no. 5 (5 January 1939): 14–15.

------------, "Rapporti fra gruppi sanguigni e caratteri antropologici," *La Difesa della razza* 2, no. 7 (5 February 1939): 8–10.

------------, "Costituzione e natalità," *La Difesa della razza* 2, no. 9 (5 March 1939): 13–14.

------------, "Il meticciato, morte degli imperi," *La Difesa della razza* 2, no. 13 (5 May 1939): 16–18.

Maffi, Attilio, "L'educazione fisica delle masse altissimo fattore di Eugenetica sociale," in *Atti del Primo Congresso italiano di Eugenetica sociale*, 355–62.

Maggi, Raffaello, "La costituzione degli attori dello schermo," in *Contributi del Laboratorio di Statistica. Serie IV* (Milan: Vita e Pensiero, 1936): 79–136.

Maggiore, Luigi, "L'assistenza dello Stato agli invalidi, storpi e mutilati," *Politica sociale* 4 (1932): 477–81.

Malcovati, Piero, "Discussione," in *Atti del convegno dedicato a "I recenti contributi della genetica umana alla medicina,"* 69.

Mantegazza, Paolo, *L'anno Tremila – Sogno* (2nd ed.) (Milan: Treves, 1897).

------------, *Un giorno a Madera. Una pagina dell'igiene dell'amore* (Florence: Salani, 1910).

Marchesini, Giovanni, "Il fattore psicologico nel dominio dell'eugenica," in *Atti della Società Italiana di Genetica ed Eugenica*, 21–29.

Marchiori, Giovanni, "Propaganda eugenica o misure coercitive?," *La Difesa della razza* 3, no. 18 (20 July 1940): 17–23.

Marchitto, Nicola, "Il meticciato e la Francia," *La Difesa della razza* 2, no. 10 (20 March 1939): 38–40.

Marcozzi, Aldo, "Voci diverse," *Riflessi* 12, no. 3 (September 1960): 71.

Maroi, Lanfranco, "Alcoolismo ed Eugenetica," in *Atti del Primo Congresso italiano di Eugenetica sociale*, 121–38.

Marro, Giovanni, "Il primato della razza italiana," *La Stampa* (30 July 1938).

------------, "La razza italiana e l'ambiente," *La Stampa* (5 August 1938).

------------, "Il problema delle origini della razza italiana studiato attraverso il materiale raccolto nel Museo di Antropologia," *La Stampa* (12 August 1938).

------------, "La razza italiana e il suo linguaggio," *La Stampa* (24 August 1938).
------------, "Egiziani, Fenici, Ebrei nella civiltà mediterranea," *La Stampa* (17 August 1939; 30 August 1939 and 22 October 1939).
------------, "La razza italiana e il suo ambiente naturale," *La Stampa* (23 February 1940).
------------, *Primato della razza italiana: Confronti di morfologia, biologia, antropologia e di civiltà* (Milan–Messina: Giuseppe Principato, 1940).
------------, "Un allarme per il razzismo italiano," *La Vita Italiana*, 29, no. 336 (March 1941): 237–51.
------------, "Nuovi orientamenti nella scienza razziale," *La Vita Italiana*, 29, no. 341 (August 1941): 135–44.
------------, *Caratteri fisici e spirituali della razza italiana* (Rome: Istituto Nazionale di Cultura Fascista, 1939).
------------, "Dell'armonia fra razza ed ambiente naturale in Italia," *Razza e civiltà* 1, no. 2 (April 1940): 165–82.
------------, "La razza e l'ambiente nella civiltà. I," *Razza e civiltà* 2, no. 2 (April 1941): 224–26.
------------, "La razza e l'ambiente nella civiltà. II," *Razza e civiltà* 2, no. 3 (May–July 1941): 438–41.
Marta, Felice, *Neomalthusianesimo medico. Quando e come non bisogna aver figli* (1915; repr., Milan: Società Anonima Editoriale, 1920).
------------, "Eugenetica e neo-malthusianismo," in *Atti del Primo Congresso italiano di Eugenetica sociale*, 455–58.
Martial, René, "Métissage et immigration," in Fédération Internationale Latine des Sociétés d'Eugénique, *Ier Congrès Latin d'Eugénique*, 16–39.
Massarotti, Vito, "La profilassi del suicidio in rapporto all'Eugenica," in *Atti del Primo Congresso italiano di Eugenetica sociale*, 435–38.
Matarrese, Fortunato, "Leon Battista Alberti, studioso di problemi razziali," *La Difesa della razza* 3, no. 9 (5 March 1940): 37–41.
------------, "Demografia ed eugenica di Tommaso Campanella," *La Difesa della razza* 3, no. 15 (5 June 1940): 40–41;
Medea, Eugenio, "L'igiene mentale e la scuola," *L'igiene mentale* 13, no. 3 (1933): 10–12.
------------, "Le malattie nervose e mentali in rapporto all'Eugenetica," in *Atti del Primo Congresso italiano di Eugenetica sociale*, 139–44.
Medugno, Domenico, "L'azione dello Stato e l'Eugenetica," in *Atti del Primo Congresso italiano di Eugenetica sociale*, 145–50.
Mendicini, Antonio, "I centri neurologici nella mostra nazionale delle opere d'assistenza nell'Esercito," *Quaderni di psichiatria* 5, no. 9–10 (September–October 1918): 229–34.
Mengarelli, Carlo, "I caratteri costituzionali delle aristocrazie italiane," in *Contributi del Laboratorio di Statistica. Serie IV* (Milan: Vita e Pensiero, 1936): 157–82.
------------, "Su i caratteri fisici della nobiltà," in *Contributi del Laboratorio di Statistica. Serie IV* (Milan: Vita e Pensiero, 1936): 239–72.
Messina, Salvatore, "Le donne violentate in guerra e il diritto all'aborto," *La Scuola Positiva* 6, no. 4 (April 1915): 289–94.
Michels, Roberto, "Eugenics in Party Organisation," in *Problems in Eugenics*, 232–37.
------------, *Problemi di sociologia applicata* (Turin: Bocca, 1919).
------------, "Sulla teoria e sulla pratica dell'Eugenica," *Echi e Commenti* 3, no. 27 (1922): 14.
------------, "Taluni effetti dell'emigrazione nei suoi rapporti coll'Eugenica," in *Atti del Primo Congresso italiano di Eugenetica sociale*, 199–201.
------------, "Nei primordi della scienza eugenetica. Le utopie di Tommaso Campanella," *Rivista internazionale di filosofia del diritto* 10, no. 25 (1930) : 8–9, offprint.
A. M. [Mieli, Aldo], "Il certificato sanitario prematrimoniale," *Rassegna di studi sessuali* 2, no. 6 (November–December 1922): 357–58.
Mieli, Aldo, "Proposte pratiche," *Il Resto del Carlino* (9 February 1927).
Mingazzini, Giovanni, "Sul certificato prematrimoniale," *Difesa sociale* 6, no. 3 (March 1927): 2–3.
Mino, Prospero, "Sulle malattie ereditarie e sulla loro etiologia (continuazione e fine)," *Archivio di antropologia criminale, psichiatria e medicina legale* 43 (1923): 1–37.

Minkowska, Franziska, "Eugénique et Généalogie," in Fédération Internationale Latine des Sociétés d'Eugénique, *Ier Congrès Latin d'Eugénique*, 341–50.

Mjøen, Jon Alfred, "Delinquenza e genio alla luce della biologia," in *Atti del Primo Congresso italiano di Eugenetica sociale*, 169–78.

------------, "Biologische und biochemische Untersuchungen bei Rassenmischung," in Gini, ed., *Atti del Congresso internazionale per gli Studi sulla Popolazione*, vol. 3, 199–202.

Mjøen, J. A. and Jon Bø, "The Norwegian System for Identification and Protection of the Individual," *Atti del Primo Congresso italiano di Eugenetica sociale*, 179–84.

Modena Camporini, Emma, "Eugenetica ed istruzione igienico-sessuale della donna," in *Atti del Primo Congresso italiano di Eugenetica sociale*, 363–64.

Modena, Gustavo, "L'organizzazione dei Centri neurologici in Francia," *Rivista sperimentale di freniatria* 43 (August 1917): 344–55.

Modica, Aldo, "Il certificato prematrimoniale," *La Difesa della razza* 5, no. 4 (20 December 1941): 28–30.

Modica, E., "Gedda bocciato," *L'Unità* (14 April 1954).

Montalenti, Giuseppe, "Utopie," *Rivista di psicologia* 35 (1939): 197–99.

------------, "I recenti studi sul problema della determinazione del sesso e dei caratteri sessuali secondari negli animali," *Genus* 3, no. 3–4 (June 1939): 193–214.

------------, "Genetica umana ed eugenica," in *Atti del convegno dedicato a "I recenti contributi della genetica umana alla medicina,"* 1–6.

------------, "L'VIII Congresso internazionale di Genetica (Stoccolma, 7–14 luglio 1948)," *La Ricerca Scientifica*, 19 (1949): 130–31.

Montandon, Georges, "Vita e opere di Vacher de Lapouge," *La Difesa della razza* 4, no. 12 (20 April 1941): 24–26.

------------, "La formazione delle razze umane," *La Difesa della razza* 4, no. 22 (20 September 1941): 9–12.

Montesano, Giuseppe, "Il certificato prematrimoniale," *Difesa Sociale* 6, no. 4 (April 1927): 3.

Montesano, Vincenzo, "Il certificato prematrimoniale e la profilassi sociale della sifilide," *Rassegna di studi sessuali* 2, no. 6 (November–December 1922): 358–61.

------------, "A proposito di certificato matrimoniale e di abolizionismo," *Rassegna di studi sessuali* 3, no. 2(March–April 1923): 119–22.

------------, "Risposta negativa," *Il Resto del Carlino* (2 February 1927).

Morselli, Arturo, "Psichiatria di guerra," *Quaderni di psichiatria* 3, no. 3–4 (March–April 1916): 65–72.

------------, "La neuropsichiatria castrense in Francia," *Quaderni di psichiatria* 3, no. 5–6 (May–June 1916): 131.

Morselli, Enrico, "La lotta per l'etnarchia," *Nuova Antologia* 151, no. 938 (1911): 231–46.

------------, *Antropologia generale. L'uomo secondo le teorie dell'evoluzione* (Turin: Un. Tip. Ed., 1911).

------------, "Progresso sociale ed evoluzione," *Rivista italiana di Sociologia* 15, no. 5 (September–October 1911): 526ff.

------------, "La psicologia etnica e la scienza eugenistica," *Rivista di psicologia* 8, no. 4 (July–August 1912): 289–93.

------------, "Problemi di psicopatologia applicata. È socialmente utile l'educazione dei frenastenici?," *Quaderni di Psichiatria* 2, no. 5, (May 1915): 223–31.

------------, "L'eugenica e le previsioni sull'eredità neuro-psicopatologica," *Quaderni di Psichiatria* 2, no. 7–8 (July–August 1915): 321–32.

------------, "La rivendicazione delle leggi di Morel," *Quaderni di Psichiatria* 3, no. 11–12, (November–December 1916): 272–78.

------------, "La funzione sociale del Manicomio," *Quaderni di psichiatria* 7, no. 5–6 (May–June 1920): 133–38.

------------, *L'uccisione pietosa (l'eutanasia) in rapporto alla medicina, alla morale e all'eugenica* (Turin: Bocca, 1923).

Muggia, Giuseppe, "Per l'avvenire della Psichiatria e dell'assistenza psichiatrica," *Quaderni di psichiatria* 9, no. 9–10 (September–October 1922): 189–94.

Mukerjee, Radhakamal, *Le migrazioni asiatiche* (Rome: CISP, 1936).
Niceforo, Alfredo, *Antropologia delle classi povere* (Milan: Vallardi, 1910).
------------, "The cause of the inferiority of physical and mental characters in the lower social classes," in *Problems in Eugenics,* 189–94.
Nicolaeff, Léon, "Les corrélations entre les caractères morphologiques des époux," *L'Anthropologie* 41, no. 1–2 (1931): 75–93.
Nix, Willi, "Eredità e destino," *La Difesa della razza* 3, no. 13 (5 May 1940): 14–16.
"Notizie. Problemi di eugenica e profilassi in un colloquio dell'on. Capasso con S.E. Mussolini," *Rassegna di studi sessuali* 3, no. 6 (November–December 1923): 438.
Nullo, Paolo, "Il razzismo nella "Città del Sole" di Tommaso Campanella," *La Difesa della razza* 4, no. 14 (20 May 1941): 13–15.
Nuzzo, Fiorella, "La diagnosi prenatale," *Sapere,* 746 (March 1972): 4–11.
Ottolenghi, Salvatore, "I rimedi legali sono insufficienti," *Il Resto del Carlino* (6 February 1927).
------------, "Sterilizzazione del delinquente in rapporto alla medicina legale," *Policlinico-Sezione Pratica* 43 (1933): 171.
Ottolenghi, S. and Mario Carrara, "Perioptometria e psicometria di uomini geniali," *Archivio di psichiatria, scienze penali ed antropologia criminale* 13 (1892): 381–408.
P. P., "La visita prematrimoniale," *Studi cattolici* 10 (January 1959): 61–63.
Pagani Cesa, Andrea, "Dati statistici sull'influenza dell'ambiente famigliare come fattore di contagio tubercolare," in *Atti del Primo Congresso italiano di Eugenetica sociale,* 293–94.
Palmieri, Vincenzo, *Denatalità. La grande insidia sociale vista da un medico* (Milan: Società Palmeritana Editrice Medica, 1935).
Pareto, Vilfredo, *Cours d'économie politique,* 1–2. (Lausanne: F. Rouge Lausanne, 1896–97) [ed. used, Turin: Bollati Boringhieri, 1961].
------------, *Les systèmes socialistes,* 1–2. (Paris: Giard & Brière, 1901–02) [ed. used, Turin: UTET, 1974].
------------, *Manuale di economia politica.* (Milan: Società Editrice Libraria, 1906) [ed. used, Milan: EGEA–Università Bocconi, 2006].
------------, "La curva delle entrate e le osservazioni del prof. Edgeworth," *Giornale degli Economisti* 13, no. 10 (1896): 439–48.
------------, "L'uomo delinquente di Cesare Lombroso e Polemica col Prof. Lombroso," in Giovanni Busino, ed., *Écrits sociologiques mineurs* (Geneva: Droz, 980), 111–25.
------------, *Epistolario. 1890–1923* (Rome: Accademia Nazionale dei Lincei, 1973).
Pasini, Agostino, "La sifilide latente nei suoi rapporti con l'Eugenica," in *Atti del Primo Congresso italiano di Eugenetica sociale,* 325–32.
Pastori, Giuseppina, "Sulla frequenza dell'eredolues nei fanciulli anormali," in *Atti del Primo Congresso italiano di Eugenetica sociale,* 425–30.
------------, "La relazione su l'eugenica e la morale cattolica," *Studium-Quaderno dei Medici. Il II Convegno dei medici cattolici (Firenze, 16–18 ottobre 1932),* suppl. no. 3 (March 1933): 65–73.
Patellani, Serafino, "Gregorio Mendel e l'opera sua," *Il Morgagni* 56 (1914), 148–54; 161–76; 201–33.
------------, "Eugenetica e guerra," *La ginecologia moderna* 8, no. 5–8 (May–August 1915): 225.
------------, *Prolegomeni di eugenetica sociale* (Milan: Cogliati, 1925).
Pearson, Karl, *The Grammar of Science* (London: A. and Ch. Black, 1900).
Pearson, K. and Alice Lee, "On the Laws of Inheritance in Man: I. Inheritance of Physical Characters," *Biometrika* 2, no. 3 (1903): 357–462.
------------, "Assortative Mating in Man: A Cooperative Study," *Biometrika* 2, no. 4 (1903), 481–98.
Pellacani, Giuseppe, "Basta la visita prematrimoniale?," *Il Resto del Carlino* (30 January 1927).
------------, "Psichiatria e psicoigiene," *L'igiene mentale* 15, no. 1 (1935): 8–10.
------------, "I servizi di profilassi neuro-mentale in Italia," *L'igiene mentale* 14, no. 1–2 (1934): 15–29.
Pende, Nicola, *Dalla medicina alla sociologia* (Palermo: Prometeo, 1921).
------------, *L'indirizzo costituzionalistico nella medicina sociale e nella politica biologica* (Genova: Le Opere e i Giorni, 1926).

------------, "Sul certificato prematrimoniale: obbligo legale od obbligo morale?," *Difesa sociale* 6, no. 3 (March 1927): 10.
------------, *Anomalie della crescenza fisica e psichica* (Bologna: Cappelli, 1929).
------------, *Bonifica umana razionale e biologia politica* (Bologna: Cappelli, 1933).
------------, "Costituzione e fecondità," in Gini, ed., *Atti del Congresso internazionale per gli Studi sulla Popolazione*, vol. 3, 86.
------------, "La scheda biotipologica individuale nella medicina preventiva e nella politica sociale," in Lucio Silla, ed., *Atti della SIPS. XXVI riunione (Venezia, 12–18 settembre 1937)* (Rome: SIPS. 1938), vol. 5, 283–86.
------------, *La scuola fascista preparatrice dell'uomo totale ed orientatrice del cittadino produttivo.* Discourse of Senator Pende in the sitting of 25 March 1938 (Rome: Tip. del Senato, 1938), offprint.
------------, *Concetto e prassi della razza nella mentalità fascista* (discourse at the Cremona section of the Institute of Fascist Culture, 15 October 1938) (Cremona: Tip. Cremona Nuova, n. d.), offprint.
------------, *Trattato di biotipologia umana individuale e sociale, con applicazioni alla medicina preventiva, alla clinica, alla politica biologica, alla sociologia* (Milan: Vallardi, 1939).
------------, "La profilassi delle malattie e anomalie ereditarie," in Lucio Silla, ed., *Atti della SIPS. XXVII riunione (Bologna, 4–11 Settembre1938)* (Rome: SIPS, 1939), 63–73.
------------, "La scienza dell'ortogenesi. Principi e finalità," *La ricerca scientifica* 10, no. 4 (April 1939): 1–6, offprint.
------------, *La scienza dell'ortogenesi* (Rome: CNR, 1939).
------------, "Il principio biotipologico unitario," *Gerarchia* 11 (November 1940): 569–72.
------------, *Corpo e anima* (Rome: SAET, 1947).
------------, *Il medico di fronte al Vangelo* (Milan: Il Giorno, 1948).
------------, *Medicina e sacerdozio alleati per la bonifica morale della società* (Ancona: Tip. Flamini, n. d.).
Pensabene, Giuseppe, "L'evoluzione e la razza. Cinquant'anni di polemiche ne 'La Civiltà Cattolica,'" *La Difesa della razza* 1, no. 2 (20 August 1938): 31–33.
------------, "Il meticciato delitto contro Dio," *La Difesa della razza* 4, no. 8 (20 February 1941): 26–27.
------------, "Le due cause maggiori del meticciato nel mondo," *La Difesa della razza* 4, no. 10 (20 March 1941):10–12.
Perico, Giacomo, "Aspetti medico-sociali della 'visita prematrimoniale,'" *La Civiltà Cattolica* 2983 (October 1974): 57–61.
------------, "Visita e certificato prematrimoniali," *Aggiornamenti sociali* 12, no. 1 (January 1961): 1–18.
------------, "Visita e certificato prematrimoniali (continuazione)," *Aggiornamenti sociali* 12, no. 2 (February 1961): 65–84.
Pestalozza, Camillo, "La natimortalità nei diversi periodi della vita italiana e milanese," in *Atti del Primo Congresso italiano di Eugenetica sociale*, 191–98; 251–52.
Pestalozza, Ernesto, "Visita prematrimoniale obbligatoria?," *Difesa sociale* 6, no. 4 (April 1927): 4–5.
------------, "Le indicazioni operatorie in rapporto all'Eugenica," in *Atti del Primo Congresso italiano di Eugenetica sociale*, 81–87.
------------, "Sterilizzazioni coattive," in *Atti del Secondo Congresso italiano di Genetica ed Eugenica*, 79–87.
Petrò, Francesco, "Un reparto psichiatrico avanzato d'Ospedale da campo nel suo primo anno di funzionamento," *Quaderni di psichiatria* 4, no. 3–4 (March–April 1917): 71–78.
Piccinini, Prassitele, "Le fonti d'Italia," in *Atti del Primo Congresso italiano di Eugenetica sociale*, 419–22.
Pieraccini, Gaetano, *La difesa della società dalle malattie trasmissibili* (Turin: Bocca, 1895).
------------, *La stirpe dei Medici di Cafaggiolo. Saggio di ricerche sulla trasmissione ereditaria dei caratteri biologici* (Florence: Vallecchi, 1924).
Pietravalle, Michele, "Per un Ministero della Sanità ed Assistenza Pubblica in Italia," *Nuova Antologia* 1131 (1919): 103–17.
Pighini, Giacomo, "Per la eliminazione dei degenerati psichici dall'esercito combattente," *Giornale di medicina militare* 1 (1918): 978–96.

Pintus, Giuseppe, "Metodologia delle ricerche di Genetica in Psichiatria," *Atti della Lega italiana di igiene e profilassi mentale* (1938): 51–105.
Pius XII, *Discorsi ai medici* (Rome: Orizzonte Medico, 1959).
Polastri, A., "I piccoli tedeschi," *Giornale di Sicilia* (21–22 February 1915).
Pope, Ernest G, Karl Pearson and Ethel M. Elderton, *A Second Study of the Statistics of Pulmonary Tuberculosis: Marital Infection* (London: Dulau and Co., 1908).
Porteus, Stanley D., "Race Crossing in Hawaii," in Gini, ed., *Atti del Congresso internazionale per gli Studi sulla Popolazione*, vol. 3, 203–12.
Preziosi, Giovanni, "Parlo di pier Nicola Gregoraci," *La Vita Italiana* 23, no. 263 (February 1935):147–48.
------------, "Fatti e commenti. Il caso Gregoraci," *La Vita Italiana* 23, no. 265 (April 1935): 515–18.
------------, "Il caso Gregoraci," *La Vita Italiana* 23, no. 267 (June 1935): 788–92.
------------, "Fatti e commenti. Scienza nuova?!...," *La Vita Italiana* 23, no. 268 (July 1935): 99–100.
------------, "Fatti e commenti. Probità scientifica," *La Vita Italiana* 23, no. 268 (July 1935): 100–01.
------------, "Per la serietà degli studi razziali in Italia (dedicato al camerata Giacomo Acerbo)," *La Vita Italiana* 28, no. 328 (July 1940): 74–75.
Putnam, Carleton, *Race and Reason: A Yankee View* (Washington: Public Affairs Press, 1961).
Raineri, Roberto, "Il problema razziale brasiliano," *La Difesa della razza* 3, no. 7 (5 February 1940): 39–42.
Ratto, Lorenzo, "La sterilizzazone coattiva in Germania," *Avvenire Sanitario* 50 (1934): 1.
Reed, Sheldon C., *Consulenza in Genetica medica* (Rome: Edizioni dell'Istituto Gregorio Mendel, 1959), 12–13.
Ricci, Marcello, "Le leggi di Mendel," *La Difesa della razza* 1, no. 2 (20 August 1938): 16–17.
------------, "Il mendelismo nell'uomo," *La Difesa della razza* 1, no. 3 (5 September, 1938): 18–19.
------------, "Ereditarietà ed eugenica," *La Difesa della razza* 1, no. 5 (5 October 1938): 29–31.
------------, "Eugenica e razzismo," *La Difesa della razza* 2, no. 6 (20 January 1939): 22–23.
Richet, Charles, *La sélection humaine* (Paris: Alcan, 1919).
Righetti, R., "Le basi scientifiche del movimento eugenico," *Difesa sociale* 4, no. 12 (December 1925): 10–14.
Rigobello, Guido, "L'ereditarietà nella tubercolosi," in *Atti del Primo Congresso italiano di Eugenetica sociale*, 319–24.
Riva, Emilio, "Il Centro psichiatrico militare di I raccolta," *Rivista sperimentale di freniatria* 45 (May 1919): 308–24.
------------, "Un anno di servizio presso il centro Psichiatrico Militare della Zona di guerra," *Rivista sperimentale di freniatria* 45 (May 1919): 443–59.
Rocchi, Lorenzo, "Razzismo nel Nord-America," *La Difesa della razza* 3, no. 8 (20 February 1940): 29–33.
Rosiński, R.P. Bolesław, "Charakterystyka antropologiczna ludności pow. Pułtuskiego," *Kosmos* 48 (1923).
------------, "Antropogenetische Auslese," *Antropologischer Anzeiger* 6, no. 1 (1929): 49–64.
Rossi-Doria, Tullio, *L'eredità delle malattie* (Milan: Vallardi, 1893).
------------, *Medicina sociale e socialismo. Scritti per l'educazione politica e igienica dei lavoratori* (Rome: Mongini, 1904).
------------, "Congresso di ostetricia sociale," *Il Policlinico – Sezione pratica* 3 (19 January 1919): 79–82.
------------, "Congresso di ostetricia sociale," *Il Policlinico – Sezione pratica* 4 (26 January 1919): 113–15.
Roussy, Gustave and Huguenin, René, "Vues sur le rôle de l'hérédité dans le cancer humain," in Fédération Internationale Latine des Sociétés d'Eugénique, *Ier Congrès Latin d'Eugénique*, 171–86.
Ruggles Gates, Reginald, "Il Meticciato di Guerra," *The Mankind Quarterly* 1, no. 2 (October 1960): 218.

------------, "Records of Y-inherited Hairy Ears in India," *Acta geneticae medicae et gemellologiae* 6, no. 1 (January 1957): 103–08.

------------, "Inheritance of Racial and Sub-racial Traits," in *Proceedings of the Second International Congress of Human Genetics*, 1, 369–70.

Sacchetti, Alfredo, "Sport ed evoluzione dei popoli," *Razza e civiltà* 1, no. 2 (April 1940): 229–38.

Saffiotti, Umberto, "Eugenica e anormali," *L'infanzia anormale* 5–6 (1920): 62–82.

Savorgnan, Franco, *La guerra e la popolazione* (Bologna: Zanichelli, 1917).

------------, *La scelta matrimoniale. Studi statistici* (Ferrara: Casa Editrice Taddei, 1924).

------------, "I problemi della razza e l'opportunità di un'inchiesta antropometrica sulla popolazione italiana," *La Difesa della razza* 1, no. 1 (5 August 1938): 18.

Schaeuble, Johann, "Il meticciato nell'America del Sud," *La Difesa della razza* 3, no. 10 (20 March 1940): 46–49.

Schreiber, Georges, "Allocations familiales et Eugénique," in Fédération Internationale Latine des Sociétés d'Eugénique, *Ier Congrès Latin d'Eugénique*, 91–100.

Semizzi, Renato, "Eugenica e terapia razziale," *Critica medico-sociale* no. 7–8–9 (July–September 1940): 34–39.

------------, "La medicina sociale attraverso i tempi e le idee," *La Difesa della razza* 4, no. 11 (5 April 1941): 21–26.

------------, "La medicina della masse," *La Difesa della razza* 4, no. 9 (5 March 1941): 13–15.

------------, "L'influenza della civiltà sui popoli," *La Difesa della razza* 4, no. 14 (20 May 1941): 10–12.

Sergi, Giuseppe, *Le degenerazioni umane* (Milan: Fratelli Dumolard, 1889).

------------, *Problemi di scienza contemporanea*. (Milan–Palermo–Naples: Remo Sandron Editore, 1904).

------------, "Francis Galton," *Rivista di Antropologia* 41, no. 1 (1911): 179–81.

------------, "Variation and Heredity in Man," in *Problems in Eugenics*, 16–22.

------------, "L'eugenica. Dalla biologia alla sociologia," *Rivista italiana di Sociologia* 18, no. 5–6 (September–December 1914): 605–33.

------------, "L'eugenica e la guerra," *Nuova Antologia* 51, no. 1064 (1916): 129–39.

------------, "L'eugenica e la decadenza delle nazioni," in Vincenzo Reina, ed., *Atti della SIPS. VIII riunione (Roma, 1–6 marzo 1916)* (Rome: SIPS, 1917), 181–99.

------------, "La guerra e la preservazione della nostra stirpe," *Nuova Antologia* 52, no. 1099 (1917): 8–18.

Sergi, Sergio, "Antropologia di Stato. L'archivio comunale delle famiglie," *Razza e civiltà* 1, no. 2 (April 1940): 183–89.

Serono, Cesare, "L'importanza del fattore femminile nella conservazione della razza," *Razza e civiltà* 2, no. 8–12 (October 1941–February 1942): 673–84.

Shapiro, Harry L., "Race Mixture Studies in Polynesia," in Gini, ed., *Atti del Congresso internazionale per gli Studi sulla Popolazione*, vol. 3, 213–20.

Silvestri, Giulio, "Per un archivio genealogico nazionale," *La Difesa della razza* 4, no. 9 (5 March 1941): 24–27.

Skerlj, Bozo, "Correspondence. 'The Mankind Quarterly,'" *Man* 60 (November 1960): 172.

Sonnabend, Enrico Haskel, *L'espansione degli Slavi* (Rome: Failli, 1931).

------------, *Il fattore demografico nell'organizzazione sociale dei Bantu* (Rome: Arti Grafiche Zamperini e Lorenzini, 1935).

Stoppato, Alessandro, "I vantaggi e i danni," *Il Resto del Carlino* (28 January 1927).

Swan, Donald, "Genetics and Psychology," *Genus* 20, no. 1–4 (1964): 23–35.

Tallarico, Giuseppe, "Il numero è anche la qualità," *Razza e civiltà* 1, no. 3–4 (May–June 1940): 283–92.

------------, "I pregi biologici della fecondità," *Razza e civiltà* 1, no. 5–6–7 (July–September 1940): 479–88.

------------, "L'alimento e la prolificità umana" *Razza e civiltà* 2, no. 1 (23 March 1941): 81–92.

------------, "Il problema degli incroci," *Razza e civiltà* 3, no. 1 (January–March 1943).

Tao, Yun Kuei, "Incroci fra cinesi ed europee," *La Difesa della razza* 3, no. 10 (20 March 1940): 33–38.

Tauro, Giacomo, "La transmigration des classes sociales par l'éducation," in Fédération Internationale Latine des Sociétés d'Eugénique, *Ier Congrès Latin d'Eugénique,* 320–21.

------------, "Eugénique et pédagogie," in Fédération Internationale Latine des Sociétés d'Eugénique, *Ier Congrès Latin d'Eugénique,* 379–80.

Thooris, Alfred, "Considérations ethnologiques et démographiques sur la population française," in Fédération Internationale Latine des Sociétés d'Eugénique, *Ier Congrès Latin d'Eugénique,* 214–27.

Trombetta, Edmondo, "Gli epilettici in zona di guerra (nota critica)," *Giornale di medicina militare* 1 (1918): 54–58.

Turpin, Raymond and Alexandre Caratzali, *Influence de l'âge maternel sur la mortinatalité des jumeaux,* in Fédération Internationale Latine des Sociétés d'Eugénique, *Ier Congrès Latin d'Eugénique,* 283–85.

Turpin, R., A. Caratzali and H. Rogier, "Étude étiologique de 104 cas de mongolisme et considerations sur la pathogénie de cette maladie," in Fédération Internationale Latine des Sociétés d'Eugénique, *Ier Congrès Latin d'Eugénique,* 154–64.

Turpin, R., A. Caratzali and M. Gorny, "Contributions à l'étude de l'influence de l'âge et de l'état de santé des procréateurs, du rang et du nombre des naissances, sur les caractères de la progéniture," in Fédération Internationale Latine des Sociétés d'Eugénique, *Ier Congrès Latin d'Eugénique,* 240–61.

Turpin, R., Caratzali A. and Nicholas Georgescu-Roegen, "Influence de l'âge maternel, du rang de naissance et de l'ordre de naissance sur la mortinatalité," in Fédération Internationale Latine des Sociétés d'Eugénique, *Ier Congrès Latin d'Eugénique,* 271–77.

Uggé, Albino, "Sulla rassomiglianza fra coniugi per alcuni caratteri somatici," in *Contributi del Laboratorio di Statistica. Serie III* (Milan: Vita e Pensiero, 1934), 117–208.

"Una cattedra universitaria per il prof. Luigi Gedda," *Paese Sera* (3–4 May 1960).

Valenziani, Carlo, *Il problema demografico dell'Africa equatoriale* (Rome: Tip. C. Colombo, 1929).

Verrotti, Guido, "Il certificato medico prematrimoniale," *Rassegna di studi sessuali* 3, no. 3 (May–June 1923): 329–34.

Verschuer, Otmar von, "Die Erbanlage als bestimmende Kraft auf dem Lebenswege," in Luigi Gedda, ed., *Genetica Medica,* 132–52.

------------, "Ein altes und ein neues Problem der Zwillingsforschung," *Acta geneticae medicae et gemellologiae* 1, no. 2 (May 1952): 180–90.

------------, "Die Häufigkeit von krankhaften Erbmerkmalen beim Menschen," in *Proceedings of the Second International Congress of Human Genetics (Rome, September 6–12, 1961)* (Rome: Istituto Gregorio Mendel, 1963), 1, 168–75.

Vignes, Henri, "De l'influence de l'intoxication alcoolique des procréateurs sur leur progéniture," in Fédération Internationale Latine des Sociétés d'Eugénique, *Ier Congrès Latin d'Eugénique,* 165–70.

Weil-Hallé, Benjamin and M. Meyer, "La survie des enfants dans les familles nombreuses et restreintes," in Fédération Internationale Latine des Sociétés d'Eugénique, *Ier Congrès Latin d'Eugénique,* 270.

Zavattari, Edoardo, "Le basi biologiche di fascismo," *Critica medico-sociale* no. 6 (June 1937): 21–28.

------------, "Politica ed etica razziale," *Vita Universitaria* (5 October 1938): 3.

------------, "Ambiente naturale e caratteri razziali," *La Difesa della razza* 3, no. 8 (20 February 1940): 6–11.

------------, "Ambiente naturale e caratteri razziali (continuazione)," *La Difesa della razza* 3, no. 9 (5 March 1940): 29–33.

------------, "Fauna e fenomeni sociale. II," *Razza e civiltà* 2, no. 3 (May–July 1941), 463–72.

------------, "La dottrina degli spazi vitali dal punto di vista biologico," *Scientia* 71, no. 361–62 (May–June 1942): 173–79.

------------, "Leggi biolgiche e spazi vitali," *Il Giornale d'Italia* (13 May 1943).

Zuccarelli, Angelo, "Asessualizzazione o sterilizzazione dei degenerati," *L'anomalo,* 8, no. 6 (1898–99): 186–89.

------------, "Per la sterilizzazione della donna come mezzo per limitare o impedire la riproduzione dei maggiormente degenerati," *Bollettino della Società Ginecologica di Napoli* 1 (February–March–April 1901): 3.

-----------, "La proposta della "sterilizzazione" dei più anormali quale misura profilattica sociale contro la degenerazione," *L'anomalo* (1909):16–17, offprint.
-----------, "Al professor Ettore Levi, membro del Consiglio superiore di sanità," *Il pensiero sanitario* 18 (1922): 3–4.
-----------, *Il problema capitale della "eugenica"* (Ferrara: Industrie Grafiche Italiane, 1924).
Zippari Garola, Aristide, "Ancora sul certificato matrimoniale," *Rassegna di studi sessuali* 3, no. 4 (July–August 1923): 323–29.

Secondary sources

Adams, Mark B., ed., *The Wellborn Science: Eugenics in Germany, France, Brazil and Russia* (New York: Oxford University Press, 1990).
Alippi Cappelletti, Maurizia, "Jucci Carlo," in *Dizionario Biografico degli Italiani* (Rome: Istituto della Enciclopedia Italiana, 2004), vol. 62, 687–90.
Allen, Garland E., "The Misuse of Biological Hierarchies: the American Eugenics Movement, 1900–1940," *History and Philosophy of the Life Sciences* 2, no. 5 (1983): 105–128.
-----------, "The Eugenics Record Office at Cold Spring Harbor, 1910–1940. An Essay in Institutional History," *Osiris*, 2nd series 2 (1986): 225–64;
Álvarez Peláez, Raquel, ed., "Dossier: Estudios sobre eugenesia," special issue, *Asclepio* 51, no. 2 (1999): 5–148.
Audoin-Rouzeau, Stéphane, *L'enfant de l'ennemi, 1914–1918* (Paris: Aubier, 1995).
Baader, Gerhard, Veronika Hofer and Thomas Mayer, eds., *Eugenik in Österreich: Biopolitischer Methoden und Strukturen vor 1900–1945* (Vienna: Czernin Verlag, 2007).
Barkan, Elazar, *The Retreat of Scientific Racism. Changing Concepts of Race in Britain and in the United States between the World Wars* (Cambridge: Cambridge University Press, 1992).
Barker, David, "The Biology of Stupidity: Genetics, Eugenics and Mental Deficiency in the Inter-War Years," *The British Journal for the History of Science* 22, no. 3 (September 1989): 347–75.
Bashford, Alison and Philippa Levine, eds., *The Oxford Handbook of the History of Eugenics* (New York: Oxford University Press, 2010).
Bernardini Mosconi, Paola, ed., *Carlo Jucci nel centenario della sua nascita. Testimonianze e documenti*, (Milan: Cisalpino, 2000).
Bianco Silvestroni, Ida, *Storia della microcitemia in Italia* (Rome: Giovanni Fioriti editore, 2002).
Billig, Michael, *Psychology, Racism and Fascism* (Birmingham: Searchlight, 1979).
Black, Edwin, *War against the Weak. Eugenics and America's Campaign to Create a Master Race* (New York: Four Walls Eight Windows, 2003).
Block, Ned J. and Gerald Dworkin, eds., *The IQ Controversy* (New York: Pantheon, 1976).
Bocci, Maria, *Agostino Gemelli rettore e francescano. Chiesa, regime, democrazia* (Brescia: Morcelliana, 2003).
Bock, Gisela, *Zwangssterilisation im Nationalsozialismus* (Opladen: Westdeutscher Verlag, 1986).
-----------, "Nationalsozialistische Sterilisationpolitik," in Klaus-Dietmar Henke, ed., *Tödliche Medizin im Nationalsozialismus. Von der Rassenhygiene zum Massenmord* (Cologne: Böhlau, 2008): 85–99.
Bonetta, Gaetano, *Corpo e nazione. L'educazione ginnastica, igienica e sessuale nell'Italia liberale* (Milan: Franco Angeli, 1990).
Bowler, Peter J., "E. W. MacBride's Lamarckian Eugenics and Its Implications for the Social Construction of Scientific Knowledge," *Annals of Science* 41, no. 3 (May 1984): 245–60.
Broberg, Gunnar and Nils Roll-Hansen, eds., *Eugenics and the Welfare State: Sterilization Policy in Denmark, Sweden, Norway and Finland* (East Lansing, Mich.: Michigan State University Press, 2005).
Bucur, Maria, *Eugenics and Modernization in Interwar Romania* (Pittsburgh: Pittsburgh University Press, 2002).
Burgio, Alberto, ed., *Nel nome della razza. Il razzismo nella storia d'Italia, 1870–1945*, (Bologna: il Mulino, 1999).
Burgio, A. and Luciano Casali, eds., *Studi sul razzismo italiano* (Bologna: Clueb, 1996).

Burleigh, Michael and Wolfgang Wipperman, *The Racial State: Germany 1933–1945* (Cambridge: Cambridge University Press, 1991).

Calvesi, Maurizio, Enrico Guidoni and Simonetta Lux, eds., *E42. Utopia e scenario del regime. 2: Urbanistica, architettura e decorazione* (Venice: Marsilio, 1987).

Canali, Stefano, "La Biologia," in Raffaella Simili and Giovanni Paoloni, eds., *Per una Storia del Consiglio Nazionale delle Ricerche* (Bari–Roma: Laterza, 2001) 1, 510–48.

Canali, S. and Gilberto Corbellini, "Lessons from Anti-Thalassemia Campaigns in Italy, before Prenatal Diagnosis," *Medicina nei secoli* 14, no. 3 (2002): 739–71.

Capocci, Mauro and Gilberto Corbellini, "Il contesto culturale della ricerca biomedica in Italia nel secondo dopoguerra," *Nuova Civiltà delle Macchine* 19 (2001): 29–41.

------------, "Adriano Buzzati-Traverso and the foundation of the International Laboratory of Genetics and Biophysics in Naples (1962–1969)," *Studies in History and Philosophy of Science Part C: Biological and Biomedical Sciences* 33, 3 (2002): 489–513.

Casella, Antonio, ed., *Una difficile modernità. Tradizioni di ricerca e comunità scientifiche in Italia, 1890–1940* (Pavia: Università degli Studi di Pavia, 2000).

Cassata, Francesco, *A destra del fascismo. Profilo politico di Julius Evola* (Turin: Bollati Boringhieri, 2003).

------------, "Cronaca di un'epurazione mancata (luglio 1944–dicembre 1945)," *Popolazione e Storia* no. 2 (2004): 89–119.

------------, *Il fascismo razionale. Corrado Gini fra scienza e politica* (Rome: Carocci, 2006).

------------, *Le due scienze. Il "caso Lysenko" in Italia* (Turin: Bollati Boringhieri, 2008).

------------, *"La Difesa della razza." Politica, ideologia e immagine del razzismo fascista* (Turin: Einaudi, 2008).

------------, "Dall'Uomo di genio all'eugenica," in Montaldo and Tappero, eds., *Cesare Lombroso cento anni dopo*, 175–84.

Chase, Allan, *The Legacy of Malthus. The Social Costs of the New Scientific Racism* (New York: Knopf, 1977).

Chiozzi, Paolo, "Autoritratto del razzismo: le fotografie antropologiche di Lidio Cipriani," in Centro Studi F. Jesi, ed., *La menzogna della razza. Documenti e immagini del razzismo e dell'antisemitismo fascista* (Bologna: Grafis, 1994): 91–95.

Ciceri, Massimo, *Origini controllate. La prima eugenetica in Italia (1900–1924)* (Rome: Prospettiva Editrice, 2009).

Cleminson, Richard, *Anarchism, Science and Sex: Eugenics in Eastern Spain, 1900–1937* (Oxford/Bern: Peter Lang 2000).

------------, "'A century of civilization under the influence of eugenics' Dr. Enrique Diego Madrazo, socialism and scientific progress," *Dynamis* 26 (2006): 221–51.

Collotti, Enzo, ed., *Razza e fascismo. Le persecuzioni contro gli ebrei in Toscana (1938–1943)* (Rome: Carocci, 1999).

Comfort, Nathaniel, "'Polyhybrid Heterogeneous Bastards': Promoting Medical Genetics in America in the 1930s and 1940s," *Journal of the History of Medicine and Allied Sciences* 61, no. 4 (October 2006): 415–55.

Cosmacini, Giorgio, "Medicina, ideologie, filosofie nel pensiero dei clinici tra Ottocento e Novecento," in Corrado Vivanti, ed., *Storia d'Italia. Annali*, vol. 4, *Intellettuali e potere* (Turin: Einaudi, 1981), 1159–94.

------------, "Scienza e ideologia nella medicina del Novecento: dalla scienza egemone alla scienza ancillare," in Franco Della Peruta, ed., *Storia d'Italia. Annali*, vol. 7, *Malattia e medicina* (Turin: Einaudi, 1984), 1223–67.

De Felice, Renzo, *Storia degli ebrei italiani sotto il fascismo* (Turin: Einaudi, 1961).

Degl'Innocenti, Maurizio, *Gaetano Pieraccini. Socialismo, medicina sociale e previdenza obbligatoria* (Manduria–Rome–Bari: Lacaita, 2003).

De Grazia, Victoria, *How Fascism ruled women: Italy, 1922–1945* (Berkeley: University of California Press, 1992).

De Sio, Fabio and Mauro Capocci, "Southern genes. Genetics and its institutions in the Italian South, 1930s–1970s," *Medicina nei Secoli* 20, no. 3 (2008): 791–826.

Detti, Tommaso, "Stato, guerra e tubercolosi (1915–1922)," in Franco Della Peruta, ed., *Storia d'Italia. Annali*, vol. 7, *Malattia e medicina* (Turin: Einaudi, 1984): 879–954.

Di Figlia, Matteo, *Farinacci: il radicalismo fascista al potere* (Rome: Donzelli, 2007).

Dikötter, Frank, *Imperfect Conceptions: Medical Knowledge, Birth Defects and Eugenics in China* (New York: Columbia University Press, 1998).

-----------, "Race Culture: Recent Perspectives on the History of Eugenics," *The American Historical Review* 103, no. 2 (April 1998): 467–78.

Dowbiggin, Ian Robert, *Keeping America Sane: Psychiatry and Eugenics in the US and Canada, 1880–1940* (Ithaca, NY: Cornell University Press, 1997).

-----------, *The Sterilization Movement and Global Fertility in the Twentieth Century* (Oxford and New York: Oxford University Press, 2008).

Drouard, Alain, "Concerning Eugenics in Scandinavia. An Evaluation of Recent Research and Publications," *Population: an English Selection* 11 (1999): 261–70.

Eberhard, Heinz, Neugebauer, Wolfgang, eds., *Vorreiter der Vernichtung? Eugenic, Rassenhygiene und Euthanasie in der österreichischen Discussion vor 1938* (Vienna: Böhlau Verlag, 2005)

Eraso, Yolanda, "Biotypology, endocrinology, and sterilization: the practice of eugenics in the treatment of Argentinian women during the 1930s," *Bulletin of the History of Medicine* 81, no. 4 (2007): 793–822.

Fantini, Bernardino, "Buzzati-Traverso Adriano," in *Dizionario Biografico degli Italiani* (Rome: Istituto della Enciclopedia Italiana, 1988), vol. 34, 563–67.

Farolfi, Bernardino, "Antropometria militare e antropologia della devianza (1876–1906)," in Franco Della Peruta, ed., *Storia d'Italia. Annali*, vol. 7, *Malattia e medicina* (Turin: Einaudi, 1984), 1181–1222.

Farrall, Lindsay A., *The Origins and Growth of the English Eugenics Movement 1865–1925* (New York: Garland Pub., 1965).

Fish, Jefferson M., ed., *Race and Intelligence: Separating Science from Myth* (Mahwah, NJ: Lawrence Erlbaum, 2002).

Franzinelli, Mimmo and Poggio, Pier Paolo, *Storia di un giudice italiano. Vita di Adolfo Beria di Argentine* (Milan: Rizzoli, 2004).

Freeden, Michael, "Eugenics and Progressive Thought: a Study in Ideological Affinity," *Historical Journal* 22 (1979): 645–71.

Friedländer, Henry, *The Origins of Nazi Genocide: From Euthanasia to the Final Solution* (Chapel Hill: University of North Carolina Press, 1997).

Frühstück, Sabine, *Die Politik der Sexualwissenschaft: Zur Produktion und Popularisierung sexologischen Wissens in Japan 1908–1941* (Vienna: Institut für Ostasienwissenschaften, University at Wien, 1997).

Fuchs, Brigitte, *'Rasse', 'Folk', 'Geschlecht'. Anthropologische Diskurse in Österreich, 1850–1960* (Frankfurt a.-M.: Campus Verlag, 2003).

Gabrielli, Gianluca, "Prime ricognizioni sui fondamenti teorici della politica fascista contro i meticci," in Burgio and Casali, eds., *Studi sul razzismo italiano*, 61–88.

García González, Armando and Raquel Álvarez Peláez, *En busca de la raza perfecta. Eugenesia e higiene en Cuba (1898–1958)* (Madrid: CSIC, 1998).

Gayon, Jean and Daniel Jacobi, eds., *L'éternel retour de l'eugénisme* (Paris: Presses Universitaires de France, 2006).

Gawin, Magdalena, *Rasa i nowczesność. Historia polskiego ruchu eugenicznego, 1880–1952* (Warsaw: Wydawnicwo Neriton, 2003).

Gentile, Emilio, "The Myth of National Regeneration in Italy: From Modernist Avant-Garde to Fascism," in Matthew Affron and Mark Antliff, eds., *Art and Ideology in France and Italy* (Princeton: Princeton University Press, 1997), 25–45.

-----------, *La Grande Italia: The Rise and Fall of the Myth of the Nation in the Twentieth Century* (Madison: University of Wisconsin Press, 2009).

Gerodetti, Natalia, "From science to social technology: eugenics and politics in twentieth-century Switzerland," *Social Politics: International Studies in Gender, State and Society*, 13, no. 1 (2006): 59–88.

Gibelli, Antonio, "Guerra e follia. Potere psichiatrico e patologia del rifiuto nella Grande Guerra," *Movimento operaio e socialista* 4 (1980): 441–64.

------------, "La guerra laboratorio: eserciti e igiene sociale verso la guerra totale," *Movimento operaio e socialista* 5 (1982): 335–49.
------------, *L'officina della guerra. La grande guerra e la trasformazione del mondo mentale* (Turin: Bollati Boringhieri, 1991).
Gibson, Mary, *Born to Crime. Cesare Lombroso and the Origins of Biological Criminology* (Wesport–London: Praeger, 2002).
Gillette, Aaron, *Racial Theories in Fascist Italy* (London–New York: Routledge, 2002).
------------, "The origins of the 'Manifesto of the racial scientists,'" *Journal of Modern Italian Studies* 6, no. 3 (2001), 305–23.
Gormley, Melinda, "Scientific Discrimination and the Activist Scientist: L.C. Dunn and the Professionalization of Genetics and Human Genetics in the United States," *Journal of the History of Biology* 42 (2009): 33–72..
Goglia, Luigi, "Una diversa politica razziale coloniale in un documento inedito di Alberto Pollera del 1937," *Storia contemporanea* 16, no. 5–6 (December 1985): 1071–92.
------------, "Note sul razzismo coloniale fascista," *Storia contemporanea*, 19, no. 6 (December 1988): 1223–66.
Gori, Gigliola, *Italian Fascism and the Female Body: Sport, Submissive Women and Strong Mothers* (New York: Routledge, 2004).
Graham, Loren R., "Science and Values: the Eugenic Movement in Germany and Russia in 1920s," *American Historical Review* 82, no. 5 (December 1977): 1133–1964.
Griffin, Roger, "Tunnel Visions and Mysterious Trees: Modernist Projects of National and Racial Regeneration, 1880–1939," in Turda and Weindling, eds., *Blood and Homeland*, 417–56.
Haller, Mark H., *Eugenics: Hereditarian Attitudes in American Thought* (New Brunswick, NJ: Rutgers University Press, 1984).
Harwood, Jonathan, "Geneticists and the Evolutionary Synthesis in Interwar Germany," *Annals of Science* 42, no. 3 (May 1985): 279–301.
Hau, Michael, *The Cult of Health and Beauty in Germany: A Social History, 1890–1930* (Chicago: University of Chicago Press, 2003).
Holmes, Samuel J., *A Bibliography of Eugenics* (Berkeley: University of California Press, 1924).
Horn, David, *Social Bodies. Science, Reproduction and Italian Modernity* (Princeton: Princeton University Press, 1994).
Ipsen, Carl, *Dictating Demography: The Problem of Population in Fascist Italy* (Cambridge: Cambridge University Press, 1996).
Jackson Jr., John P. "'In Ways Unacademical': The Reception of Carleton Coon's *The Origin of Races*," *Journal of the History of Biology* 34 (2001), 247–85.
------------, *Science for Segregation. Race, Law and the Case against Brown v. Board of Education* (New York–London: New York University Press, 2005).
Kallis, Aristotle A., "Racial Politics and Biomedical Totalitarianism in Interwar Europe," in Turda and Weindling, eds., *Blood and Homeland*, 389–415.
Kevles, Daniel J., *In the Name of Eugenics. Genetics and the Uses of Human Heredity* (Rev. ed. Cambridge, MA: Harvard University Press, 1995).
Kline, Wendy, *Building a Better Race: Gender, Sexuality, and Eugenics from the Turn of the Century to the Baby Boom* (Berkeley and London: University of California Press, 2001).
Krementsov, Nikolai, "From 'Beastly Philosophy' to Medical Genetics: Eugenics in Russia and the Soviet Union," *Annals of Science* 68, no. 1 (January 2011): 61– 92.
Kröner, Hans-Peter, *Von der Rassenhygiene zur Humangenetik* (Munich: Urban & Fischer, 1998).
Kühl, Stefan, *The Nazi Connection. Eugenics, American Racism and German National Socialism* (New York: Oxford University Press, 1994).
------------, *Die Internationale der Rassisten. Aufstieg und Niedergang der internationalen Bewegung für Eugenik und Rassenhygiene im 20. Jahrhundert* (Frankfurt a. M.: Campus, 1997).
Largent, Mark A., *Breeding Contempt: the History of Coerced Sterilization in the United States* (New Brunswick: Rutgers University Press, 2008).
Larizza Lolli, Mirella, *Le scienze politiche e sociali*, in *Storia di Milano. Il Novecento*, vol. 18, (Rome: Istituto della Enciclopedia Italiana, 1995).

Larson, Edward J., *Sex, Race, and Science: Eugenics in the Deep South* (Baltimore: Johns Hopkins University Press, 1995).

Lawrence, Cristopher, and George Weisz, eds., *Greater than the Parts: Holism in Biomedicine, 1920–1950* (New York: Oxford University Press, 1998).

Leon, Sharon M., "'Hopelessly Entangled in Nordic Pre-suppositions': Catholic Participation in the American Eugenics Society in the 1920s," *Journal of the History of Medicine and Allied Science* 59, no. 1 (2004): 3–49.

Locorotondo, Giuseppe, "Boldrini, Marcello," in *Dizionario Biografico degli Italiani* (Rome: Istituto della Enciclopedia Italiana, 1988), vol. 34, 465–67.

Löscher, Monika, "Eugenics and Catholicism in Interwar Austria," in Turda and Weindling, eds., *Blood and Homeland*, 299–316.

Lyons, Andrew P., "The Neotenic Career of M. F. Ashley Montagu," in Larry T. Reynolds and Leonard Lieberman, eds., *Race and Other Misadventures. Essays in Honor of Ashley Montagu in His Ninetieth Year* (Dix Hills NY: General Hall, 1996).

Maccabelli, Terenzio, "Social Anthropology in Economic Literature at the End of the 19th Century. Eugenic and Racial Explanations of Inequality," *American Journal of Economics and Sociology* 67, no. 3 (July 2008): 481–527.

Maiocchi, Roberto, *Scienza italiana e razzismo fascista* (Florence: La Nuova Italia, 1999).

-----------, "Agostino Gemelli critico dell''eugenica' tedesca," *Vita e Pensiero* 83, no. 2 (2000): 150–69.

Mantovani, Claudia, *Rigenerare la società. L'eugenetica in Italia dalle origini ottocentesche agli anni Trenta* (Soveria Mannelli: Rubbettino, 2004).

Marletti, Carlo and Emanuele Bruzzone eds., *Teoria, società e storia. Scritti in onore di Filippo Barbano* (Milan: Franco Angeli, 2000).

Masutti, Monica, "La rivista 'Razza e civiltà': un aspetto del razzismo fascista," *Sociologia*, I (2002): 83–100.

Mazumdar, Pauline M. H., *Eugenics, Human Genetics and Human Failings: the Eugenics Society, its Source and its Critics in Britain* (London–New York: Routledge, 1992).

McGinn, Patrick, "'Quality not quantity tells': The Eugenics Movement in India," unpublished manuscript.

Mehler, Barry, "Foundations for Fascism: The New Eugenics Movement in the United States," *Patterns of Prejudice*, 23 (1989): 17–25.

Mendelsohn, Andrew J., "Medicine and the Making of Bodily Inequality in Twentieth-Century Europe," in Jean-Paul Gaudillière and Ilana Löwy, eds., *Heredity and Infection. The History of Disease Trasmission* (London and New York: Routledge, 2001), 21–80.

Michaelis, Meir, "Mussolini's unofficial mouthpiece: Telesio Interlandi, Il Tevere and the evolution of Mussolini's anti-Semitism," *Journal of Modern Italian Studies* 3, no. 3 (1998): 217–40.

Mignemi, Adolfo, "Profilassi sanitaria e politiche sociali del regime per la 'tutela della stirpe.' La 'mise en scène' dell'orgoglio di razza," in Enzo Collotti, ed., *La menzogna della razza. Documenti e immagini del razzismo e dell'antisemitismo fascista* (Centro di Studi F. Jesi) (Bologna: Grafis Edizioni, 1994), 65–72.

Miranda, Marisa and Gustavo Vallejo, eds., *Darwinismo social y eugenesia en el mundo latino* (Buenos Aires: Siglo Veintiuno De Argentina Editores, 2005).

Montaldo, Silvano and Paolo Tappero, eds., *Cesare Lombroso cento anni dopo*, (Turin: UTET, 2009).

Motti, Lucia and Marilena Rossi Caponeri, eds., *Accademiste a Orvieto: donne ed educazione fisica nell'Italia fascista, 1932–1943* (Orvieto: Quattroemme, 1996).

Müller-Hill, Benno, *Murderous Science. Elimination by Scientific Selection of Jews, Gypsies, and Others in Germany, 1933–1945* (Cold Spring Harbor: Cold Spring Harbor Laboratory Press, 1988).

Munro Prescott, Heather, "I was a Teenage Dwarf: The Social Construction of 'Normal' Adolescent Growth and Development in United States," in Alexandra Minna Stern and Howard Markel, eds., *Formative Years: Children's Health in the United States, 1880–2000* (Ann Arbor: University of Michigan Press, 2002), 153–82.

Murray, Charles and Herrnstein, Richard J., *The Bell Curve. Intelligence and Class Structure in American Life* (New York: Free Press, 1994).

Newby, Idus A., *Challenge to the Court: Social Scientists and the Defense of Segregation, 1954–1966* (Baton Rouge: Louisiana State University Press, 1967).
Nisot, Marie-Thérèse, *La Question Eugénique dans les divers pays*, 2 vols., (Brussels: Librairie Falk Fils, 1927 and 1929).
Paul, Diane B., "Eugenics and the Left," *Journal of the History of Ideas* 45 (1984): 567–590.
------------, *Controlling Human Heredity: 1865 to the Present* (Atlantic Highlands, NJ: Humanities Press, 1995).
------------, *The Politics of Heredity. Essays on Eugenics, Biomedicine, and the Nature–Nurture Debate* (Albany: State University of New York, 1998).
------------, "Eugenic Anxieties, Social Realities, and Political Choices," *Social Research* 59, no. 3 (1992): 663–83.
Petraccone, Claudia, *Le due Italie: la questione meridionale tra realtà e rappresentazione* (Rome–Bari: Laterza, 2005).
Pick, Daniel, *Faces of Degeneration: A European Disorder, c.1848–c.1918* (Cambridge: Cambridge University Press, 1989).
Pickens, Donald K., *Eugenics and the Progressives* (Nashville: Vanderbildt University Press, 1968).
Pogliano, Claudio, "Aldo Mieli, storico della scienza," *Belfagor* 5 (1983): 537–57.
------------, "Eugenisti, ma con giudizio," in Alberto Burgio, ed., *Nel nome della razza. Il razzismo nella storia d'Italia, 1870–1945* (Bologna: il Mulino, 1999), 423–42.
------------, "Bachi, polli e grani. Appunti sulla ricezione della genetica in Italia," *Nuncius. Annali di Storia della Scienza* 14, no. 1 (1999): 133–68.
------------, "Le scienze biomediche," in Antonio Casella, ed., *Una difficile modernità.*, 257–86.
------------, *L'ossessione della razza. Antropologia e genetica nel XX secolo* (Pisa: Edizioni della Normale, 2005).
Porter, Dorothy, "Eugenics and the sterilization debate in Sweden and Britain before World War II," *Scandinavian Journal of History* 24, no. 2 (1999): 145–62.
Prévost, Jean-Guy, *A Total Science. Statistics in Liberal and Fascist Italy* (Montréal: McGill-Queen University Press, 2009).
Proctor, Robert, *Racial Hygiene: Medicine under the Nazis* (Cambridge, MA: Harvard University Press, Cambridge 1988).
Promitzer, Christian, Sevasti Trubeta and Marius Turda, eds., *Health, Hygiene and Eugenics in Southeastern Europe to 1945* (Budapest–New York: Central European University Press, 2011).
Quine, Maria Sophia, *Population Politics in Twentieth Century Europe: Fascist Dictatorships and Liberal democracies* (London: Routledge, 1996).
Ramsden, Edmund, "Carving up Population Science: Eugenics, Demography and the Controversy over the 'Biological Law' of Population Growth," *Social Studies of Science* 32, no. 5–6 (Oct.–Dec. 2002): 857–99.
------------, "Social Demography and Eugenics in the Interwar United States," *Population and Development Review* 29, no. 4 (December 2003): 547–93.
------------, "Eugenics from the New Deal to the Great Society: genetics, demography and population quality," *Studies in History and Philosophy of Biological and Biomedical Sciences* 39 (2008): 391–406.
Raspanti, Mauro, "I razzismi del fascismo," in Centro Studi F. Jesi, ed. *La menzogna della razza. Documenti e immagini del razzismo e dell'antisemitismo fascista* (Bologna: Grafis 1994), 73–89.
Regal, Brian, "Amateur versus Professional: the Search for Bigfoot," *Endeavour* 32, no. 2 (June 2008): 53–57.
Reggiani, Andrés Horacio, "La ecología institucional de la eugenesia: repensando las relaciones entre biomedicina y política en la Argentina de entreguerras," in Miranda and Vallejo, eds., *Darwinismo social y eugenesia en el mundo latino*, 273–309.
------------, *God's Eugenicist. Alexis Carrel and the Sociobiology of Decline* (New York–Oxford: Berghahn Books, 2007).
Reilly, Philip R., *The Surgical Solution: a History of Involuntary Sterilization in the U. S.* (Baltimore: Johns Hopkins University Press, 1991).

Repp, Kevin, "'More Corporeal, More Concrete': Liberal Humanism, Eugenics and German Progressives at the Last Fin de Siècle," *Journal of Modern History* 72 (2000): 683–730.

Rifelli, Giorgio, *Per una storia dell'educazione sessuale* (Florence: La Nuova Italia, 1991).

Roll-Hansen, Nils, "Geneticists and the Eugenics Movement in Scandinavia," *The British Journal for the History of Science* 22, no. 3 (September 1989): 335–46.

Sale, Giorgio, *Hitler, la Santa Sede e gli ebrei* (Milan: Jaca Book, 2004)

Sapp, Jan, "The Struggle for Authority in the Field of Heredity, 1900–1932: New Perspectives on the Rise of Genetics," *Journal of the History of Biology* 16, no. 3 (1983): 311–42.

Sarfatti, Michele, ed., *La Repubblica sociale italiana a Desenzano: Giovanni Preziosi e l'Ispettorato generale per la razza* (Milan: Giuntina, 2008).

Scartabellati, Andrea, *Intellettuali nel conflitto. Alienisti e patologie attraverso la Grande Guerra (1909–1921)* (Bagnaria Arsa: Edizioni Goliardiche, 2003).

Scarzanella, Eugenia, *Italiani malagente. Immigrazione, criminalità, razzismo in Argentina, 1890–1940* (Milan: Franco Angeli, 1999).

Schmuhl, Hans-Walter, *The Kaiser Wilhelm Institute for Anthropology, Human Heredity and Eugenics, 1927–1945. Crossing Boundaries* ([Dordrecht]: Springer, 2008).

Schneider, William H., *Quality and Quantity. The Quest for Biological Degeneration in Twentieth-Century France* (Cambridge: Cambridge University Press, 1990).

Schwartz, Michael, *Sozialistische Eugenik. Eugenische Sozialtechnologien in Debatten und Politik der deutschen Sozialdemokratie, 1890–1993* (Bonn: Dietz, 1995).

Soloway, Richard A., *Demography and Degeneration: Eugenics and the Declining Birthrate in Twentieth-Century Britain* (Chapell Hill: North Carolina University Press, 1990).

Sonn, Richard, "'Your body is Yours': Anarchism, Birth Control, and Eugenics in Interwar France," *Journal of the History of Sexuality* 14, no. 4 (October 2005): 415–32.

Sòrgoni, Barbara, *Etnografia e colonialismo. L'Eritrea e l'Etiopia di Alberto Pollera 1873–1939* (Turin: Bollati Boringhieri, 2001).

Stepan, Nancy Lays, *"The Hour of Eugenics": Race, Gender, and Nation in Latin America* (Ithaca, NY: Cornell University Press, 1991).

Stern, Alexandra Minna, *Eugenic Nation: Faults and Frontiers of Better Breeding in Modern America* (Berkeley: University of California Press, 2005).

------------, "From Mestizophilia to Biotypology: Racialization and Science in Mexico, 1920–1960," in Nancy Applebaum, Anne S. MacPherson, and Karin Alejandra Rosemblatt, eds., *Race and Nation in Modern Latin America* (Chapel Hill: University of North Carolina Press, 2003), 187–210.

Strassoldo, Raimondo, "La sociologia austriaca e la sua ricezione in Italia. La mediazione di Franco Savorgnan," in Marletti and Bruzzone, eds., *Teoria, società e storia*, 403–21.

Tedesco, Luca, "'For a healthy, peace-loving and hardworking race': anthropology and eugenics in the writings of Giuseppe Sergi," *Modern Italy* 16, 1 (February 2011): 51–65.

Teti, Vito, *La razza maledetta: origini del pregiudizio antimeridionale* (Rome: Manifestolibri, 1993).

Tomeo, Vincenzo, *Il Centro nazionale di prevenzione e difesa sociale. Un caso di ricerca sociale e di azione sui centri di decisione politica* (Milan: Giuffrè, 1961).

Tracy, Sarah W., "George Draper and American Constitutional Medicine, 1916–1946: Reinventing the Sick Man," *Bulletin of the History of Medicine* 66, no. 1 (1992): 53–89.

Treves, Anna, *Le nascite e la politica nell'Italia del Novecento* (Milan: LED, 2001).

Tucker, William H., *The Science and Politics of Racial Research* (Urbana: University of Illinois Press, 1994).

------------, *The Funding of Scientific Racism. Wickliffe Draper and the Pioneer Fund* (Urbana–Chicago: University of Illinois Press, 2002).

Turda, Marius, "New Perspective on Race and Eugenics," *Historical Journal* 51, 4 (2008): 1115–24.

------------, "The Biology of War: Eugenics in Hungary, 1914–1918," *Austrian History Yearbook* 40 (2009): 1–27.

------------, "'To End the Degeneration of a Nation': Debates on Eugenics Sterilization in Inter-war Romania," *Medical History* 53, 1 (2009): 77–104.

------------, *Modernism and Eugenics* (Basingstoke: Palgrave Macmillan, 2010).

Turda, M. and Paul J. Weindling, eds., *Blood and Homeland: Eugenics and Racial Nationalism in Central and Southeast Europe 1900–1940* (Budapest–New York: Central European University Press, 2007).

Uzarczyk, Kamila, *Podstawy ideologiczne higieny ras i ich realizacja na przykładzie Śląska w latach 1924–1944* (Toruń: Wydawnictwo Adam Marszałek, 2003).

Vallejo, Gustavo, "Males y remedios de la ciudad moderna: perspectivas ambientales de la eugenesia argentina de entreguerras," *Asclepio* 59, no. 1 (January–June 2007): 203–38.

Ventrone, Angelo, *La seduzione totalitaria. Guerra, modernità, violenza politica 1914–1918* (Rome: Donzelli, 2003).

Volpone, Alessandro, *Gli inizi della genetica in Italia.* (Bari: Cacucci, 2008).

Wanrooij, Bruno, *Storia del pudore. La questione sessuale in Italia* (Venice: Marsilio, 1990);

Weiler, Bernd, "Ludwig Gumplowicz (1838–1909) e il suo allievo triestino Franco Savorgnan (1879–1963). Analisi del rapporto fra la sociologia austriaca e quella italiana," *Sociologia* 1 (2003): 9–41.

Weindling, Paul J., "Weimar Eugenics: The Kaiser Wilhelm Institute for Anthropology, Human Heredity and Eugenics in Social Context," *Annals of Science* 42, no. 3 (May 1985): 303–18.

------------, *Health, Race and German Politics between National Unification and Nazism, 1870–1945* (Cambridge: Cambridge University Press, 1989).

------------, "The 'Sonderweg' of German Eugenics: Nationalism and Scientific Internationalism," *The British Journal for the History of Science* 22, no. 3 (September 1989): 321–33.

Weingart, Peter, "Science and Political Culture: Eugenics in Comparative Perspective," *Scandinavian Journal of History* 24, no. 2 (June 1999): 163–77.

Weingart, P., Jürgen Kroll, and Kurt Bayertz, *Rasse Blut und Gene: Geschichte der Eugenik und Rassenhygiene in Deutschland* (Frankfurt a. M.: Suhrkamp, 1988).

Widmann, Giovanni, "Pionieri della medicina genetica preventiva in Italia. Luisa Gianferrari e l'esperienza dei consultori genetici prematrimoniali," in *Atti della Accademia Roveretana degli Agiati,* 3, no. B (2003): 35–66.

Witkowski, Jan A. and John R. Inglis, *Davenport's Dream. 21st-Century Reflections on Heredity and Eugenics* (Cold Spring Harbor, NY: Cold Spring Harbor Laboratory Press, 2008).

Electronic resources:

Institute for the Study of Academic Racism, Ferris State University: www.ferris.edu/isar/
ISAR Archives, Barry Mehler's Archives.

INDEX OF NAMES